Tourism, Environment and Ecology in the Mediterranean Region

Tourism, Environment and Ecology in the Mediterranean Region

Edited by

Recep Efe and Münir Öztürk

CAMBRIDGE
SCHOLARS
PUBLISHING

Tourism, Environment and Ecology in the Mediterranean Region,
Edited by Recep Efe and Münir Öztürk

This book first published 2014

Cambridge Scholars Publishing

12 Back Chapman Street, Newcastle upon Tyne, NE6 2XX, UK

British Library Cataloguing in Publication Data
A catalogue record for this book is available from the British Library

ISBN (10): 1-4438-6218-5, ISBN (13): 978-1-4438-6218-9

TABLE OF CONTENTS

PREFACE

Tourism has different appeal and significance to people of contrasting cultural backgrounds, sociohistory as well as geographical location in the Mediterranean countries. After industrial revolution they have become good homes for long-term investment but their environment is threatened. We need fundamental changes in basic values and practices, otherwise we will destroy the diversity and beauty of the region, as well as its ability to support diverse human cultures. Current investments in tourism are leading towards a biodiversity loss, extinction of species and destruction of habitat.

More our knowledge of "ecology" increases more we get astonished at the unity of life and its diversity. Ecological perspective means thinking about soils, waters, air, plants, animals, and all their relationships. Major aims being balancing of physical, biological and human dimensions. Time and space act as limiting factors at all levels in our environment. When any ecosystem is changed by us, it becomes fragile and there is a collapse. An ecosystem is not a bottomless pit, it is a depletable asset. People are dependent on ecosystem resources for tourism as well as environment.

There is a need for an emphasis on "ecolasy" and "ecowisdom". Ecowisdom solution to our problems depends on ecologically viable, economically feasible and socially acceptable principals. The principles of "ecowisdom" are well-being and flourishing of human and nonhuman life together with the richness and diversity of life forms and realization of their values. We have no right to reduce this richness and diversity except to satisfy vital human needs. The flourishing of nonhuman life requires a substantial decrease of human population as the present human interference with the nonhuman world is excessive. The policies followed are affecting basic economic, technological and ideological structures.

This book with 472 pages includes 31 chapters from Bosnia, Turkey, Bulgaria, Kosova, Italy, Egypt, Georgia; covering such aspects as tourism, environment and ecology. The aim is to examine both natural and cultural environment in the Mediterranean Basin. Thirty-one reviewed chapters were accepted for publication. The editors hope that this book will upgrade the importance of awareness among public for protecting the fragile Mediterranean environment.

We would like to express their gratitude to the team at Cambridge Scholars Publishing for their collaborative efforts in preparing this book for publication.

The Editors

CHAPTER ONE

SMES NETWORKS ON DESTINATIONS: GEOGRAPHICAL AND ECONOMIC IMPACTS FOR NEW FORMS OF TOURISM

CARMEN BIZZARRI
AND MARGHERITA PEDRANA

Introduction

Enterprise Networks and Network Contracts in Tourism

In tourism, as in other economic areas, enterprise networks can assume the form of enterprise as an aggregation of business - for example, franchising, unions, confederations, and business associations - aiming to a greater bargaining power than others and making profits, creating incentives and facilitating harmonization of the existing provisions. Essentially, enterprise networks, more recently considered a form of lobbing in an outer environment, today could assume a new form of contract.

Therefore, in the present scenario enterprise networks are also considered as a new form of contract, the legally recognized, so-called network contract (law decree n.5/2009 and following), that brings about a new form of collaboration and cooperation between affiliated enterprises so as to reach a specific objective, that is a network project. To access said contract, enterprises must already exist and operate on the territory. In reality, this kind of enterprise network was defined as a "plot of relations that joins entities institutionally different, without damaging formal independence and without a unitary management and control, an organization based on cooperation and coordination between enterprises or other interdependent organizations" (Soda 1998, 66).

The new form of contract was established to develop all small and medium enterprises (SMEs) presently in crisis. Particularly in tourism, a time of crisis such as this is greatly felt, since global competition and technology have actually reduced profit margins and the power of attraction.

In reality, globalization induces SMEs to compete with large enterprises and multinational and international corporations, which by nature can benefit from scale economies, agglomeration economies, reduction of cost transactions, credit legibility and external effects deriving from the various phases of production. On the contrary, without these benefits SMEs proved their insufficiency in international competition and in a short time saw their sales reduced.

The main causes for this can be found in the modest financing power, poor innovation and the inefficient spread of know-how. The first cause is the main obstacle to dimensional growth and competitiveness of the entire production system, since this element limits leading companies and consequently all enterprises associated with them. The second and third causes represent important breaks to competitiveness and the overcoming of barriers with the international market. As a matter of fact, innovation promotes entering into the international market, facilitating access to new markets.

The realization of innovation policies, therefore, is essential to developing Italian tourism enterprises, which at present are blocked by financial obligations due to the limited guarantee given. Moreover, in tourism generally innovation does not derive from investments of large assets, but from ideas which on one hand are easily copied, since they are economical and not particularly risky, and on the other cannot be reproduced since they are associated with the cultural and environmental context in which they were conceived. Mainly, in the area of tourism, innovation doesn't produce a new tangible product, but a new action that changes the method of fruition of resources and the service offered.

The solution to these limits can be found in the network contract that make the whole network a guarantee, since it is the project being financed. Therefore, funding is subordinate to the evaluation of the project which must integrate the goal and the object of a network contract.

The decision to fund a project is actually oriented to appraising the material and immaterial assets of the single SMEs, sharing the risk connected to the project, though maintaining management autonomy and ownership of the associated companies.

Advantages and disadvantages of tourism enterprises in network contracts

In tourism, enterprises can greatly benefit from network contracts since aggregation processes are functional to the increase of business and the realization of new investments. In reality, in tourism the latter are essential elements for developing and increasing the offer; a tourist that perceives innovation as a new product, in discovering how unique and irreproducible the good is, would be willing to pay a higher price, giving up part of the income and spreading, by word of mouth, the experience encountered.

Moreover, it seems very beneficial to activate this type of network contract for the entire tourism division due to the many advantages deriving from it.

Besides the feasible innovation and credit eligibility, we can find:

- greater efficiency relative to the specialization of the different enterprises
- reduction of cost transactions
- scale economy—for example tourism SMEs with such a contract can share costs by cutting supply expenses
- spreading of know-how
- enlarging supplier networks.

The last factor produces an important development in enterprises since, by increasing the number of supporters, both horizontally (all enterprises operating in the system) and vertically (of other economic areas, for example agricultural and food), it activates a collaboration grid that enables a greater specialization and consequent cost reduction.

Moreover, new relations allow for the creation of new tourism products, also increasing the offer in new markets.

Moving into different territories could cause many difficulties, such as:

- access barriers
- legal restrictions
- information costs.

Enterprises operating in such a territory can assist other enterprises participating in the network by sharing their own organizational model, supporting the intake in the market and thus making the network international.

An international network, that is a network composed of several enterprises operating in different regions, is much more complex than a national network since there are legal restrictions both in the contract and the company, especially in networks operating in Eastern European countries and in developing countries. Instead, in national networks enterprises are characterized only by contractual ties, even when representing several sectors that belong to the same region.

Therefore, a network contract has many advantages for corporate enterprises, but certain critical issues determined by the network governance and particularly management costs, and decision making and possible conflicts of interest that could arise should also be considered.

To realize the project, enterprises establish a Common Assets Fund, i.e. an ad hoc fund destined to the realization of a network program and entrusted to a common managing body. Essentially, a network contract limits the responsibilities, risks and obligations of single participants that are bound only for the part destined to the network.

Therefore, the tasks of the common agent are very delicate and complex, particularly if there is a predominant enterprise in the market wanting to be a leader. However, the presence of all the enterprises in said agent should guarantee a wide sharing of strategic goals of networks and a certain facility in the organization. In fact, the network government system was conceived to be simple and fast in making choices, precisely so as to activate the spread of know-how among enterprises.

In conclusion, this kind of networking determines many advantages for SMEs. But, at the same time, some limits must be mentioned, made up for the major part by interdependence and conflict risks deriving from the more predominant enterprises on the market, forming the network. Its expansion could comprise a further element of network instability, when if, from a starting group of a few, more join by increasing the potential of the common fund, the latter could reduce the trust among the original members due to the different needs that could arise.

Network contract for developing a tourism destination

The projected aim of a network contract could cause various impacts on the territory where enterprises involved in the network operate. Actually, if enterprises in tourism destinations establish networks to develop new attractions, deriving from the different use of local resources, the project would obviously be an element of change in the territory. Being an explosive project, a thorough, efficient and acceptable planning and political program aiming to organize both the resources involved and the welcoming services, the first damaged by new visitors in tourism, would be necessary. The introduction of new resources and the resulting new tourism flows spread positive and negative effects very rapidly both inside and outside of a specific area, since innovation, either of process or product, changes travel practices. This transformation has an effect "from the cradle to the grave" of the trip, i.e. from the beginning to the end.

In fact, in the usual locations where tourism flows start from, generally there are no suitable services to reach innovation attractions. New connections and new infrastructures will be needed, not only as regards to transport but also for planning the trip, both from the standpoint of equipment as well as the new know-how for new attractions.

For a new tourism product to succeed it would be advisable to consider the participation of enterprises and territories from where new tourism flows start. In reality, such an expansion would be essential to the attainment of an efficient management and would be optimal for the network. The arrangement of welcoming services and infrastructures ad hoc are essential for acknowledging the good outcome of the project on the part of the tourists. In fact, communication and word of mouth among tourists constitute the only vehicles to confirm the innovation.

Furthermore, if it is true that networks arise to improve business performance, without an initial and evident participation of the territory, it is also true that in tourism, a network contract project in most cases includes the use of resources favouring access and availability of joint local goods. At this point we must consider the problems and environmental and social impacts deriving from the increased flow of tourists to the destination from where the network started and the likely subsequent welcome overflow in the area produced by the crowding of both traditional and new tourism flows.

In order to determine costs and benefits of a network on the territory it would be suitable to carry out a SWOT analysis to check the strengths and weakness, opportunities and threats of a network contract project.

We can certainly state that the key strength is the innovation and originality of the project, both:

(a) internally among enterprises participating in the contract—the spreading of the new know-how enables a different use of resources, developing agglomeration economies and new know-how among enterprises.
(b) external to the network with a double effect—the first consists in attracting new tourism flows that cooperate, wherever existing, with the traditional circuits interested in innovation, while the second consists in spreading innovation, not only in the destination but also the entire region, providing a permanent development in tourism.

In this virtuosity the weak point is reported outside the network, i.e. when enterprises do not inform the territory and all stakeholders about the new established network and the new know-how and the relative project, triggering a vicious circle introducing outer diseconomies such as congestion, difficulty with waste disposal, traffic, etc. that force new tourists to run from the destination.

Also within the network the weak point could be the fragility with which the network is managed. As mentioned, since the network is made up of intersectorial enterprises, often these do not have the same risks or results, in the sense that for some enterprises it may be very convenient, while for others it is not so important to participate.

Hence the greatest threat for a network contract is the internal governance where an enterprise, either a leading one, or several that form a coalition, could adopt a ruling position and would be inclined to an unlawful behaviour that would harm enterprises participating in the network. If enterprises were to act in such a way the network would break up and the consequences for the territory would be cascading distrust towards all enterprises, and consequently the entire tourism system would deteriorate.

At the same time, the weakness of this network fluidity could become a great opportunity since its being easily accessible could not only constantly renew the network by developing new projects and consequently create other contracts with the same and other enterprises, but could also easily expand and be absorbed into another existing network, if it held the same goal.

Therefore, a network contract is a great opportunity for enterprises, which together can both participate in public competitions requesting a series of services and goods that an enterprise alone cannot produce, and also have incentives for appraising and protecting resources that could be part of the project.

Therefore, cooperation with local agents is essential for the success of tourism networks, since tourists will appreciate and return to a territory where the regional organization provided for the realization of a holiday as a good experience.

Such public-private synergy can also take place in regions and territories different to those where networks start. In fact, a public body could endorse some networks for the attraction and strategies that they hold, offering incentives and support. In tourism, in reality, some services which are not necessarily connected to the resources of a locality could relocate since they do not need big infrastructures or large capital. Therefore, if a network does not have a local cooperative body, it could choose a different area from where it started, and afterwards also involve the other enterprises of the locality.

Territorial economic policies in SME networks

In the field of complexity and differences among policies and projects related to enterprise networks there are many examples, especially for what concerns SMEs. Firstly, we have to notice that many regions have development plans in which public announcements evaluate policies and contracts which aim at creating a pool of competences and skills—both public and private—whose purpose should be to stimulate territorial growth and development. In particular, tourism is the sector that had more benefits from these public interventions.

As an example, the Tuscany region of Italy proclaims to select industrial and development experimental researches in the field of sustainable and competitive tourism. Also, the Lombardy region gives public contributions to the sector defined as CTS (*commercio*, *turismo* and *servizi*—trade, tourism and services). The main purposes of these public investments are connected with the increase of the competitiveness of aggregated enterprises, managing costs, improving and modernize spaces, and supporting the development of informal networks also aimed at knowledge exchange and innovation diffusion. The regional government aims at the financial support of old and new enterprise networks.

Why do regional and territorial institutions try to finance enterprise networks? Which are the advantages for the development of territorial areas? In particular, why finance enterprise networks in the tourism sector? The main purpose of this chapter is to answer to these questions through a case study.

Tourism is a sector that cannot ignore the territory and its future development. The territory is the area in which the enterprise works and to which tourists come. Therefore, the future deterioration of the area and the lack of attention to the themes of sustainability, both from environmental and a socio-cultural points of view, may lead to losing the heart of the tourism sector, which is the territory and its attractiveness.

Initiatives regarding enterprise networks are many, both vertical and horizontal, in particular considering their applications in the tourist sector. For the purpose of this chapter it is interesting to analyse some specific features which, in some of the analysed case studies, lead to the evaluation of the territory development with a strong tourist propensity. Among the different case studies, the initiative of enterprise networks usually starts from institutions which know enterprise organization and all interested actors in the promotion. More often enterprise networks are born on specific projects and develop on permanent enterprise networks in the territory, with long-run purposes for the promotion of the tourism sector and the cultural tissue of the territory itself.

Destination management and management issues

Economic and territorial policies, in the tourism sector, have developed through the tourism destination concept, defined as an organized system of actors in a territory whose purpose is the organization and systemic-territorial management of the tourism activities. This paragraph aims at the evaluation of problems related to the management at consortia and network contract levels.

Globalization underlines some issues at the territorial and local levels, creating the phenomenon known as "glocalization."[1] Tourist destinations concern problems of globalization on local and territorial entities. The need to define what we understand by tourism destination is essential for the definition of policies and network contracts in a specific territory. Tourism destination is defined as the "place objective of a journey" (Franch 2010). The concept is linked to the local system of tourist supply (in Italian SLOT[2]), defined as all activities and factors of attractiveness organized in a system of tourism hospitality (Della Corte 2000).

The development of enterprise networks in a tourism destination, especially if supported by territorial policies, may lead to the strengthening of tourism organization. In enterprise networks and in tourism destinations, public interventions are often fundamental, especially at the local level.

The main problems of the management of tourism destinations, also through the destination management organization, start from the complex heterogeneity of actors in the local area. Actually, there are different types of enterprises, some of which may be dedicated exclusively to the tourism sector. The strategic-territorial vision may be efficient if we take into consideration three important factors.

First is the different strategic vision of the tourism destination, which has to widen the strategic vision of the specific actor and specific category of operators. The second is the involvement of the actors that deal with all the different aspect of the holiday, which are different according to the territorial systems. The third and last factor is the long-term planning quality of every strategy concerning the tourism destination, in comparison with the short-term strategies of the individual actors.

An association of enterprises in a territory aims at the improvement of efficiency, decrease of costs, increase of the efficiency of the promotion

---

[1] "Glocalization" comes from sociology, and is a term created in juxtaposition to the excessive use of "globalization."

[2] From the Italian "*sistema locale di offerta turistica.*"

and distribution of the tourism activities and level of internal competences, and influences the competitiveness of enterprises. All these objectives are long term and aim at the valorisation of the association in comparison with the individual interests linked to the kinds of enterprises or individual actors.

The consortium contract is defined as a common organization among entrepreneurs "for the management of some phases of the own enterprises."[3] In particular, the hotel consortium often has support functions to the connected enterprises in order to provide for the problems connected to SMEs. The consortium has some specific objectives:

- reduction of the service costs for enterprises
- economies of scale, both operative and management
- valorisation of the contract power both on national and international markets
- development of individual enterprise skills thanks to network economies and cooperation between enterprises and consortia.

The main activity provided by a consortium may concern the centralization of purchases, the central management of bookings, consulting services, learning and promotion. The nature of the consortium participant is very important because they can be also public institutions, which should grant general interest towards the interests of individual actors.

Another important fact linked to the tourism destination is the concept of industrial district. Starting from the first definitions, there are some peculiar characteristics for the local development of industrial districts, especially in specific areas. These features are the spatial concentration, the productive specialization, the small dimensions of enterprises and the strong linkages with the territory. These indications are also in the tourist sector which, although it is not an industrial sector, presents some characteristics of industrial districts. In particular, the kinds of cooperation and competition typical of industrial districts may be strong assets in tourist destinations organized at the central level. The organization and territorial policies may be potential inputs important for the area and the economic development, especially if inspired by governance principles of tourism destination in a systemic-organizational view.

Destination management is defined as the management of the tourism destination, understood as a set of different activities, sometimes not in the tourism sector, which have to be organized for tourism purposes. The organization and management of enterprises belonging to different sectors may lead to some problems, especially for different time strategies and

[3] Article 2602 of the Italian civic code.

objectives that a systemic organization has in comparison with the individual enterprises.

Obviously, problems related to organization of destination management are those typical of consortia and enterprise networks where, especially for what concerns the hospitality sector, the double role of competition-cooperation is often an obstacle to the tourism system, investing the government, organizational, management and reporting of kind of marketing, supplying and tourism production (Ruisi 1999).

As stated, the governments of such associations should follow objectives which may also be different compared to the individual actors. In particular, many difficulties of management and organization of innovation may emerge from the individual interests, which may transform into interests of the whole organization, leading the purposes of the organization to be the same for just one or a few members. In the ideal model of management of the destination, the governance should evaluate general objectives in the long term to get common interest results. These problems of management may lead to abuse of interests of some in general, imposing on the global government individual interests.

For instance, the definition of the market target for the destination may be linked to the needs of specialized travel agencies in a specific market with long-term perspectives, which have perhaps not yet been explored. There may also be problems in terms of management of destination and its perspectives, which may lead to a long period of stagnation or the decline of the tourism destination. Moreover, the centralized booking system may not find the potentiality of new solutions leading to a monopoly of the supply for the whole destination and the management system, with lock-in effects. All these evaluations have to be done in the planning phase of the tourism destination to avoid further management problems, and first of all the objectives and functions of enterprise associations have to be analysed.

In the next paragraph we will analyse a case study of an enterprises network concerning both the industrial and tourism sectors. Therefore, the industrial tourism through the enterprises network has a complexity of organization, which has to be managed to be effective in both the industrial and tourism sectors.

Industrial tourism in Friuli-Venezia Giulia[4]

The case study concerns industrial tourism in the Friuli-Venezia Giulia region, and in particular Udine. Confindustria Udine[5] elaborated a project

[4] Friuli-Venezia Giulia an Italian region located in the north east of the country.

for the promotion of the territory, which concerns industrial tourism, defined as the valorisation of the industrial heritage through architecture, infrastructure, enterprise museums, historical archives and industrial villages. This kind of tourism allows for the presence of firms in a territory and am enterprise network, which may support it.

Industrial tourism had a great development from the association of SMEs that, especially at local level,[6] allowed for the use of the economic tissue already present through associations and enterprise networks. Industrial tourism is a part of cultural tourism based on the presence of enterprises, allowing visitors to visit firms and find out the history of some sectors and enterprises. The fields of visit range from industrial archaeology to enterprise museums, visits to the firms of products and services in industrial districts, and industrial villages to enterprise shops.

At local and territorial levels, industrial tourism may have a threefold objective. The first purpose concerns the knowledge and promotion of the territory, including not just the natural and historical-cultural attraction, but also the cultural tissue at enterprise level and enterprise culture spread throughout the territory, exploring the economic part of the area. The second purpose, linked to corporative and enterprise associations which often promote this kind of tourism, concerns the promotion of the enterprises through tourism and the knowledge spread on techniques and production processes specific to this kind of sector. The third objective concerns the industrial tourism dedicated to traditional museums and the re-qualification of areas with a reduction and closing of important enterprises for the territory and the re-qualification of the history of such enterprises through industrial museums and exhibitions. This last purpose may also be linked to the project of eco-museums and the re-evaluation of former industrial areas.

The specific kind of tourism, also defined as "economic discovery tourism," and the project which involved the enterprises of Friuli-Venezia Giulia and the entrepreneurs, usually not directly involved in the tourism sector, through an enterprise network, lead to an enterprise network which is both horizontal (the enterprises become tourism destinations) and vertical (involving tourism activities and the tourism supply chain which is in the territory).

The first objective of this project of enterprise network for industrial tourism concerns the promotion of territory. Therefore, industrial tourism

---

[5] Confindustria is the association of Italian Industry. The source is the Confindustria Udine and the project is "*Industria e turismo. Andata e ritorno*" ("Industry and Tourism. A Return Trip").

[6] The level in this case is usually the regional or provincial one.

communicates the territory value, especially in a territory with a high rate of enterprises, allowing for knowledge of the product, the productive process and the entrepreneur. As already said, this kind of tourism is linked to the territory and productive tissue promotion, promoting the excellence of the territory. In fact, industrial tourism in this case is articulated through three different products, following three different routes. The first concerns enterprises in the design sector in the region. The second concerns the food and promotion of excellence in the food industry, such as San Daniele ham and the Sauris area. The third route concerns so-called industrial archaeology, which is enterprise history both with industrial museums and tourism in enterprises which are closed, but that were very important in the past.

An important characteristic for industrial tourism is the de-seasonality of tourism. Therefore, the product may promote the territory and modify the typical seasonality of tourism, leading to the use of the territory by interested tourists who would visit in different periods from the usual ones. This is also important for the tourism supply chain in the period characterised by the low season.

Factors which push an enterprise to participate are usually of a sales-based nature. Industrial tourism allows for direct promotion and sale of products. There are also communication and industrial factors which lead to promoting excellence and enterprise culture communicating a positive image, also through an integrated communication and diversifying of the relationship instruments. Also at the internal communication level, industrial tourism involves employees who may be motivated by a new project, and communication with all stakeholders and shareholders, in order to exploit the enterprise at the economic-financial level. Finally, industrial tourism may create new meeting and exchange opportunities between professionals and enterprises, leading to a promotion and a communication linked with business to business.

From the other point of view, tourism is stimulated by visiting enterprises through a project of wider tourism and territory knowledge. There is an untapped tourist demand which wants to discover the territory through enterprise culture and knowledge, especially from foreign tourists who want to evaluate Italian fineries and places where they are produced. There are many opportunities at the enterprise level to encourage this kind of tourism, through many initiatives and projects in different Italian areas, such as publicising the enterprise and its culture through the territory promotion. The most important strengths in such a kind of enterprise network linked to this kind of tourism concern the former presence of enterprise associations, which may have this opportunity to increase the seasonality of tourism and help in the promotion of the territory through integrated communication.

The most important challenge concerns the opportunity to promote this kind of tourism at the national level, both for the re-qualification of the area with many enterprises and the economic tissue and product promotion that help in the de-seasonality of tourism. At the national level, obviously, industrial tourism may be promoted through enterprise associations and networks. The most important weaknesses concern the coordination of the needs of enterprises and tourism actors and try to use these initiatives in an integrated promotion of the territory, which may not be out of the evaluation of all possible kinds of tourism, in a project for the sustainable use of the territory. The coordination and organization of this kind of tourism have to also consider the economic tissue and tourism seasonality and its composition.

In conclusion, industrial tourism and this use of enterprise networks already on the territory may lead to a better coordination of productive activities in a tourism destination, also helping in the de-seasonality of the tourism product.

Conclusions

In the tourism sector, enterprise networks are an element of valorisation of both economic activities and territorial resources. The contract of enterprise networks, in particular, may be considered a strong factor of innovation for the development of new projects and may influence the tourist and territorial systems because of its positive effects on the tourism life cycle. The SWOT analysis helped to positively valuate the added value of the enterprise network contract for the local entrepreneur tissue. It is very important to consider the instability and possible conflicts that network governance may create both inside and outside the network itself.

In conclusion, the phenomenon of enterprise networks is positive as the case study on industrial tourism in Friuli-Venezia Giulia shows. Enterprises have created an ad hoc enterprise network for the valorisation of entrepreneur and cultural local resources. The success of the enterprise network comes both from the coordination of involved enterprises and tourism activities and public institutions in the territory.

References

AA.VV. (2012). Linee Guida per i contratti di rete, RetImpresa - Confindustria e Comitato interregionale dei consigli notarili delle Tre Venezie, Marzo 2012- www.reteimpresa.it consultato il mese di Marzo 2012.
Cafaggi F. e AA.VV. (2009). Le reti di imprese e i contratti di rete, Giappichelli.
Calabrò, A., (2003). Turismo industriale in Italia, Touring Editore.
Della Corte, V. (2000). La gestioen dei sistemi locali di offerta turistica, CEDAM.
Franch, M. (2010). Marketing delle destinazioni turistiche. Metodi, approcci e strumenti, McGraw- Hill.
Promoveneto, (2006). Marketing per il turismo d'impresa, ed. Teamwork.
Ruisi, M. (1999). I consorzi alberghieri. Problemi economico-aziendali, Giuffrè.
Soda, G. (1998). Reti tra imprese. Modelli e prospettive per una teoria del coordinamento tra imprese, Roma, pag.66 e ss.

Web sites

http://www.industriaeturismofvg.it/
http://www.turismoindustrialevicenza.it

CHAPTER TWO

BRANDING POLICIES FOR MARGINAL AREA COMPLEXITY AND MULTIDIMENSIONALITY IN A REGIONAL CASE STUDY

DANIELA LA FORESTA

Introduction

Technological, economic and political changes have occurred in recent years, sometimes altering relations between nations and sometimes the meaning of words of territorial identities, geographies, economies and alliances between various places. These processes have been determined by globalization that has shrunk the world, led to new opportunities and new problems as well as new forms of competition, and has attracted attention to new spaces. The competition now operates through different channels, using innovative tools, to assure the constant availability of products and services in the global market.

The first challenge for the territories, then, consists in the ability to ensure its presence in new spaces created by globalization; innovation, inclination and the capability to exploit local potential are fundamental in the management of territory.

Territorial competitiveness is the most important and most difficult challenge for every country in the world, because it requires effort and constant attention, as a dynamic concept in continuous change.[1]

An area becomes competitive when it is able to compete in the market while ensuring an environmental economic, social and cultural sustainability, based on network organization and inter-territorial relations.

To compete territories have therefore grown and developed, using the intangible resources deeply embedded within them, such as skills, relationships between people, and trust in products, services and people.

---

[1] The reference to the concept of territorial competitiveness, however, is not unique in the scientific literature.

Local governments might therefore face the difficult challenge of aggregating the various stakeholders in the local context around a specific mission (territorial competitiveness). The operation is centred around the need to integrate all the resources and activities of the area into one system and create a high-density experiential offer for the chosen target. To this end it is essential to initiate territorial strategic management based on "a shared vision and a shared cultural model able to promote the construction of ongoing working relationships between all stakeholders both internal and external to the territory" (Sicca 2000).

Competitiveness, however, is measured in a comparative perspective that needs specific operational strategies aimed at increasing the internal and external visibility. To this end, specific promotion policies have been made in territories and widely analyzed by scientific literature; social and cultural transformations, however, now require a review of the promotional models traditionally used and the continuous search for new means of communication and promotion.

The aim of this work is to promote a critical reflection on the effectiveness of traditional policies of territorial branding through the analysis of promotional choices made in the Italian regional context based on the most recent developments in scientific thinking.

Trends and challenges in branding policies

The watchword for each territory becomes "differentiation" and the primary goal is to raise awareness and appreciation of the managed area.

Differentiation allows those who do not live in the territory concerned to immediately recognize the cultural, historical, geographical, social and economic features of a place. To obtain recognition of a place, however, it isn't enough to increase the advertising business, because people are subject to a daily chaos of information, meaning that the message would not have the necessary resonance, reducing the usefulness of the investments and efforts. The regional managers, therefore, have sought new tools to place their areas in people's minds and, in the wake of corporate literature searches, identified the geographical brand as a powerful tool to positively influence people.

Brand management, the potential of which has been widely discussed in the literature of the field,[2] is known as the application of marketing

---

[2] Within companies one of the most widely used management tools to improve the competitiveness of enterprises and impose their presence in world markets is the brand. According to the American Marketing Association, "brand is a name, term, sign, symbol, or any other characteristic that is intended to identify the goods or services of one seller and to distinguish them from those of other sellers." For

techniques to a specific product or a product line. The brand name may perform the functions of identification, warranty, guidance and symbolism as well as protection from imitations, competitive positioning, strengthening of market power and influence in the internal and external relations (van Gelder 2003). The aim is to increase the perceived value by consumers of a product or an offering which determines, as a consequence, an increase in brand equity (brand value).

The brand is characterized, therefore, as a great tool for the differentiation of tenders, and a fundamental element to create a preference for consumer trust, because it is the only resource that is not imitable by competitors.

The brand, in its generic sense, is an intangible instrument of great impact in fantasy and in the opinions of people because it creates familiarity and confidence in the products, charges them with meanings and expectations, and gives them an identity and "life."

It allows for the establishment of "a lasting relationship with customers, generating an amount of value that, at least potentially, implies functional, emotional or self-expression benefits and giving credibility to the guaranteed products."[3] Its purpose is to make sure that the brand can last over time, communicating consistent messages that are realistic, positive and relevant. It is precisely this set of aspects and features that have led the branding policies from consumer products to territories.

Kotler & Gertner (2002) were the first to say that the brand "is the primary tool to transfer the value proposition of territory because branding strategies create involvement, evoke emotions and influence the behavior of people." The territorial brand is, therefore, "the result of a continuous and dynamic process of construction in the mind of the user of the territory

---

Kotler (2002), brand "is a name, term, sign, symbol or design or combination thereof that is used to identify products or services of one seller or group of sellers and to differentiate them from those of their competitors." Aaker (1991) argues that a brand "is a set of assets (or liabilities) linked to a distinctive mark (brand, name, logo) that you add (or subtract) to the value generated by a product or service."

Pratesi & Mattia (2006) state that "the brand has its own expressive event, so it is a set of signs and symbols, tangible and intangible, which connote character and personality, as is the case for an individual." According Fiocca, Marino & Testori (2007), "the brand is for businesses and consumers a moment of attraction and connection between what the company is able to offer and what consumers perceive and desire."

[3] Aaker & Joachimstaler (2001).

which, therefore, is influenced by the experiences, memories and opinions expressed by other users with whom they come into contact.[4]

The territorial brand derives, therefore, from specific attention to identity factors and distinctive features of a territory that allow a clear strategic positioning versus competitors reorganizing the offer to the needs and/or interest of the potential beneficiary of the promotion (Kotler 2002).

The brand, directly related to the concept of region or regional reputation, is characterized by a multi-identity dimension (economic, political, socio-cultural, environmental, etc.) and a positioning aimed to highlight one or more "positive" aspects that can create reasons of attention/attraction.

The development of an institutional brand as a priority vehicle for a modern and homogeneous communication of tourism offer allows, when well designed, both short- and long-term benefits. Associating the multifaceted system of the offering to a distinctive graphic sign becomes the glue that holds together the different tourist offers, ensuring identity and common purpose, requiring a strategic and integrated vision developed into a geographic perspective of overcoming the institutional constraints that may impede a common and effective strategy. The first action therefore concerns the increasing of the reputation and the placement of a "new" proposal in the imagination of tourists.

In particular, the main problems that must be taken into account for the effective management of branding policies relate to administrative and political management, the multidimensionality of the factors of communication, effectiveness of the scale to which they relate and, finally, to the ability to respond to the new challenges posed by contemporary society.

The identification of the boundaries of the tourist destination area to which the brand reports is inevitably related to the identification of the governing body responsible for the branding activities and the resulting co-existence of different levels of land management, each with different goals and philosophies of branding. This condition in multi-brand territorial management overlaps with the need for coexistence between public and private brands. The cooperation between different sectors, both public and private, is one of the strongest needs for defining competitive identity, as it provides a reversal of thought - identity management is not a practice that stands alongside the normal practice of government, but must

[4] The literature coined the terms "Place Brand," "Country Brand," Nation Brand" and "Brand City" to define specific fields of action of the various policies of territorial branding. This document refers to regional branding policies with the objective of promoting tourism purposes.

be an integral part of this. This objective can be realised only if institutions, entrepreneurs and civil society are involved in a joint effort.

Territorial multiplicity is also reflected in the presence of multiple stakeholders (with not always compatible expectations and for which the tourist development of area is not necessarily a priority) and the difficulty of synthesizing, through a single brand, a portfolio of very heterogeneous territorial activities. Frequently, the use of these territorial options is characterized by their experiential nature and for the existing territorial hierarchy that can be marked by the coexistence of geographical and historical brands consolidated alongside the newest products and tourist destinations.

As for Italy, the promotion of Rome and its region Lazio is emblematic. For some years, these territories were promoted as one, trying to realize a single promotional activity that exploited the driving effect of the reputation of the city of Rome to launch the "Lazio brand" on domestic and, especially, international markets. It has now returned to a model of communication that considers the two territories as separate, and which required the adoption of completely different models of communication.

The geographical configuration of the reference scale is a matter of considerable importance. In a regional branding policy, in fact, the geographical contexts of proximity are without doubt the most receptive. Physical proximity determines neighbourly relations and, therefore, of knowledge, solving the issues of accessibility and travel times that are among the main factors influencing the choice of a journey, effectively promoting it.

The brand strategy declined in these terms, therefore, favours the short-range approach, considering the contiguity not only as a spatial dimension but also, and especially, a temporal one. The infrastructure links that reduce travel times, and particularly air links (point to point), define a new horizon of contiguity of accessibility in a short time, determining a new geography of relationships.

In the long term this correlation is dramatically reduced and regional brands are less powerful. As a result, on domestic markets a localistic promotional approach, based on regional singularities, has good value and a considerable impact; on international markets, however, the visibility must be achieved by connecting to a well-known and accredited institutional brand.

The branding policies, then, cannot avoid considering the contemporary dimension of communication, more and more characterized by the specific gravity assumed by the technological component and with the processes of fragmentation and convergence of information simultaneously appearing.

The progressive approval of the message and the uniqueness of evocative contents risk producing the opposite effects to that intended, such as homogeneity rather than differentiation, and approval rather than innovation.

Under the pressure of these factors, promotion of the territories has to rethink the classical setting of communication, founded on unidirectional and hierarchical dynamics, exploring the new opportunities offered by the digital society but constantly holding the difficulty of capturing the attention in a landscape populated by a multiplicity of overlapping ideals.

In this respect, the technique of guerrilla marketing that attempts, with unconventional marketing actions, to capture the attention on identity, values and offers of the territories gains some interest

Basilicata. What a discovery!

The Basilicata region, a new entry on the touristic market, is of some interest in terms of brand positioning. The market, though not particularly large, is steadily growing, with 517 million arrivals in 2012, representing a slow but progressive growth index of increasing attractiveness even in a time of deep economic and financial crisis on a global scale.

However, the growing interest in this place of culture, nature and adventure, proposed as a destination for the "good life" and "well-being," a place for those who want to take care of themselves, has not yet been translated into real and effective travel choices.

These transformations take place only thanks to attractive and competitive offers and through the quick adoption, by the private system supported by the public, of strategies and advanced tools in the promotional marketing.

The growing market shares are, furthermore, those of the domestic market related to the Italian regions - international tourism comprises only 5% of the total, a dimension which reflects a regional infrastructural deficit (only the eastern part, and in particular Matera, are able to attract a significant share of foreigners), and the difficulty of a marginal and little-known brand to succeed in international markets or geographically farest.

Considering the Italian case, it is easy to detect that mass schooling has contributed to spreading a geographical and historical culture. So, a region like Basilicata, perceived as small and fair, is not completely absent in the imagination of Italians, though scarcely present in terms of its cultural, historical, economic and social status. However, even though it enjoys some notoriety in the neighbouring regions of Puglia, Campania and Calabria, further away, in other regional contexts, this is not the case.

This awareness is present in the regional institutions responsible for its promotion and deduced by the plans of activities that, since 2007, have devoted priority attention to neighbouring regions (which comprise 50% of the current tourist flow).[5]

The increase in the number of tourists was not enough to compensate for the losses resulting from a lower number of overnight stays (-4.2%). Overnight stays, in fact, amounted to about 1,881,000 in 2012. The significant reduction in the number of days of holidays is a phenomenon observed in recent years. However, whereas some years ago an increment of short breaks during the year was noticed, today the number of holidays is shrinking, in many cases, to one holiday a year (for those who can afford it).

Overnight stays decrease and the decline in tourist spending caused by disposable income abatement for expenditure on luxuries has had an inevitable negative impact especially on the tourism industry which has suffered the most recently (visible through restaurants, bars, activities recreational, tourist services, etc.). It must be remembered that the companies operating in the Basilicata region in the context of the activities attributable to tourism are more than five thousand, according to the latest national report on tourism, of which only 705 in the hospitality sector are marked by growing competition from new hotels.

Table 2.1. Arrivals and overnight stays in Basilicata (absolute values and annual changes)

Year	Arrivals	%	Overnight stays	%
2003	427,286	9.2	1,761,876	3.8
2004	438,263	2.6	1,779,592	1
2005	467,030	6.6	1,953,150	9.8
2006	451,056	-3.4	1,743,528	-10.7
2005	448,546	-0.6	1,856,789	6.5
2008	466,280	4	1,862,373	0.3
2006	467,284	0.2	1,888,696	1.4
2010	493,828	5.7	1,890,108	0.1
2011	511,677	3.6	1,963,474	3.9
2012	517,022	1.0	1,881,120	-4.2

Source: processing on APT data

Data show, in the areas where the number of tourists stays the same or increases, that some structures improve their performance with significant

[5] Apt Basilicata, Piano di attività di comunicazione e marketing, 2007, 2008, 2009, 2010, 2011, 2012, 2013.

increments, while others lose market share. This is a scenario that is much more selective than in the past, where the most enterprising operators were successful, their ability to act creating a balance between quality of services and affordability, compared to those who are less attentive to promotion and to quality management factors.

The brand policies of Touristic Basilicata are made on an annual investment of around two million Euros (the fact that in the last four years Calabria has invested 32 million Euros for its promotion makes for an easy comparison). This is important as it is one of the fundamental conditions of branding policy that relates to the ability to increase awareness about some market in a non-sporadic but intense and recurrent form.

There must be in a particularly "talkative" and "noisy" market such as tourism, in which thousands of "sirens" make their voices heard, strongly influenced by the spending amount, as every mass market promotional campaign highlights. This refers to the guerrilla marketing (or "predatory," to evoke a theme often associated with the Basilicata, at least in Italy) approach that was periodically used to communicate the specificity offered by this region.

The unconventional marketing identifies ways and means to promote the brand with respect to "direct" and generalist campaigns. It favours indirect forms of narration (as in the case of support for the movie *Basilicata Coast to Coast* [2010]), unusual areas such as town squares or outlets ("*Basilicata in tir*" or "Basilicata indoor") where proposals are presented in intense and rhapsodic ways, or unusual promotion such as the construction of the artistic crib at Saint Peter's Square, in Rome, that contextualizes the Nativity in the Sassi of Matera.

In addition to these offline actions an intense online campaign unfolds which on direct contact (face to face or one to one) with non-conventional marketing adds research and methods of the digital world using an integrated strategy between websites and social networks (see basilicataturistica.com) and innovative actions to capture the attention which is the nodal point of each communication. The "Digital Diary of Basilicata" project—audiovisual stories for the web created by seven young artists from around the world, with very significant viral results on the web (more than a million views) through multiple sources (Youtube and Vimeo, several Facebook pages, etc.)—is a demonstration as evidenced by the special mention given by the Italian Ministry of Public Administration in 2012 and the Social Hospitality award for the best integrated strategy for social media marketing at Florence BTO (the most important event dedicated to tourism online in Italy). This strategy also encompasses the experiences of mini websites abroad where, aware of the "nationalist" nature of search engines and the need for a cultural mediation in the proposition of the regional offer and the Basilicata brand, it has been chosen to acquire partners in the target countries like England with the task of translating and promoting the image of south Italy Basilicata (See Discover Basilicata) with UK domains, and content generation making use of this mediation and developing the site in different languages. In this case the choices of England, Germany, Netherlands and Russia define a limited area of target countries connected with the logistics hubs of Bari and, therefore, Matera or Naples (considered the two-part hub of the region), and also linked with cultural ties arising from the presence of the Lucan community abroad or with cultural and scientific relations, as in the case of Russia.

The system of logistical infrastructure that draws new relatedness based on the temporal rather than spatial dimension makes for an example, as it is easier to get to Matera from London through Bari (the distance between the airport and the centre of Matera is about 65 kilometres, a one hour journey) than from some regions of north-central Italy.

This condition determines opportunities for promotional activity targeting specific cities/destinations starting from the correlation between dissemination and reputation policies of a destination and the concrete accessibility of the proposed travel suggested by the brand.

Table 2.2. Basilicata Turistica online

Site	Feb-Sept 2011	Feb-Sept 2012
Basilicata travel / blog	40,084 visits	50,619 visits
Basilicata.turistica.com	---	38,174 visits
Total	40,084	88,793
Facebook pagina apt.basilicata travel	Not calculable	5,600 fans
Twitter.com	Not calculable	1,348 followers
Altri social network	---	350 contacts
Youtube.com	6,248	10,304

Source: APT data

The case of the United Kingdom, however, shows that even if it is growing the capacity of contacts and therefore the Basilicata brand's fame, in the short term this has not led to a significant increase in overnight stays (between 2011 and 2012 overnight stays passed from 5,809 to 5,939). The latter aspect is leading to a rethink of policies of regional branding, as the plan sets out this year. Rather than insisting on positioning strategies, independent from the Italian brand, on new and more distant markets, it should maybe work itself onto those markets for which the Italian brand has already had a consolidated position and contribute to enriching the specific content of the region. This experience is emblematic in defining the relationship between the Italian brand and its regional variations.

If it is acceptable that the Friuli region in the north of Italy, close to the national border, can be promoted directly on the Austrian market, being considered "proximity region," would a similar promotion policy implemented on more distant markets, such as, for example, the Chinese, be justifiable?

On the domestic market, thus, the geographical and historical regionalist culture determines certain preconditions to strengthen its image inside the country (even within the competitive framework in which they operate, active in all national regions at different intensities).

On the European market, the same cultural preconditions that can justify a similar strategy do not exist, except, perhaps, for those countries with which there have always been relationships, where the internal history has been intertwined with the Italian (e.g. France, Spain, Austria, Germany, England) and where there still exists an imaginary deriving from the tradition of the Grand Tour. On these markets beyond the Alps a strong image of some cultural destinations exists, such as Rome, Venice, Florence and Tuscany, Naples, the Amalfi Coast, Rimini or Sicily and Calabria or, more recently, Trentino and Puglia-Salento (in this case, thanks to aggressive promotional campaigns and maturation of local tourist systems).

In the long term, the Italy brand as a whole has international recognition and only rarely is the declination of this made in Italy in its regionalist versions or destination is sufficiently known. Paradoxically, this evidence is neglected in internationalization and the national policies of brand reputation, such as in the case of ENIT, the Italian National Agency of Tourism.

ENIT has the task to promote Italian touristic image abroad. The most recent data in our possession show that the participation of Italian regions in a unified manner in ENIT-proposed events (international fairs, workshops) sees a maximum of thirteen regions involved (Fair London, Berlin, Moscow). In all the others comes down between 4 and 6. This does not mean that other regions do not participate but that some of them, such as Veneto, Friuli, Piedmont, Calabria and Sicily, exist on their own, as if their brands do not need the more general connection to the Italian brand.

Beyond the debate on the usefulness and effectiveness of the "Fair model" to promote countries in the digital era, this approach is a demonstration of "regionalist instinct" and political conditioning that, in order to give an account to local political systems of promotion policies and costs, creates a paradoxical effect of "pre-unification."

The promotion of Italian regions abroad, without the support of the Italian umbrella brand, obtains unconvincing results. Just consider the fall of appreciation of Italy in the international market or the inability of the southern regions to increase their overall attractiveness in recent years. In fact, only 2 in every 10 foreigners choosing Italy come to the south of the country, despite the generally positive numbers of Italian tourism until 2011, as shown below in Table 2.3.

Table 2-3. Tourism demand in Italy

Anno 2012	Italians		Foreigners		Total	
	Value	Var. % 12/11	Value	Var. % 12/11	Value	Var. % 12/11
Arrivals	50,705,768	-9.9%	47,394,537	-0.1%	98,100,305	-5.4%
Overnight stays	186,850,624	-11.2%	175,198,164	-0.7%	362,048,788	-6.4%

Anno 2011						
Arrivals	56,263,060	2.3%	47,460,809	8.4%	103,723,869	5.0%
Overnight stays	210,420,670	0.0%	176,474,062	6.8%	386,894,732	3.0%

Source: ONT

The Italian model, then, struggles a "system" even in the tourism sector in which it could have a better chance. Paradoxically, the connotation of "thousand Italy, one homeland," typical of this country, is not translatable into consequential policies, despite studies and research highlighting the positive perception of the world about made in Italy and opportunity that this perception could be used effectively in the touristic dimension. However, the “Made in Italy” brand (which refers to the lifestyle, fashion, art and culture, the great industrial brands and the image conveyed by cinema) does not find support in a localist approach.

Data about overnight stays in Italian accommodation in the period 2000–2010 are positive. Despite the lack of coordination of tourism policies, in fact, there has been an increase of 10.8% over the decade, with a record 386,895,000 overnight stays from about 46% of foreigners in 2011.

Data are unfortunately not confirmed for 2012 and 2013 due to a significant decline in the domestic market resulting from the economic crisis. However, if the regional competitiveness derived from an expansion of market share and, therefore, from a growth of Italian tourism as a whole, other considerations should be made. The mutual subtraction seems to be a weakening of the Italian brand without any significant compensative strengthening of regional brands.

As a result, although some Italian regions, more than others, can commit substantial investments in promotional campaigns, these policies' effects are not significant on the national and international markets.

Basilicata, South Italy or Made in Italy?

As stated previously, south Italy is a “potential category,” and in fact is well known as a place of leisure (it is a “temple” of sun, sea, hospitality, gastronomy, cultural and artistic heritage) and sin (organized crime, lack of services and corruption have deep roots in this area). Furthermore, an integrated strategy among the regions of south Italy does not work in a common and co-operative way.

In south Italy, sharing a policy of local development is difficult for many reasons; in fact, there are different administrative units which promote different products without having a common idea of brand equity.

It happens that people who do not well know the reality of south Italy are not able to distinguish, on a global market, the specific characteristics of the “south Italy brand.”

It could be helpful, for this reason, to imagine a territorial and corporate brand made up of different and connected “thematic brands.” These thematic brands have not only specific characteristics, but are also linked with emotions, feelings and theoretical values.

Building a marketing strategy in south Italy offers two solutions: we could promote an unique territorial brand, which is an artificial entity and connects different geographical realities; or we could also promote thematic brands, which are able to develop interregional topics.

Nowadays, the traveller could follow a “thematic way” which doesn’t consider the administrative and specific reality of a place. For example, they could visit castles or choose options of ecotourism, practise sport or embark on adventures. The tourist framework in south Italy offers a lot of options both for local organizers and foreign travellers, and “personalizing” travel is really easy in this area because it is so rich, that our administrative boundaries are not able to pen in a great historical heritage. This idea is particularly true in Basilicata and Lucania, whose boundaries have never changed (for example, in the past the area of “Vulture Alto Brandano” belonged to ancient Apulia, and “Vallo di Diano,” which is now in the administrative boundaries of Salerno, was part of Lucania).

For these reasons, cultural tourism in south Italy must be an interregional activity, and local institutions must promote this great cultural continuum. For example, between the territories of Puglia and Basilicata it is possible to discover King Federico's itinerary, which requires a marketing policy not limited to episodic interventions of promotion.

This new marketing policy has been expressed in some activities of rural development (the "Leader," for example) and in some environmental projects (National and Interregional Parks, "Parco del Pollino" in Calabria and Lucania).

On the other hand, the European Union has often enhanced the interventions on local territory, as Cecil Douglass North (2005, 38) also noticed, underlining that it is still necessary to develop partnerships and co-operation agreements to promote south Italy.

In fact, natural and cultural heritage, sustainable development and marketing policy are keywords in the European projects and guidelines. In these fields the national and regional policy, supported by Structural Funds for Underdeveloped Ares, is necessary linked with the European agenda.

In the perspective of a unique regional policy that is national and European at the same time, the institutional co-operation in the activities of safeguarding and promoting cultural heritage will became an international priority, which must work in all regions of south Italy.

A concrete tool of this strategy is the "Operative Interregional Programme," which is clearly linked with all the programmes promoted by each region in the fields of Culture, Arts, Tourism and Environment. Furthermore, the surplus value of this programme is defining centres of excellence (for example, museums and parks), also building, around them, activities of sustainable tourism and local development which must help the territorial economy.

In fact, the programme is a synergic tool which is naturally linked with the regional agenda of local tourism. It is also aimed at overcoming its geographical and administrative limits to work in an interregional framework and use European, national and local funds.

So, we can consider south Italy as a co-operative and global brand; institutions and private managers should co-operate to develop this great and unique product in the international market of tourism. As stated previously, this synergy is also a great opportunity for the economy of the Mediterranean area of Italy. For example, co-operation is necessary to develop and promote the coast from Taranto to Calabria and Basilicata (the Ionic seaside), which could easily be represented by a unique brand on the international market of tourism. Similarly, the other side of the coast (Mar Tirreno) of Basilicata and Maratea requires the same integrated marketing policy.

Recently, in south Italy there has been a new interest in the development of tourism shown during different meetings organized by regional institutions. As stated recently, "tourism has been changing and it is necessary to revolutionize the approach to its dimension, rethinking the Italian strategy and leadership in the international market of tourism" (Ambrosetti 2007, 257).

On the international market of tourism, Italy (and, in particular, south Italy) needs new strategies and models which should use modern technologies and ways of communication to promote itself; otherwise, on the internal market, Italian regions are adopting marketing policies which are often aggressive, as happened in Campania, Sicilia and Puglia where the promotion of local brand has been very expensive, costing millions of Euros drawn from local funds.

It is also useful to verify if these marketing policies are able to express a real praxis of interregional co-operation, because an abstract model sometimes doesn't fit the reality.

In Basilicata, whose territory could easily find points of convergence with other regions, the strategy of co-operation identifies great advantages for the local economy.

First of all, it is necessary to strengthen the local brand to be more competitive in the national market of tourism, which might have a new acceleration in the future. It is important to emphasize the territorial identity, which is based on nature, history, art, culture and gastronomy, and this activity of promotion could suggest that Basilicata is a centre of excellence in south Italy. Basilicata has another great resource, which is represented by the development of the real estate industry after the earthquake. This resource, which is undervalued nowadays, should be linked with a policy of development in the field of tourist infrastructure.

In Basilicata, tourism could be nourished by the local emotions and atmosphere, which are still uncontaminated in the small, quiet villages, where travellers can enjoy their free time.

Conclusions

An aggregated local offer, geared towards different segments (mass market, independent travellers, low cost tourism, innovative niches, countries and regions of origin of flows) and structured on composite regional realities can be a resource for those regions whose tourism is placed on the edge of the main flows and which fail to urge the driving effect of the resources of excellence.

In Basilicata in particular, development and implementation of an intangible and inter-regional district, based on the achievement of quality and integrated touristic projects, able to meet many needs, can represent an opportunity for redemption from a condition of marginality and a lack of infrastructure.

Those conditions which have held back the promotion and affirmation of Basilicata can instead become a development opportunity. The authenticity of wild places, the absence of stereotypes and genuine nature of the products of this land represent a concrete response to contemporary tourism that is always looking for authenticity, participation and emotion.

In a context of programming and planning of actions it is however necessary to continue the action of macro-segmentation on a geographical basis of the potential market identifying possible targets, outlining product policies and defining the operational phases.

In the short to medium term policies should be put in place focused on a known and accessible target. In the medium to long term a new and creative product addressed to innovative and experimental market segments is needed. So, over a period of more than five years, products that refer to a south Italy brand, or indeed the "Italy" brand, can be offered, identifying new themes of fruition for the global marketplace.

The development of tourism in south Italy requires, however, not only extraordinary and ordinary activities of communication and promotion, but also a new model of governance which could express the local and cultural background.

Nowadays, south Italy still has to cope with decades of inactivity in the public governance and administration, trying at the same time to solve structural problems and enhance institutional programmes.

References

Aaker, D.; Joachimsthaler, E. (2000). Brand Leadership. Milano.

Anholt, S. (2007). Competitive Identity. London: Palgrave Macmillan.

Anholt, S. and GMI (2004). Anholt-GMI Nation Brands Index.

Apt Basilicata, Piano di attività di comunicazione e marketing, 2007, 2008, 2009, 2010, 2011,2012, 2013.

Ashworth, G.; Kavaratzis, M. (ed.), (2010). Towards Effective Place Brand Management: Branding European Cities and Regions, Cheltenham, Edward Elgar Publishing Inc.

Dinnie, K. (2008). Nation Branding: concepts, issues, practice. Oxford: Butterworth-Heinemann.

Dinnie, K. (2011). City Branding: Theory and cases, London: Palgrave Macmillan.
Douglass North C. (2005). Capire il processo di cambiamento economico, Bologna, Il Mulino.
Ferrucci, L. (2000), Strategie competitive e processi di crescita dell'impresa, Milano.
Fiocca, R.; Marino, A.; Testori, M. (2007). Brand management. Valori e relazioni nella gestione della marca. Milano.
Gilmore, F. (2002). A country - can it be repositioned? Spain - the success story of country branding. Journal of Brand Management Journal of Brand Management 9, 281–293.
Gnoth, J. (2002). Leveraging export brands through a tourism destination brand. Journal of Brand Management Journal of Brand Management 9, 262–280.
Govers, R. (2013), International Place Branding Yearbook 2012: Managing Smart Growth and sustainability, London: Palgrave Macmillan.
Gudjonsson, H. (2005). Nation branding, Place Branding, 1:3, 283-298
Hatch, M.J. and Schultz, M. (1997), Relations between organisational culture, identity and image. European Journal of Marketing, 31, 356-365.
Hankinson, G. (2004). Relational network brands: towards a conceptual model of place brands. Journal of Vacation Marketing; Mar 2004; 10, 2.
Hildred, J. (2006). The Saffron City Brand Barometer. Revealing which cities get the brands they deserve. London.
IMD (2005). The World Competitiveness Yearbook 2005. Geneva: The International Institute for Management Development.
Kapferer, J.N. (2004). Strategic Brand Management. London.
Kavaratzis, M.; Hatch, M.J. (2013). The dynamics of place brands: An identity-based approach to place branding theory. Marketing Theory March 1, 13: 69-86.
Kotler, P. and Gertner, D. (2002). Country as brand, product, and beyond: a place marketing and brand management perspective. Journal of Brand Management, 9:4/5, 249-261.
Kotler, P.; Haider, D.H. and Rein, I.J. (1993). Marketing Places: attracting investment, industry, and tourism to cities, states and nations, Free Press, NY.
Krugman, P.R. (1996). Making sense of the competitiveness debate. Oxford Review of Economic Policy.
Lee, K.M.; Cho, D.S. and De Boer, S. (2006). Measuring the performance in nation branding: capability, stakeholders and identity-oriented views. The European International Business Academy (EIBA) 2006 Conference.
Mazzalovo, G.; Chevalier, M. (2003). Pro logo. Le marche come fattori di progresso, Milano
Olins, W. (1999). Trading identities: why countries and companies are taking on each other's roles. London: Foreign Policy Center.
O'Shaughenssy, J. and O'Shaughenssy, N.J. (2000). Testing the nation as a brand: some neglected issues. Journal of Macromarketing, 20:1, 56-64.
Pasquinelli, C. (2010). The Limits of Place Branding for Local Development: The Case of Tuscany and the Arnovalley Brand. Local Economy: The Journal of the Local Economy Policy Unit September 1, 2010 25: 558-572.

Pastore, A.; Vernuccio, M. (2008). Impresa e comunicazione. Principi e strumenti per il management. Milano.

Pratesi, C.A.; Mattia, G. (2006). Branding. Strategia, organizzazione, comunicazione e ricerche per la marca. Milano.

Progetto Marchi d'Area (2005), Analisi comparata delle valutazioni socio economiche delle aree interessate e per un modello di Marchio d'Area. Volume 1.

—. (2007). Relazione finale progetto. Parco Nazionale del Cilento e Vallo di Diano.

Risitano, M.(2008). Il ruolo della marca nella gestione delle destinazioni turistiche, Napoli.

Scott, B.R. and Lodge, G.C. (1985). U.S. competitiveness in the World Economy. Boston: Harvard Business School Press.

Sicca (2000), Prefazione in Napolitano, M.R. (2000), Dal marketing territoriale alla gestione competitiva del territorio, Napoli.

The European House, Ambrosetti, (2007). Sistema Turismo Italia, Proposte per essere vincenti. Milano.

Van Gelder S. (2003), Global brand Strategy, Kogan Page.

Walsh, G. and Wiedmann, K.P. (2008). Branding Germany-managing internal and external country reputation. In: Dinnie, K. (2008). Nation Branding: concepts, issues, practice. Oxford: Butterworth-Heinemann, p.154-158.

Wilkin-Armbrister, E. (2008). Brand Nevis-the role of the financial service sector. In: Dinnie, K. (2008). Nation Branding: concepts, issues, practice. Oxford: Butterworth- Heinemann.

World Economic Forum (1996). Global Competitiveness Report 1996. Geneva: World Economic Forum.

CHAPTER THREE

USING REMOTE SENSING AND GIS TECHNIQUES TO MONITOR THE GEOENVIRONMENTAL CHANGES AT ABYDOS AREA, SOHAG, EGYPT

ELSAYED ZAGHLOUL AND SALWA ELBEIH

Introduction

The application of satellite image analysis to archaeology has emerged alongside other uses, but archaeologists are only now beginning to exploit the broad range of analytical tools available for assessing the satellite image data of the earth's surface and sub-surface more fully. The evolution of satellite image technology is also enabling the manipulation of a greater range of data contained in increasing types of satellite images (e.g. Aster, Corona, Landsat TM, etc.). Archaeologists can now examine a broad spectrum of reflectivity signatures and bands within and between archaeological sites, including both surface and sub-surface features.

One of the historical sites is Abydos at the western bank of El-Baliana city, Sohag Governorate, Egypt. The site of Abydos is huge, and includes a number of ruins and mounds around the edge of the desert. Much of the information we know about early dynastic Egypt comes from the hieroglyphic symbols carved into Abydos's walls. Most visitors, however, stick to the grand Temple of Seti I (Frankfort et al. 1933).

Material and Method

Abydous is one of the most ancient cities of Upper Egypt (Figs. 3.1 and 3.2), located about 11 kilometres west of the Nile at latitude 26° 10' N (near the town of Al-Balyana about 45 km south west of Sohag), and it is considered one of the most important archaeological sites of Ancient Egypt (Fig. 3.3). The sacred city of Abydos was the site of many ancient temples, including Umm El-Qa'ab, a royal necropolis where early pharaohs were entombed, the place in which the sacred head of Osiris was preserved. It is said that at night you can hear the spirits congregating to pay homage to Osiris (Brand Peter 2000), and all spirits must travel to Abydos as part of their journey in the afterlife.

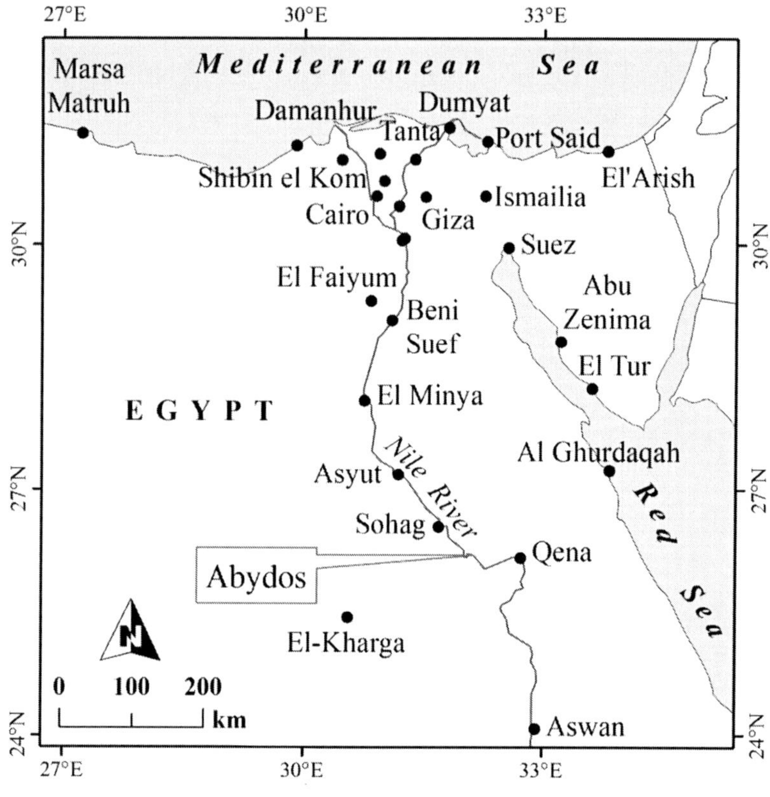

Fig. 3.1. Location map of the study area

- Digital image processing of an ETM+ satellite image (1997) was executed using ENVI 4.7 software. Image processing included image calibration to reflectance, colour enhancement, rectification and sub-setting.
- Digital Elevation Model (DEM): A digital elevation model of the study area was generated from the elevation points and the vector contour lines (using topographic map scale 1:50,000) with an accuracy of 89 %, ESRI ArcMap V.9.3 software

- Landsat ETM+ image and digital elevation model (DEM) were used in ENVI 4.7 software to produce the physiographic map of the studied area.
- The hydrogeological map (scale 1:100,000 RIGW, 1990) was used to determine the groundwater flow and the groundwater characteristics.
- A series of change detection maps was produced based on the variation of the land cover from years 1980s, 1997 and 2009 respectively.

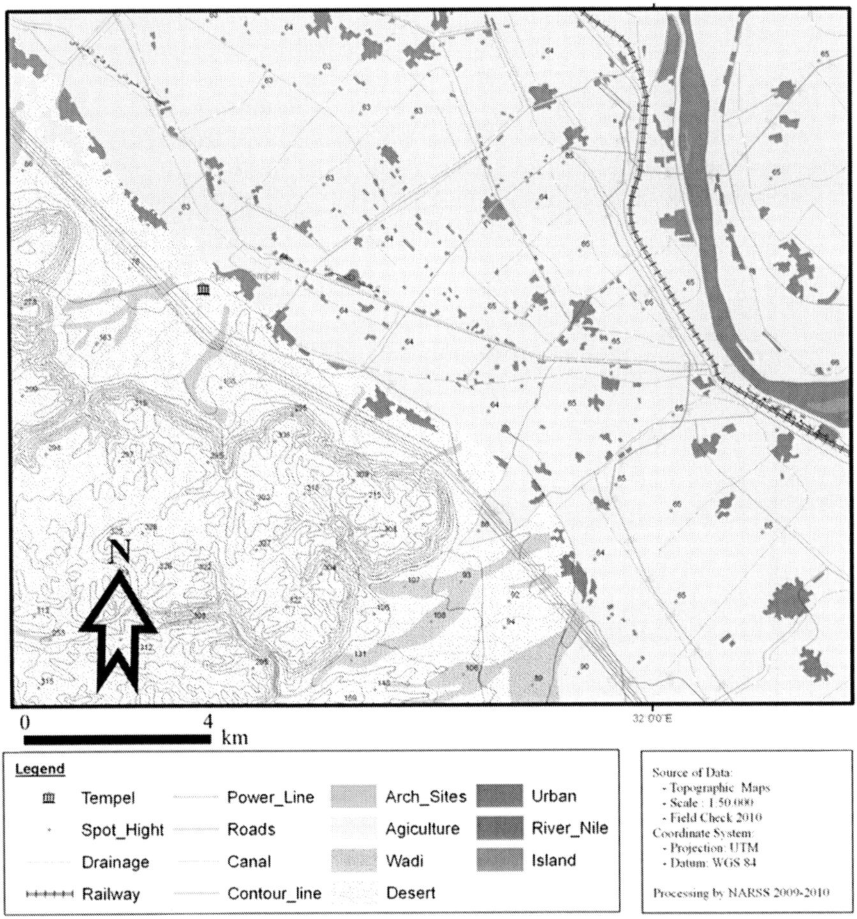

Fig. 3.2. Topographic map

Results

In the area of study, as shown in the location map (Fig. 3.1) and confirmed by field investigations, the limestone scarp that bounds the area from the western side does not form a straight line-scarp, as in most places west and east of the old cultivated lands along the Nile Valley. The scarp in the region forms two embayments, one located west of Nag-Hammadi and the other west of El-Baliana. The two embayments form between a promontory of limestone located near to the temples area (less than 1.0 km to the west). The detailed hydrogeologic investigations show that this limestone promontory acts as a hydrogeologic barrier for groundwater that flows from south to north in the unconfined aquifer. As the Paleocene Esna shale and the Lower Eocene marly limestone in the area are completely impervious (Abdel Ghani & Zaghloul 1973; Said 1981), water returns back towards the east passing through the area of the temple; thus, water rises constantly in the region (Fig. 3.4).

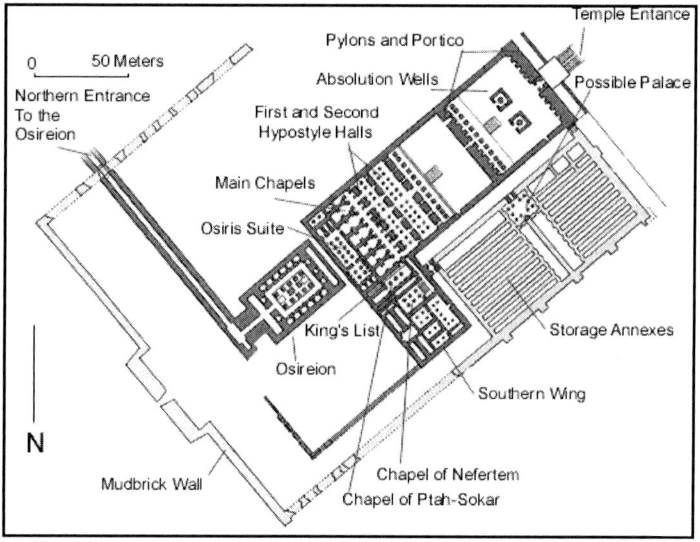

Fig. 3.3. Layout of Abydos Temple and Osirion

It is important to mention that the fluctuation of the Nile water has no direct effect on the groundwater flow in the study area. This is confirmed by measuring the Nile water levels at El-Baliana city and groundwater table levels at the Roman well located at the entrance of Seti-I of the First Temple. As shown in Fig. 3.5 below, there is no hydraulic relation between the River Nile and the groundwater in the temple area. The figure also shows that the difference in head between the Nile water and the ground water table is about 7 m. This excludes the contribution of the Nile water to the groundwater rising in the area. This study also shows that the main irrigation canals in the area control groundwater head in the study area considerably (Fig.3.4). The minor changes in the ground water regime are probably due to local seasonal changes (e.g. irrigation rotation in the old and new lands, as well as the climatic changes).

It is believed that for the safety of the temples it is important to control water seepage towards the area. This can be done by constructing a set of pumping stations in the upstream direction of water flow (south of the temples) in order to control the volume of water that flows to the area from the south. Also, constructing some observation wells to monitor the water level changes in the region will help to better understand the present problem. The number of wells, their locations and the quantity of water needed to be pumped are processed by developing a numerical simulation model for the area.

Discussion

Groundwater Flow: The flow of ground water in the area takes two directions. According to Abdel-Moneim (1992, 1999) the main direction is from the south to the north-north east, following the slope of the topography and the ground surface. The other groundwater flow which dominates in the semi-confined part (cultivated land) is the east west direction. This flow direction is based on the difference in head between the water level in the aquifer and that of the main irrigation canals, drains and the Nile River. In the unconfined part of the aquifer (Nile fringes), the most predominant flow direction is from the southwest to the northeast. Also, water flows from the cultivated land to the fringes (i.e. westward) due to the seepage of subsurface water percolating from the cultivated lands. This understanding is in accordance with the earlier hydrogeological studies carried out in the Nile Valley (Attia 1974; 1985; Barber & Carre 1981; Brooks & Issawi 1993) (Fig. 3.6).

As one of the main purposes of this study is to clarify the groundwater conditions in the area around the temples, a water table map has been constructed and three hydraulic profiles presented by Abdel Moneim (1988) in and around the study area (Figs. 3.5 and 3.6).

The cross- section (C-C') which passes by the temple is given in Fig. 3.7. It shows that the general flow direction is also northeast (i.e. toward the old cultivated lands). Thus, one can say that there is a general trend of groundwater in the opposite direction to the prevailing flow trend with a hydraulic gradient of about 30 cm/100 m.

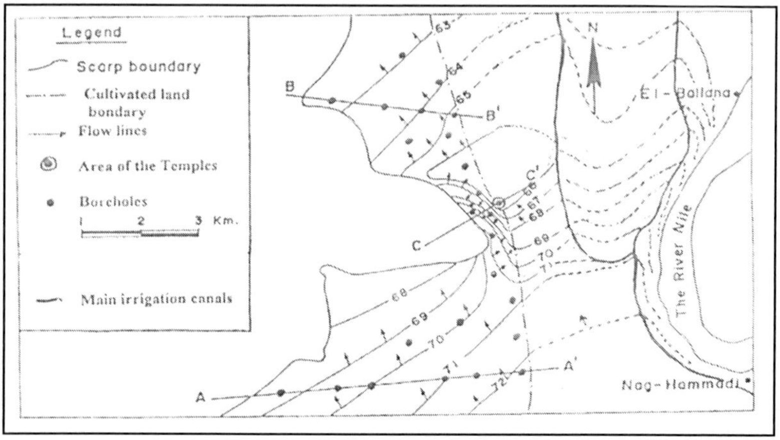

Fig. 3.5. Water table contour map of the area showing the directions of the groundwater flow (after Abdel Moneim 1999).

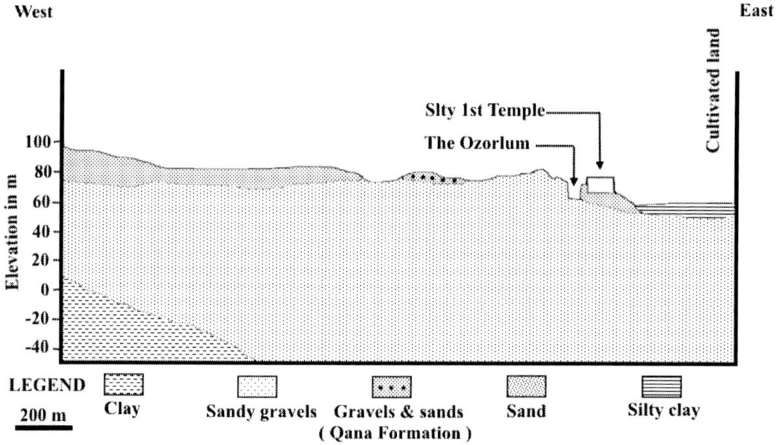

Fig. 3.6. Lithostratigraphic section crossing the Seti-I Temple at Abydos (modified after Brooks & Issawi 1993).

Therefore, a continuous rising of the groundwater table has been recorded around the temples causing serious damage to such historic places. The water table rising and the local reverse flow direction of water in this area are attributed to the following:

The excess water of the drainage agriculture
The sewage water discharging from the urban areas encroaching the archaeological site
New reclaimed areas located south and southwest of the study area.

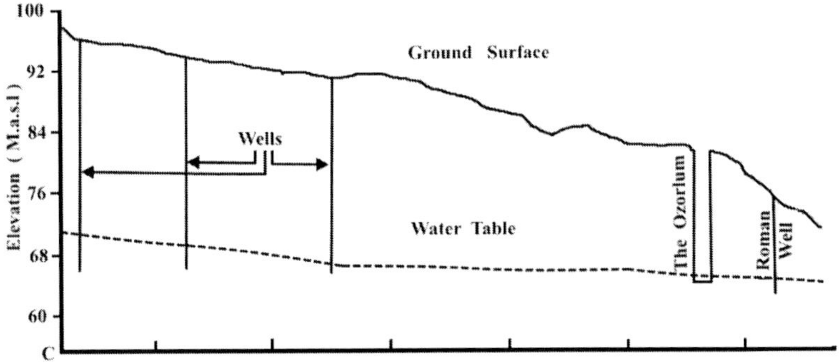

Fig. 3.7. Hydraulic section crossing Sity 1st Temple (after Abdel Moneim 1999)

Fig. 3.8. Osirion groundwater deterioration

Fig. 3.9. Urban encroachment around the temple

Fig. 3.10. Digital Elevation Model (DEM) obtained from ASTER image

Conclusions

Groundwater deterioration: It is worth mentioning that water has never been successfully drained before as it is attached to the water table (Fig. 3.8). The dewatering project carried out before was not effective because:

(1) The temple is located below the present water table.

(2) The groundwater flow is from south and southwest towards the area of the Osirion temple.

(3) The continuous dewatering processes will cause the re-arrangement of the soil particles and this may cause the increasing of the fracture density, fracture widening, tilting of the temple construction and collapse of the foundation blocks.

(4) The groundwater extraction from the temple will increase the groundwater flow towards the temple.

It is believed that, for the safety of the temple, it is important to control water seepage towards the area.

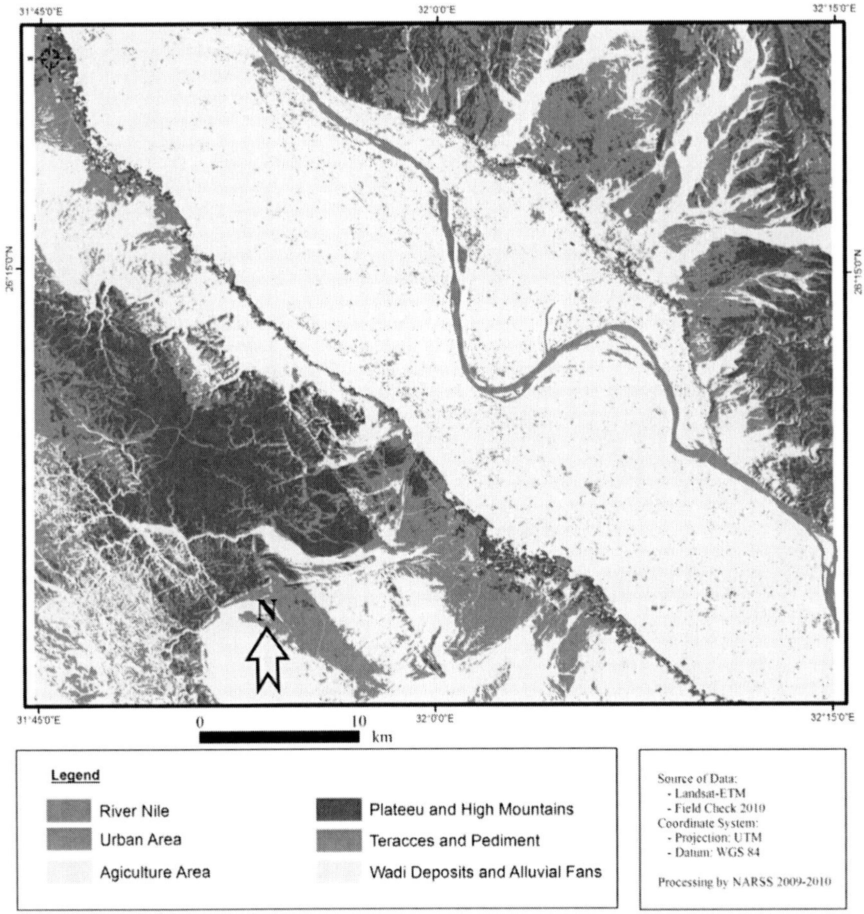

Fig. 3.11. Land covers classification at 1980's

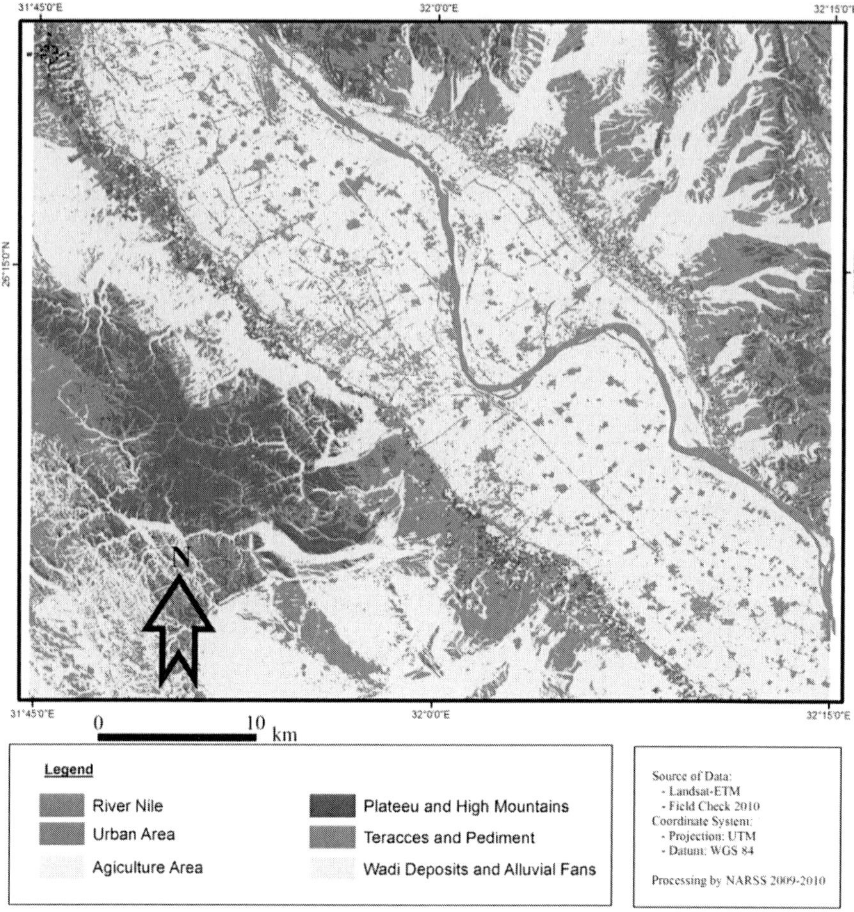

Fig. 3.12. Land Cover Classification at 1997

This can be done by constructing a set of pumping stations in the upstream direction of water flow (south and southeast of the temples) in order to control the volume of water that flows to the area from the south.

Also, constructing some observation wells to monitor the water level changes in the region will help achieve a better understanding of the present problem. The number of wells, their locations (south and northwest of the temple) and the quantity of water needed to be pumped/extracted are being processed by developing a numerical simulation model for the area.

Urban Encroachments: In fact, the unplanned urban encroachments around the temple (Fig. 3.9) have a negative impact on the temple area due to the following reasons:

(1) The sightseeing in the area has become very limited and almost invisible.

(2) There are no sewage systems, so the untreated domestic sewage water drains directly to the ground which causes the increase of the sub-soil water level flowing toward the temple low area.

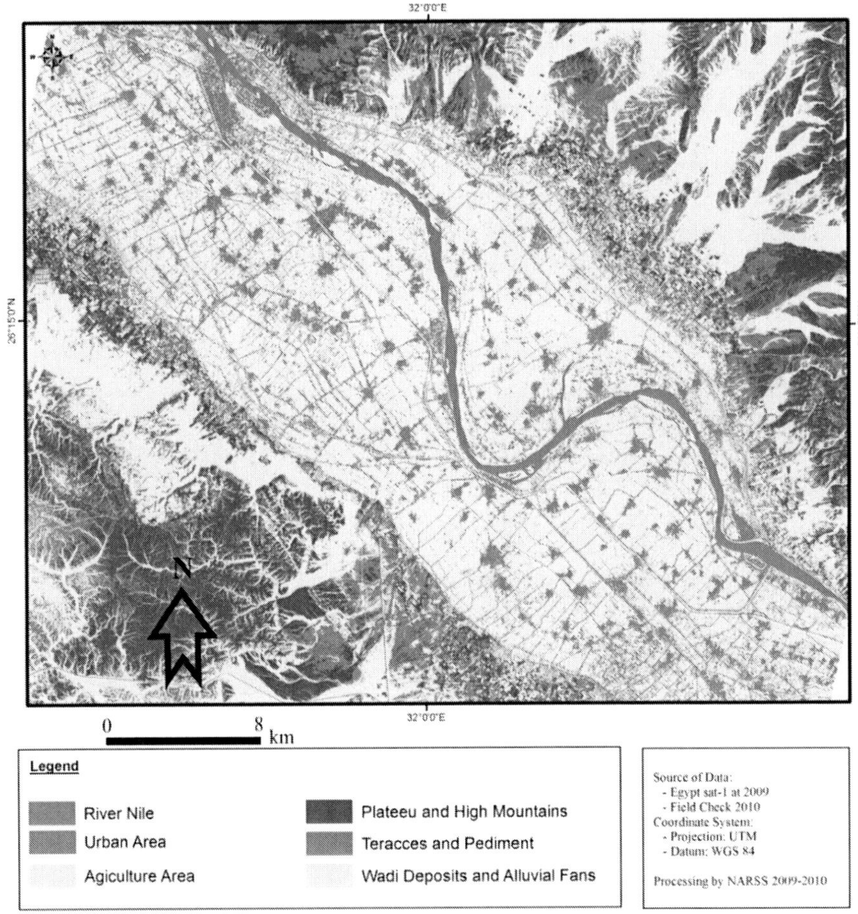

Fig. 3.13. Land cover classification at 2009

Agricultural encroachments: After the construction of Aswan High dam new reclaimed areas were added to the cultivated lands. The farmers use excess water to irrigate the sugarcane crop which consumes more water to grow. The sub-soil water level increase (Fig. 3.8) causes deterioration and salinization of soil around the temple area and causes the aggressive deterioration of the sandstone rock foundation.

Capillary water levels within the foundation stones are sustained at positive elevations with respect to the excavated ground surface. These capillary waters evaporate on the newly exposed surface of the foundation stones. Pressure developed during the crystallization and hydration of these salts promotes exfoliation of the outer layers of the stone foundations allowing more water to infiltrate the blocks and hasten their disintegration by wind and other physical processes (Lewin 1982; Siebert et al. 1984; Duttlinger & Knofel 1993; Wust & Schluuchter 2000).

In the study area, as well as in all temples located at the fringe of the cultivated lands, an effective fundamental engineering work should be carried out to develop methods to prevent deterioration from sub-soil water, wind erosion, urban encroachment and traffic parking impacts to those temples.

Management System: It is believed that the right approach to handling the situation is the application of the concept of "biosphere reserves" as proposed and adopted by the UNESCO Man and Biosphere Program (MAB). This objective can be met by a "Zonation System" that applies different management policies to different zones. In the central, most significant "core" area, human intervention is kept to the minimum.

The core is to be surrounded by one or more "buffer zones" in which supplementary research and training activities may be carried out. The "inner" buffer zone may contain limited research and training facilities that are in harmony with the overall site landscape. In the "outer" buffer zone touristic facilities and some traditional local activities may be allowed. The sizes and shapes of the core and buffer zones may vary widely.

References

Abdel Moneim, A.A. (1988). Hydrogeology of the Nile Valley at Sohag Province. M.Sc. Thesis, Faculty of Science, Sohag, Assiut University, ISO p.

—. (1992). Numerical simulation and groundwater management of the Sohag aquifer, the Nile Valley, Egypt. Ph.D Thesis, Civil Eng. Dept.,Univ. of Strathclyde, Glasgow, U.K. 464 p.

—. (1999). Groundwater studies in and around Abydos Temples, El-Baliana, Sohag, Egypt. Annals Geol. Surv. Egypt. V. XXII, pp. 357-368.

Attia F. (1974). Characteristics of groundwater reservoir in Upper Egypt . M. Sc. Thesis, Faculty of Engineering , Cairo University . 150 p.

—. (1985), Management of water system in Upper Egypt . Ph. D. Thesis Fac. Eng., Cairo University , 207 p.

Barber, W. and D.P. Carre (1981). Water management capabilities of the alluvial aquifer system of the Nile Valley, Upper Egypt . Technical Report No. 11, Ministry of Irrigation , Egypt. 145 p.

Brand Peter, J. (2000). The monument of Seti 1, Epigraphic, historical and historical analysis. ISBN 978- 90 - 04 11770-9 p.

Brooks J. and B. Issawi (1993). Groundwater in Abydos area . Proc. Inter. Conf. 30 years Cooper. Geol. Surv. Egypt. pp. 297-308

Frankfort, H.; De Buck and Gunn, B. (1933). Cenotoph of Seti 1 at Abydos, 2 Vol. Memoir of the Egypt Exploration Society , London

RIGW, (1990). Hydrogeological map of Luxor area, Scale 1:100,000, Explanatory note, p.4.

Said, R. (1981). The geologic evolution of the River Nile; Springer Verlag, New York.

CHAPTER FOUR

SMART CITIES: RANKING THE ITALIAN EXPERIENCE

FRANCA MIANI AND VALENTINA ALBANESE

Introduction

The enormous economic and technological changes caused by globalization have important implications for the way we experience the city. Today, cities are facing the dual challenge of combining competitiveness with a sustainable urban development. There is therefore an urgent need to rethink the city.

The concept of Smart City, if used carefully and concretely, appears to be an appropriate response. There is no single definition of Smart City; instead, there are various perceptions reflecting the time periods of research. The term "smart" was coined by multinationals IBM and CISCO for marketing their products and services for an ideal city using ICT infrastructure as the keystone of urban intelligence. The concept subsequently came to mean a city characterized by widespread and intelligent use of digital technology, made "smart" by accessible and available information.

In reality, this vision is likely to create confusion, as a Smart City embraces much more than technology. Subsequent definitions note that if innovation technologies are not contextualized in a systemic overall vision of the city and its future they will remain fragments of a mosaic of which the design is unreadable (Mochi Sismondi 2012).

In the current literature, the first definition that generated a shift in the perspective was offered by the University of Vienna in 2007. Six main axes were identified: smart economy, smart mobility, smart environment, smart people, smart living and smart governance. Seventy middle-size European cities were ranked along these axes, their rankings based on theories of regional competitiveness, transport and ICT economics, natural resources, human and social capital, quality of life, and participation of societies in cities (Giffinger et al. 2007). This research constituted the first attempt to measure the degree of "smartness" of a city, and to attract the attention of governments and policy makers towards different levers which could be used. In this interpretation, "smartness" is no longer related only to the digital infrastructure, as human and relational capital are also key factors in urban development.

The main objective is to transform the city in order to improve the citizens' quality of life. This means identifying benefits as economic, social and environmental or all of these, which is the task of those who create the vision and who run planning and development policies. The vision is the key to success of Smart Cities, and the city needs to capitalize on its strengths to succeed. Cities that hope to capture long-term benefits from commitment to Smart City development will need to create and effectively communicate an intuitive and compelling vision. As a process necessary for urban revival, Smart City is a project that needs to be shared and developed jointly.

The concept of Smart City maximizes technological applications to improve the functional efficiency of the social, economic and physical systems. A Smart City includes safety, health and social inclusiveness as essential characteristics for a well-designed urban environment. It is also important to involve the community's intelligence for innovation, and essential to take account of the history of a city as a value in the processes of cultural regeneration. The history of each area or city offers natural and cultural opportunities, which can be assets in urban redevelopment (Miani 2011).

Some research points to the role of a creative class, human capital and sustainable urban development in identifying a Smart City. A crucial point for long-term urban development is the role of high-tech and creative industries (Caragliu et al. 2009). This is the central thesis of "The Creative City" by Richard Florida (2002), in which he argues that a successful city needs to invest in creativity and a high level of human skills. Smart City inhabitants need to be able to benefit from and use technology. For Florida, the key strategies of a Smart City are environmental and social sustainability.

Supranational institutions such as the European Commission have also identified the Smart City as a concrete and virtuous solution for future wellbeing and development. For this reason, the Smart City is a key priority in the Digital Agenda for Europe which "aims to reboot Europe's economy and help Europe's citizens and businesses to get the most out of digital technologies. It is the first of seven flagships initiatives under Europe 2020, the EU's strategy to deliver smart sustainable and inclusive growth" (European Commission 2010).

In the numerous definitions of Smart City, which vary according to the point of view and the stakeholder, the only common aspect is environmental sustainability. This is, in fact, a reflection of current trends, as there is increasing global awareness of the importance of efficient use of resources for future generations. It is important in a world where resources are scarce, and where cities are increasingly basing their development and wealth on tourism and natural resources. Exploitation must guarantee the safe and renewable use of natural heritage (Miani 2011).

An alternative approach focuses on the role of social and relational capital in urban development. Here, a Smart City is where the "community has learned to learn, adapt and innovate" (Coe et al. 2001). This approach can emphasize the social inclusion of urban residents in public services and citizen participation in co-design. Quality of life, like other social dimensions, is mentioned explicitly only in academic definitions of Smart City. European definitions are in general more restrictive. They tend to focus on network infrastructure and assume a "wired" model of urban development, where connectivity is considered the main source of growth in the short term. This focus on ICT in a Smart City is a common feature in several definitions, although it is clearly particularly prevalent in the marketing of these products and services. IBM and Siemens, for example, use Smart City as a marketing concept, and conduct dedicated research on the key role of ICT in making a city smarter.

There are increasing international and national initiatives aimed at spreading the concept of the Smart City. At the European level, for example, there is the "Covenant of Mayors" which commits member states to curbing their CO2 emissions by at least 20% by 2020 through the implementation of a Sustainable Energy Action Plan. Launched in January 2008, the Covenant of Mayors now counts 4,480 signatory cities, in other words about 171 million citizens, of whom about 2,000 are in Italy. The main fields of action are clean mobility, energy requalification of public and private buildings, and making citizens aware of energy consumption (eumayors.eu).

Another initiative launched on July 10, 2012 by the European Commission is the "Smart Cities and Communities Initiative." This supports cities and regions taking ambitious measures to progress by 2020 towards a 40% reduction of greenhouse gas emissions through sustainable use and production of energy (European Commission 2012).

In Italy ANCI, the association of Italian municipalities, has promoted a scheme to launch system-wide action to "allow Italian cities to free up their as-yet latent potential for development and innovation." ANCI has also recently created the Smart City National Observatory in collaboration with the Italian MIUR (Minister of Education, Universities and Research) and other ministries (Reggi 2012).

The application of smart solutions to the city is currently attracting a great deal of interest in Italy. Only slightly behind European counterparts, many Italian cities are today approaching the Smart City concept with concrete projects. They are increasingly involved in the reset of core elements of urban life, such as mobility, efficiency, energy saving, building and construction, public lighting, waste and public parks. It is a challenge that Italian mayors are frequently choosing to meet by focusing on innovation and taking an intelligent system revolving around individual needs in respect of the environment as an organizational model—the Smart City of the future.

A true urban revolution is taking place, although in reality there are many small urban revolutions, because small-scale actions tailored to local requirements and characteristics are what gives substance to the concept of Smart City. It is true that increased urbanization means that problems faced by different towns are similar, but there is no standard formula; each city or town has to draw up a plan for its own specific needs and urban social stratification.

Currently, smart intervention tends to be focused on individual areas, and the use of technology is the common feature. Schemes run by towns all over Italy range from the smart management of energy consumption in buildings to innovative mobility solutions. They include tele-heating, tele-cooling, and energy improvement of residential buildings, in cooperation with industry and research. In sustainable mobility, schemes include computerization, traffic congestion and pollution management, facilities for cycling and electric transport. What is still lacking and what cities are striving for is an overall vision of the complexity (Miani 2010).

Smart City rankings: definitions and comparisons

The recent emergence of Smart City rankings is further evidence that the concept is booming in Italy. Although the term has not yet been unequivocally defined, there is broad agreement on indicators and measurements. The emerging rankings may however reflect the need of local authorities for a reference framework to assess progress and/or problems encountered in the transformation towards smartness.

It is well known that the most common way to define a policy problem is to measure it. In fact, the emergence of rankings may indicate a delicate phase in which the Smart City concept is no longer purely theoretical, but also practiced, and to discuss and assess the smartness of Italian cities the rankings may be a useful tool.

This contribution focuses on two rankings: The Smart City Ranking 2012, developed by FORUM PA, and the Quality of Life 2012, developed by the business daily Il Sole 24Ore.

Both rankings use objective indicators, such as economic indicators and subjective indicators, such as social indicators. This is because material well-being, as measured by GDP per head, cannot explain the broader quality of life in a city alone. Subjective indicators measure attributes that are relatively intangible and directly experienced, and which in reality characterize quality of life. It would be inadequate to measure quality of life without measuring more distinctive human attributes.

The Smart City Ranking 2012 was carried out by FORUM PA, a public relations and institutional communications company in collaboration with Bologna Fiere.[1] FORUM PA runs meetings and debates between public administration, companies and citizens on communities, research, multimedia communications and training.

The FORUM PA ranking compared cities on the basis of about one hundred indicators in six basic categories: city governance, economy, mobility, environment, social capital and quality of services.[2] Variables

[1] The Smart City Ranking 2012 was presented at a conference held in Bologna on October 29–31, 2012.

[2] To measure these dimensions, the indicators were: *Smart Mobility*—share of Euro IV and V cars of the total cars, urban public transport in province capitals per 100 square km, seats per km offered by public transport in province capitals, sustainable mobility, cycling, limited traffic zones, parking; *Smart environment*—number of pollutants detected in province capitals, Separate waste collection, density of urban green spaces (percentage of the municipal area), number of collection centres, purification capacity, maximum number of days exceeding the PM10 limit for the protection of human health, difference in volume between the water provided and the amount consumed for domestic, industrial and agricultural; *Smart people*—difference between male and female employment rate, graduates per thousand young people 25–30 years, attractiveness to foreign students, directors in office women/total directors, employment rate, university for older people, attendance at cultural events per 100 thousand inhabitants, readers on an average day per 1000 inhabitants, number of PCs per pupil, associations of social promotion per 100 thousand inhabitants; *Smart living*—capital expenditure per capita on culture and cultural heritage, capital expenditure per capita by function in the social sector, applications for enrolment in public nurseries met/applications, places available in public nurseries/potential demand, book readership, index of social inclusion; *Smart governance*—use of recycled paper in the municipal offices, separate collection within the city, participation in environmental planning, participation in decision-making, public and social services, transparent

that showed limited variation were discarded and a total of 89 variables/indicators, some of which are summary indicators, were identified as useful. The ranking is not an end in itself but reveals the current state of the process. A critical reading can bring out policy considerations useful to better exploiting the potential for optimization of urban systems.

The second ranking, Quality of Life 2012, was carried out by a research team from the newspaper *Il Sole 24Ore*. For over twenty years, the newspaper has evaluated the liveable nature of 107 Italian provinces, producing a series of statistical data and drawing up an annual ranking. It has developed a "quality of life" index based on a unique methodology that links an indicator of subjective life-satisfaction to the objective determinants of quality of life across cities and administrative provinces.[3]

---

governance, political strategies & perspectives, affordable housing schemes, ecological criteria in procurement (GPP—green public procurement) and purchase of fair trade products, energy policies; *Smart economy*—loan/deposit ratio, ISO environmental certification, net enrolment rates in the trader registers, incidence of youth enterprises on the total number of companies registered, incidence of women-owned businesses on the total number of companies registered, municipalities with universities and research institutes per 100 thousand inhabitants, employment rate, unemployment rate, attractiveness to tourists, accommodation per 1000 inhabitants, distance from the nearest airport (km), index endowment of airports, index endowment of the railway network .

[3] The *Il Sole 24Ore* research examined six dimensions, and within each dimension several indicators were identified: *Material wellbeing*—cost of house €/sq m/distance from the centre, per capita expenditure on vehicles/mobile phones etc./ pc, deposits per capita, national index of consumer prices cost of living index, average amount of pension per month in €, GDP per capita in euro; *Job and business*—unemployment rate, share of exports in GDP, entrepreneurs 18–29 years per thousand, registered businesses per 100 inhabitants, loan/deposit ratio, % share of non-performing loans; *Services and environment*—municipal kindergartens in % of users 0–3 years, difference between the hottest and coldest months, index broadband coverage, Index Legambiente Urban Ecosystem, lawsuits processed on new and pending, emigration rate hospital (in %), family life: divorce rate (per 1,000 population); *Population*—Number of inhabitants per square kilometre, divorces and separations every 10 thousand families, 2011 graduates per thousand young people 25–30 years, % share of young population, live births per thousand inhabitants, immigrants in % of population; *Security*—car theft every 100 inhabitants, extortion every 100 inhabitants, burglary every 100 inhabitants, muggings / robberies / pickpocketing / 100 thousand people, total crimes variation trend, fraud and computer fraud every 100 people; *Leisure time*—creativity index, libraries every 100 inhabitants, restaurants and bars every 100 inhabitants, no.

The Smart City ranking 2012
source: www.forumpa.it

TOP TEN	
1°	Bologna
2°	Parma
3°	Trento
4°	Firenze
5°	Milano
6°	Ravenna
7°	Genova
8°	Reggio Emilia
9°	Venezia
10°	Pisa

BOTTOM TEN	
94°	Catania
95°	Catanzaro
96°	Oristano
97°	Brindisi
98°	Agrigento
99°	Vibo Valentia
100°	Siracusa
101°	Caltanissetta
102°	Crotone
103°	Enna

The Quality of Life ranking 2012
source: Il Sole 24Ore

TOP TEN	
1°	Bolzano
2°	Siena
3°	Trento
4°	Rimini
5°	Trieste
6°	Parma
7°	Belluno
8°	Ravenna
9°	Aosta
10°	Bologna

BOTTOM TEN	
98°	Caserta
99°	Palermo
100°	Bari
101°	Foggia
102°	Vibo Valentia
103°	Trapani
104°	Brindisi
105°	Caltanissetta
106°	Napoli
107°	Taranto

Both the Smart City and the Livable City rankings are an attempt to measure "liveability." An assessment of liveability is not straightforward, because there are problems in identifying relevant indicators, measuring them, and using them in comparisons. Data on qualitative or subjective dimensions of liveability are particularly elusive, and there is a tendency to disregard them in favour of firm statistics because they are inaccessible, not because they are unimportant. Moreover, even with objective indicators, "more" may not mean "better," "for the meeting of needs may not lead to the satisfaction of wants" (Ley 1983, 328).

And neither do the rankings represent absolute truths as different indices of course yield different outputs. For example, the smartest city is Bologna, in Emilia-Romagna, and the most liveable province is Bolzano, in Trentino Alto Adige, two regions in the northern part of Italy. On the other hand, at the bottom position for smartness is Enna in Sicily, and for

shows every 100 thousand inhabitants, no. of people taking part in sport, no. of tourists per inhabitant.

quality of life Taranto in Puglia, two regions in the southern part of the country.

Generally, it is interesting to note that liveability and smartness progress negatively from north to south in Italy, confirming that Italy is moving at two different speeds. The northern regions are moving at a fast pace, and the top ten cities in both rankings are located in the north. The bottom ten cities in both rankings are located in the south. This may be a result of of several factors that negatively influence the smartness of cities in the south. The continuous scandal of waste management in Palermo and Naples is evidence of a bad resource management and bad governance. Some of the difficulties of those two cities may also be due to a high rate of criminal infiltration. Another difficulty could be poor management of cultural heritage, which leads to lower participation in the "res publica."

The clear divergence between the north and the south of Italy provides further evidence of how cities are responding to the new wave of smartness. In other words, cities in the north, for example Torino, Genova and Parma, are concretely approaching the new model by implementing projects. But cities in the south are still lacking concreteness, and are responding very slowly, and will still have a long way to go to be considered really smart.

It is also interesting to observe that Emilia Romagna as a region is on the right track for all-round smartness and liveability, with five of its cities (Bologna, Parma, Ravenna, Reggio Emilia and Rimini) in the top ten. This is not by chance, but the outcome of many years' effort and spending in different dimensions.

Bologna is found to be the smartest city in Italy. The town councillor responsible for urban planning commented that:

> the municipality of Bologna has been working in partnership with the University of Bologna, a group of important companies, and research centers of our territory, to achieve the objective of being smart. Our smart vision is that of a city that makes available the maximum level of services and technologies to the citizens. The reason why Bologna is in 1st place in the ranking is because it is one of the cities in Italy with the highest quality of life. Bologna has a high score in different dimensions. At the heart, Smart City represents an idea of urban complexity management, and of all those policies and issues that concern citizens. Success in working on different dimensions in a "good" way proves that our approach is smart and efficient.
>
> (Lepore 2012)

In July 2012, the Mayor of Bologna, the University of Bologna, and the president of ASTER (a consortium between the Region Emilia-

Romagna, Universities, Research institutes operating in the national territory—CNR and ENEA—the Regional Union of Chambers of Commerce and regional business associations), presented the Memorandum of Understanding between the City of Bologna, University of Bologna and ASTER for the establishment of the platform project "Bologna Smart City." For the first time, the City and the University acted as a single point of contact to promote innovation and represent the urban system of Bologna to investors and companies interested in "betting on" the city.

The mayor says that this is an ongoing aspect of Bologna's civic tradition, through an alliance between the world of research and universities, businesses and government agencies to develop useful solutions to address urban and social issues, putting technology at the service of people (www.iperbole2020.comune.bologna.it). The fact that the Quality of Life 2012 ranking places Bologna only tenth confirms that rankings have limitations, and in the case of Bologna there are probably differences between the city and the province.

Conversely, it is interesting to note that Trento is third in both rankings. Evidence of this success are the numerous projects undertaken by the city toward a more liveable/Smart City concerning, for example, waste management and solar panels installed on municipal buildings. There is also a high presence of well-organized urban areas and countryside, a positive feature of its geographic complexity. Another example of its success is a fibre optic backbone of 900 km, which reaches 217 municipalities in the province, complemented by a wireless network (Hiperlan and WiFi) with more than seven hundred access points (comune.trento.it).

Moreover, the rankings show the success of medium-sized cities compared to bigger cities. The medium-sized city is in fact the hub of the territorial organization, where problems of sustainability are better resolved.

In fact, the Quality of Life Ranking 2012 includes none of the big cities or provinces such as Milan, Rome and Naples. This may be due to the greater urban complexities that these metropolises are facing today.

The rankings do not diverge significantly, given that liveability is a common goal. The main problem in the rankings is the selection bias and arbitrariness in the factors selected to assess quality of life/smartness and particularly weighting different indicators. Questions are often raised about the reliability of these rankings. To argue that they lack accuracy and that better proxies would result in more truthful rankings would be to miss the point. It is however interesting to note dynamics from year to year, and identify what factors push cities up or down in the tables.

The fact that two rankings place one and the same city at opposite ends of their rankings (Miani 1988), based on the same qualities, shows how fuzzy the indicators used in the rankings are. It may in fact be that the indexes tell us little about actual reality but a lot about those who measure and order it. Another important fact is that rankings stimulate competition, as they are produced annually or even quarterly to see how the performances of individual players develop. As a result, every city wants to be considered the "most liveable" or the "smartest," a title that can attract new business and investments, boost local economies and real estate markets, and foster community involvement and pride. In other words, such rankings not only create an image of a city, but shape the way cities allocate resources and manage their identities. Free et al. (2008) make the point that this is a form of brand management. As such, these rankings make highly complex and uncertain qualities knowable and comparable. Other authors argue that "the quantification of qualities leads to the commodification of cities: rankings treat complex assemblages such as lifestyle as a good that can be managed and consumed within free markets" (Carter & Kornberger 2010, 342).

Pure quantitative statistics do not add up to the images and symbols which the new economy demands. The new "branding" leads to new qualities being introduced into the global economy and previously uncontrollable qualities become manageable as they relate to each other. This is an important aspect of rankings - they not only transform the qualitative into the quantitative, they also create hierarchical order the dynamics of improving or declining. In short, rankings represent indicators of cities' global standings.

It is interesting to note that a city brand index, the Anholt index (simonanholt.com), already exists. The logic behind this ranking is that there is no choice but to compete - in a global world, cities want to attract the same resources. Thus, the city brand becomes a crucial factor as it links a strong emotional message with rational argumentation. It seems, however, that the Smart City concept is replacing the "Liveable City" idea, and in the same way smart rankings are replacing those on quality of life, which were more fashionable in the 1980s and 1990s (Miani 1988). These aimed to describe the quality of life and the characteristics of cities which made them more or less liveable. Since the 1980s, a major concern with assessing and monitoring national levels of what has variously come to be known as liveability, or the quality of life, or today's smartness, has arisen both in governments and the academic world.

Today, city rankings and particularly quality of life rankings are thought to influence the locational decisions of corporations and city management. As the city theorist Peter Hall underlines: "since the sources of the new economic growth are so various and finally perhaps so fickle, the possibilities are endless. But one central element is quality of life. It is no accident that, as never before, rankings of cities dominate the media" (Hall 1995, 20).

In short, it is literally impossible to render the city knowable, predictable and calculable.

Being Smart: a challenge or a problem for historic Italian cities?

The theme of Smart Cities is a good opportunity for Italy if presented from the right perspective. It would be difficult for models and/or solutions developed elsewhere, in radically different circumstances, to produce the desired result if they were simply transferred to Italy. Solutions adopted in another city or country are not the answer the Italy's problems. By definition, a Smart City cannot be a standardized product to be reproduced anywhere and under the same conditions. It is the result of a delicate equilibrium of factors which, most of the time, are unique and unrepeatable. All that can be done is formulate system-wide guidelines and provide the tools needed. Nor would it be effective to simply respond to European or national calls for proposals as an easy way to alleviate the increasing financial pressure on local authorities. Cities must be able to adopt and adjust solutions that proved successful in another reality, and make them feasible regarding the local context.

In 2007 the Centre of Regional Science (SRF) of Vienna carried out a research project that elaborated a ranking of European Smart Cities. The project dealt with medium-sized cities because the vast majority of the urban population lives in them. This study was the first of its kind to focus on medium-sized cities, thereby considering a broad range of factors and indicators inherent to the concept of Smart Cities. The objective of the ranking was to compare characteristics and to identify strengths and weaknesses of medium-sized cities.[4] In the European dimension, Italy is

[4] In this ranking the Smart City is defined as a city performing well in a forward-looking way in six characteristics, constructed on the "smart" combination of endowments and activities of self-decisive, independent and aware citizens: environment, mobility, governance, economy, people and living. All indicators that describe the factors of a Smart City are derived from public and freely available data. This research project presents a current overview of medium-sized European cities, but an analysis of time-series data is beyond its scope. However, a time-series analysis would be extremely useful, as smart city development is perhaps more concerned with making progress than a city's positioning in a ranking which is inevitably a snapshot in time. Truly, Smart Cities use this city ranking as a tool

still far from the top of the ranking; the first Italian city is Trento in forty-fifth place.

The recent commitment of the government and of the single municipalities is proof of the willingness to catch up with other countries. Dozens of Italian cities like Turin, Genoa, Parma and Bologna have launched or intend to launch smart programs. This, of course, is not enough: the marked fragmentation of Smart City initiatives in Italy is inefficient and means that change will not be complex or broad-based enough to benefit the entire country. There is a lack of correct frameworks for order and organization. At times, it almost appears that governments prefer disorder, and hamper innovative initiatives (Miani 2012). Individual sector-related policies, individual projects and individual applied technologies do not make an area truly smart. Although they are fundamental, technologies, projects and policies will produce results if and only if they are at the service of a common idea that enhances potential.

An additional difficulty for Italy in smart intervention is its structure of small historical towns, rather than big cities. Italy's small towns have their origins in the early medieval period, or even earlier, as ecclesiastic or university centres, defensive strongholds, or administrative centres within regional feudal systems. The importance of Italy's historical centres poses a complex challenge based on the preservation of cultural heritage and, at the same time, on their active contribution to being a smarter place to live (Miani 2010). In a context like Italy it would be absurd to think of transforming towns steeped in a heritage going back millennia into huge futuristic cities. Historical Italian towns represent a cultural and identity heritage of undisputed charm envied by the whole world. The challenge is to valorize the heritage adequately as a strategic resource in resetting the environment and social and cultural development (Granelli 2012) as well as a unique expression of what is "Made in Italy." Even in the current period of economic downturn, small historical towns and villages may represent a competitive advantage and ensure the development of innovative processes.

They in fact provide an opportunity to formulate a distinctive and original concept of Smart City within the context of a "smarter" country. Critical mass and cohesion are essential but can only result from an approach which is both top down and bottom up. Indeed, smartness is likely to be facilitated most effectively by peer-to peer networks and sharing best practices, building at the same time on individual towns'

---

to benchmark with other cities, and draw lessons from better performing cities, perhaps resulting in policy transfer (Giffinger et al. 2007).

experiences, resources, and priorities (Knox & Mayer 2009, 178). The new technologies capable of making today's cities "smarter" are often soft, not invasive (networks, digital, widespread use of electronics, etc.), and are therefore generally appropriate to Italy's delicate urban centres, custodians of historical wealth and beauty.

It is certainly a difficult task. On the one hand there are places and buildings to preserve, and on the other there is a need to upgrade those places and buildings to meet contemporary needs. Smart innovation can make a big contribution to minimizing one of the risks of valorisation of cultural heritage—the enhancement of "a nostalgic celebration of the good old days, which tends to mummify our glorious past. The old must become a bridge to modernity" (Granelli 2011). At present, historic centres are often excluded from technological and innovative development by a strict and complex legislative and planning framework which currently treats them as museums under the guise of preserving the cultural heritage (Miani 2013). There is an urgent need for a new culture of city planning that combines the support and, where applicable, automation of urban processes proposed by the major ICT players, with instruments for the protection, management and enhancement of cultural heritage (Granelli 2011). In general, Italy's problems tend to be due to the absence of a regulatory framework, or due to a framework which is in urgent need of modernization and updating.

The exclusion of medium-sized cities from the European Directive (2002/91/EC) and national energy efficiency policies and obligations also holds them back from any dynamic development and technological innovation. But precisely because historic centres play an important role in contemporary society, they can contribute, for example by reducing carbon emissions and therefore becoming more sustainable and smarter. A big problem is the restrictions on the installation of photovoltaic panels in historical centres. Aesthetic solutions however already exist. New technologies in photovoltaic tiles such as the "integrated photovoltaic system" can help to reduce visual impact and can be an aesthetically neutral or even a visually attractive element in architecture.

Historic centres in Italy have so far been largely the opposite of smart. From the lighting system to street design and decor, from mobility to buildings, everything requires updating and innovation. The limited number of interventions are often disconnected and ineffective. A framework of order and organization is still lacking and it is necessary to take several steps for Italy to be a smarter country. First of all, the country needs to define a vision and a strategy for attaining it. It is fundamental that this be a "national vision" shared by the majority of the population.

Italy needs an organic governance framework with national jurisdiction over the multiple facets of the smart concept (energy, mobility, construction, public health, environment, etc.). Central government must be the first to demonstrate what becoming "smarter" means.

An Italian way of becoming smarter could include capitalizing on the cultural identity of cities, and promoting potential areas of excellence in smart solutions for sectors in which Italy is a world leader, such as tourism, preservation of cultural heritage and the food industry (Miani 2011). A national award scheme for the top five cities could serve to foster cities' efforts, and award-winning cities would have an economic and image return creating a positive tourism flow.

Before launching any new initiatives, completion of still incomplete ones would be a positive strategy. For decades, numerous projects in Italy have shared a common fate. They are undertaken, perhaps with great emphasis, and then abandoned along the way, due to changes in the governing party, bureaucratic rivalry, lack of funds, and so on (Granelli 2011).

To make initiatives more effective, it is important to note potential obstacles in advance. Warning signs need to be identified to facilitate the planning and implementation of actions. Particular obstacles leading to the failure of system-wide projects in Italy are the limited tendency to think in an integrated manner, and the presence of interest groups. The Italian political and cultural context is characterized by a high level of fragmentation into groups with contingent and conflicting goals, and low propensity for mediation.

Finally, the Smart City theme needs to be communicated to the country with clear emphasis on the benefits of the model, or there is a risk that only a small group of people will be concerned. It is urgent and fundamental that a national information campaign about smart cities be planned and implemented. Information is the key to both achieving consensus and spreading the benefits to all (Sanchez Chillon 2012). The information campaign needs to communicate the concept of inclusion clearly and forcefully. The government has a primary role in illustrating inclusiveness, and needs to find the best way of ensuring that everyone feels the need to be part of the project.

Conclusions

The issue of Smart Cities is undoubtedly complex, and the complexity stems from the multifaceted aspects that the term embodies. Today, the term "smart" has become trendy, a word used by specialists to denote the possibility of improving quality of life and services. But a Smart City is nothing new—the ideal cities of the Italian Renaissance were born out of the same motivation as today's smart cities, with harmony and beauty of urban architecture coupled with a farsighted perspective in governmental policy and on daily lives of the civic community (Miani 2009).

Cities have always been at the centre of change, protagonists in the economic and social development of our planet, but never before have they faced the opportunities and challenges they do today. It is clear that simply repeating what has been done in the past will not be enough. The city must be re-thought. Smart cities offer a very promising answer to this need and are an important step in the social and cultural change required. More intelligent and efficient urban systems are not an option; they are an absolutely necessity. From this point of view, Smart City is also an opportunity to re-enable the public administration, revitalizing the resources that trigger innovative processes in the relationship between public-private partnerships (Mochi Sismondi 2012). A Smart City is continuous "work in progress" and should be considered a journey, rather than an overnight transformation.

The narrow view of Smart Cities as places that make better use of ICT is still current, but a more appropriate perspective is a broad, integrated approach to improving the efficiency of city operations, quality of life for its citizens, and developing the local economy.

So the first step requires a shift in thinking and a break from the past. The Smart City concept has a strong social connotation and will revolutionize the way people live in and experience cities. It is unthinkable that a process involving such a major change in the lives of citizens be launched without their being adequately informed, prepared and motivated towards the potential benefits.

The risks are numerous. There is a widespread belief that it will be possible to create models exportable to any town, but Smart Cities are not "one size fits all." The construction of "smart" projects that simply aim to win European or national recognition is not enough. There is a great risk that the term will become simply a concept of local marketing, oriented toward the private sector and the promotion of small interest groups. It is vital to provide a definition of the term Smart City to prevent the idea from becoming a mere slogan, trivializing a concept that in fact represents a social revolution.

A better approach is necessary for the difficult times that local administrations are facing today. The current downturn could be an opportunity to help mobilize resources and find synergies between different strategic actors for the economic and urban development of the city. Until recently, the enormous wealth of cities and smaller centres scattered across Italy has often produced more fragmentation, parochialism, power conflicts than economic innovation. This has created a system of fragmented islets, isolated and not part of a network of innovation. Although local government has been involved in independent initiatives for some time, on a national level Italy has been somewhat behind and its small number of large cities would seem to be a weakness. In reality, however, this presumed weakness could become a strength. Small towns, with their historical centres, comprise a feature that makes Italy unique. They are often outstanding towns that maintain their own identity and a sociable and enjoyable way of life for their inhabitants. They often have a cultural and artistic heritage envied by all, together with the most solid foundations of social capital anywhere in the world, and these aspects should be exploited. In fact, they offer the opportunity to formulate a unique and original concept of Smart City within the context of a "smarter" country. They are essential elements for the competitiveness of the country as a whole with the well-being of its citizens. The Smart City issue needs to be seen as an opportunity that enables the dialogue between modernity and tradition and enhances the cultural heritage of urban centres (Miani 2014).

To maintain and increase positive impacts of smart interventions, small towns need to monitor and evaluate their success. Published rankings are a way to hold individuals and organizations accountable and to monitor progress (Knox & Mayer 2009) and are thus emerging as a tool to measure the smartness of a city. They also reveal the trending topic of the the Smart City phenomenon. It is true, however, that such rankings can be a tool for local marketing strategy to enhance a city's image, rather than serving as a reference tool for smart interventions.

Technologies will undoubtedly help the Italian system towards significant development, but economic, financial, human and creative resources are also required for a Smart City strategy. Urban planning and policy need to analyze new phenomena to correctly interpret a city's complexity and the opportunities it presents.

A Smart City is governance too, and there is a central role for studying local opportunities. The choice of applying innovation, growth and virtuous processes is a crucial decision.

But a Smart City is also, and especially, people, and the involvement of citizens is crucial to success. New tools and applications are making the coordination of collective action less expensive, easier and more effective than ever before, and the new trend in citizenship participation can be leveraged by the emerging technologies. But this will not happen simply by putting technological tools in people's hands. In order to start solving the Smart City puzzle, citizens must be placed in the middle of the board. The intelligent development of cities needs to be seen in human terms, and citizens must be led to self-realize the services they need. The role of city governments is to encourage and enable citizen participation, offering them more opportunities and answers, and listening to their needs. This is important because a potential risk is that if a city makes new technologies available to its citizens, it may widen the digital divide, the gap between those who have access to digital technology and those who do not. This is of particular concern regarding older people.

A further question which may arise is whether a Smart City is a Just City, a city focusing on equity considering diversity and participation so as to foster a better quality of urban life. If cultural diversity is not balanced, it will be difficult to innovate and activate processes of change (Miani 2009).

A Smart City is an opportunity to re-think the contemporary city in an innovative way. City resource management aimed at improving the quality of life and integrated strategic planning are key requirements for sustainable urban development. But if citizens of different generations and cultures are not all included in the project, the whole "castle" of smartness will blow away.

References

Caragliu, A.; Del Bo, C.; Nijkamp, P. (2009). Smart cities in Europe. Series Research Memoranda, 0048 VU University Amsterdam, Faculty of Economics, Business Administration and Econometrics.

Carter, C.; Kornberger, M. (2010). Manufacturing competition: how accounting practices shape strategy making in cities. Accounting, Auditing & Accountability Vol. 23 (3), Emerald Group Publishing Limited, 325-349.

Coe, A.; Paquet, G.; Roy, J. (2001). E-Governance and Smart Communities: A Social Learning Challenge. Social Science Computer Review, 19 (1), 80-93.

European Commission (2010). The European Strategic Energy Technology Plan. Towards a low-carbon future. Luxembourg: Publications Office of the European Union.

—. (2012). Smart cities and communities - European innovation partnership. Brussels: European Commission.

Florida, R. (2002). The Rise of The Creative Class. New York: Basic Book/Perseu.

Free, C.; Salterio, S.; Shearer, T. (2008). The construction of auditability: MBA rankings and assurance in practice. Accounting, Organizations and Society, Vol. 34 (1), 119-40.

Giffinger R.; Fertner, Christian; Kramar, Hans; Kalasek, Robert; Pichler-Milanović, Nataša; Meijers, Evert (2007). Smart Cities - Ranking of European medium-sized cities. Final report.

Granelli, A. (2011). L'anima tecnologica dei centri storici. L'innovazione (ri)disegna la tradizione. In: Fiera di Roma e Camera di Commercio di Roma, Ecopolis. La città sostenibile, March 23-25, Italy: Roma.

—. (2012). Una via italiana alle smart cities. Eidos, 2, 26-30.

Hall, P. (1995). Towards a general urban theory. In: Brotchie J. et al. (Eds). Cities in Competition, Melbourne: Longman, 1-33.

Knox, P.; Mayer, H. (2009). Small Town Sustainability. Germany: Birkhauser.

Lepore, M. (2012). Le Agende Digitali nelle città: le esperienze di Bologna ed Emilia Romagna a confronto con gli altri territori. In: Smart City Exhibition 2012, October 29-31, Bologna.

Ley, D. (1983). A social geography of the city. New York: Harper & Row.

Miani, F. (1988). Le proiezioni territoriali dei concetti di benessere e disagio sociale. In: P.Coppola (Ed.), Soggetti economici e gerarchie territoriali. L'Italia nella transizione, Bologna: Patron, 259-270.

—. (2009). Dalla insostenibilità alla qualità urbana. Politiche urbanistiche e scenari per lo sviluppo territoriale e l'equilibrio sociale. Bollettino Società Geografica Italiana, Serie XIII, vol. II (1), 99-111.

—. (2010). Cultural Landscape and Local Identity for a Sustainable Urban Development: The Italian Historical Heritage. In: Lehmann S., Al Waer H., Al-Qawasmi J. (Eds.), Sustainable Architecture & Urban Development, Amman: The Center for the Study of Architecture in Arab Region, 475-490.

—. (2011). Cultural values and sustainability in European Cities: new strategies for the future of historical urban landscapes. In: UGI Regional Geographic Conference, Chile: Santiago.

—. (2013). Il futuro è nel passato: riflessioni sull'evoluzione della città europea. Bollettino Società Geografica Italiana, Serie XIII, vol. VI (3), (in press).

—. (2014). The lost urban identity. Tradition and fusion for a new sense of place. In R. Efe (Ed.) Town, City an Urbanization in Globalizing World: a Theoretical and Practical Perspective. Sofia: Kliment Ohridski University. (in press).

Mochi Sismondi, C. (2012). Non facciamo diventare la Smart City una moda vuota. Smart Innovation, Forum PA.

Reggi, R. (2012). Il progetto-paese ANCI per le città ad alto potenziale di innovazione. In: Le Smart cities dell'ANCI, February 23, Torino.

Sanchez Chillon, P. (2012). A Smart City overall vision, from infrastructure to the role of citizens. In: Smart City Exhibition 2012, October 29-31. Bologna.

CHAPTER FIVE

BALNEOTOURISM IN BULGARIA

MARIA V. SHISHMANOVA

Introduction

Researches in spa tourism as an economic sector in Bulgaria are an important task of the country, and these rich mineral water resources should be developed as a priority. This is something that requires full coordination between all institutions at various levels of government, as well as contemporary and relevant laws and regulations.

Material and Methods

The study includes an overview of the historical development of the use of mineral waters, characteristics and locations of mineral water springs, characteristics and types of balneological resorts, a review of key documents related to the development of spa tourism, and an assessment of the main organizations involved in spa tourism and its activities. Analyzing the mentioned fields of study, the development of a separate law for spa tourism and its strategy is proposed.

For the purpose of this study we used numerous researches, legislation such as the National Strategic Reference Framework, the draft-law of tourism, The National Strategy for Sustainable Development of Tourism in Bulgaria 2009–2013, the Operational Programme "Regional Development" 2007–2013, and the National Spatial Development Concept of Bulgaria, the main organizations involved in spa tourism and its activities. Also taken into account are the many statements and opinions of specialists from various forums.

The methods used are retrospective analysis, field analysis and synthesis of the results in order to output strategy of mineral waters. The results would be useful for drawing up an adequate policy in using this precious resource to develop spa tourism as a priority economic sector in the country.

Results

Historical development

Bulgaria has had a spa tradition since the time of the Thracians, Slavs and Romans who built flourishing cities near the abundant hot springs, making them spa treatment centres for sports, culture and entertainment. Thus arose Serdica (Sofia), Pautalia (Kyustendil), Augusta (Hisar), Thermopolis (Bourgas mineral baths) Tonzos (Sliven mineral baths) and Germanea (Sapareva bath).

Thracians inhabiting the Bulgarian lands worshipped the spring waters, and the resort activities originated on the basis of the Thracian medicine known as nature healing. Ancient Greeks also widely used Bulgarian mineral springs and worshiped them, as can be seen on their coins. Swimming pools were built near spas, and balneotherapy was used. In Roman times (first to seventh centuries CE) balneological resorts were related to the care of the physical strength of the legions, the first classifications of mineral water were provided, and honoured deities were associated with the cult of water and health. The remains of baths, spa devices, axchlepions, nympheum springs can be found near Kyustendil, Sofia, Hisar, and Sapareva bath. In the Middle Ages, mineral waters were used mainly for hygienic purposes. The increasing interest in health resorts during the twentieth century attracted more foreign tourists.

Nowadays, Bulgaria still has hydrothermal resources, unique in their composition and well known for their healing properties. The climate is healthy and has a curative effect in certain regions of the country. Varied holidays, warm seas, greenery and fresh mountain air, combined with other treatments, have beneficial effects on many diseases, and health centres are located in some of the most picturesque places in Bulgaria.

Balneotourism is one of the main Bulgarian tourist products and the main way to acquire considerable resources such as hot, warm and cold mineral springs and mud, in which the country is very rich. Analyses show that interest in spa tourism in Europe is constantly growing and is largely determined by demographic trends.

Characterization and location of mineral water

There are few countries in Europe that can compete with Bulgaria in the abundance and diversity of thermal waters, with over 600 mineral water sources grouped in about 240 fields. The total capacity is about 270 million litres per day. Over 75% of them are warm and hot, with temperatures from 37 to 101 degrees Celsius, and almost all kinds of mineral waters which can be found worldwide. Chemically, the mineral waters are: carbon-acidic—containing more than 500 mg/l carbon dioxide, hydro-sulphuric—over 10 mg/l hydrogen sulfide, nitrogen—high alkalinity, low salinity and often rich in trace elements and radioactive water (radon). With regard to water temperature, the springs are: cold (hypo)—up to 20 °C; warm (subthermal)—21–37 °C; warm (thermal)—37–42 °C; and hot (hyper)—above 42 °C. Over 80% of the wells in the country have a low total mineralization of solids under 1g/l. Bulgaria has deposits of mud: inorganic (precipitation and spring)—Pomorie, Banya (Karlovo), Shabla, Marikostinovo and others; organic (plant origin)—Kyustendil, Sadovo, etc., and peat (alpin and earthy) (Cras, 2008).

A lower temperature difference is characteristic for thermal springs in the northern part of Bulgaria, where the max temperature is measured at the field of Varshets at 38 °C, and the warm and hot (hyper) waters are in Southern Bulgaria. Hyperthermal mineral waters are mainly in the valley of Struma, Mesta and the northern foot of the Western Rhodopes. The hottest mineral spring in the country is in Sapareva Banya (101.4 °C—superheated water). The country is dominated by deposits with temperatures of 37 °C to 50 °C.

Bulgaria has more than 225 deposits of mineral springs with a total capacity of 5000 l/sec. In terms of South Bulgaria they are 148, and in the North 77. In South Bulgaria natural deposits dominate, while in Northern Bulgaria it is bored water. The total annual exploitation stocks of mineral water are 109 million cubic meters (3.45 l/sec.), only half of which are used for different, mainly industrial purposes.

In South Bulgaria, 82 fields are in the form of springs and 66 are bored, while in North Bulgaria the bored dominate at 57, and only 20 are springs (Cras, 2008).

Characteristics and types balneological resorts

Varied mountains, vast sea coasts, proximity to the Mediterranean Sea and the favourable geographic location of Bulgaria all create excellent bioclimatic conditions for balneotourism. The number of sunny days in Bulgaria is 18–20% higher than that of central, northern and north western Europe, and rainfall during the spring and summer months is lower by 26–65%.

In Bulgaria there are over one hundred spa resorts, staffed with very well trained professionals. A wide range of medical tourism and recreational programs are continually updated, including exotic, unique items, and more advanced treatments. Most of the services are aimed not only at people who need treatment and prevention, but also at healthy people looking for relaxation and to rid themselves of daily stress. Parallel to the treatment and prevention various opportunities for sports are offered, such as water sports, tennis, bicycle sports, jogging, and hiking in the mountain environment.

Analyses show that interest in balneal tourism in Europe is steadily increasing, largely determined by demographic trends, and Bulgaria offers hotels of a high category (more than 3 stars) specializing in a high quality balneology product. National balneological resorts can be found in the mountain area of Sredna gora: Bankia, Hisar, Pavel Banya, Stara Zagora mineral baths, Sliven mineral baths, Gorna Banya; and in the area of Rila-Rhodope: Sapareva bath, Sandanski, Dolna Banya, Momin Prohod, Kostenets, Kyustendil, Velingrad, Narechen, Mihalkovo (near Devin), Haskovo mineral baths Merichleri, Ognyanovo (near Gotse Delchev), Marikostinovo (near Petrich). On the coest of the Black Sea there are resorts such as Albena, Golden Sands, St. Constantine and Elena, Varna, and Sunny Beach. In the Balkan area are Varshets, Shipkovo and Voneshta water (Veliko Tarnovo). Resorts with mud clinics are Pomorie are Tuzlata (near Balchik). Climatic resorts are Sandanski, Tryavna, Teteven, Shipk, Apriltsi, Elena, Kotel and Dryanovo Govedartsi village (Ordinance amending Ordinance № 30 of 2004 for establishing medical standards Physical Medicine and Rehabilitation, SG 85 from 2004)..

The National Strategy for Sustainable Tourism Development meets the objectives of the National Strategic Reference Framework, focusing on the rich and diverse natural and cultural heritage. It takes into account the contribution of tourism to the national growth, seeing tourism as a key element of local development potentials and one of the engines for regional and local development; it calls for strengthening the regional dimension of product development focusing on its role in economic diversification, especially in rural areas, and on the protection and use of natural and cultural heritage (National Strategic Reference Framework 2007 - 2013).

The operational Programme Regional Development 2007–2013 (OPRD) is aimed at the practical implementation of Priority 4 of the National Strategic Reference Framework "Balanced Territorial Development," and outlines the main areas in which Bulgaria will receive funding for its regions from the European Fund for Regional Development.

The objectives of the program are implemented through support of operations included in the five priorities, one of which is Priority 3: Sustainable Tourism Development.

Tourism is considered as one of the most important sectors of the Bulgarian economy, with its contribution to GDP, exports and foreign exchange earnings, as well as job creation demonstrating its potential. The high spatial concentration, however, combined with product uniformity and gaps in destination marketing reduce the chances of the continuation of recent growth in the medium to long term. This trend creates significant environmental, social and economic risks for already overdeveloped resorts.

The priority axis is in line with the Community Strategic Guidelines for Cohesion, where tourism is considered one of the measures for the promotion of economic growth and the preservation of historical and cultural heritage as potentials for tourism development. The guidelines also emphasize the role of tourism in development, especially in rural areas. They require an integrated approach to quality, focusing on consumer satisfaction and based on the economic, social and environmental dimensions of sustainable development.

The specific objective of the priority is to increase the potential for regional tourism development and the marketing of sustainable, diverse, specific tourist products with high added value and increased sectoral contribution to sustainable regional development.

Priority activities are concentrated into three operations:

Operation 3.1: Improvement of tourist attractions and related infrastructure

Operation 3.2: Regional Tourism Product Development and Marketing of Destinations

Operation 3.3: International Tourism Marketing (Operational Programme "Regional Development" 2007 - 2013).

National Spatial Development Concept of Bulgaria (NSDC)

The success of the "tourism" industry and its contribution to the overall progress of the country is closely linked with the need to achieve environmental, socio-cultural and economic sustainability in all aspects of the tourism industry and all territorial levels. Tourism development should recognize and support the identity, culture and interests of indigenous people.

One of the main tasks for the development of a national strategy and policy for tourism development is not yet completed. Decisions are taken on the basis of specific documents for ecotourism, cultural tourism, rural tourism, etc. The formation of cluster structures is limited without taking into account their role in pooling resources and expertise for development and diversification of the tourism product. Promoting the "development of tourism regions to overcome the disproportion in the geography of tourism in the country" can be achieved through proper measures set out in the preparation of regional development plans for the areas of level 2.

Despite the integrated approach to the management of cultural heritage the possible clusters which might include diverse cultural values should be investigated. In this field the following can be identified:

• "Sea" cluster—"Black Sea tourism cluster" range areas: Dobrich, Varna and Bourgas. It is appropriate to develop and offer products integrated with maritime, environmental, cultural and rural tourism.

• "Ski tourism" cluster—including prominent ski resorts: Bansko, Smolyan, Chepelare and Borovets. This cluster will be extended to the area "Vitosha."

• "The three rivers" cluster—the river Tundzha in the north ("Valley of Kings"), the valley of the river Maritsa (Plovdiv, Starosel and Kozi gramadi [Goats cairns], Pazardzhik, Asenovgrad), and the river Arda in south, Perperikon, Tatul.

• "Spa and Wellness" centres cluster—Kyustendil, Velingrad, Varshets, Devin, Pavel Banya, town of Banya, municipality of Banite (Smolyan District), Gorna Banya, Lozenec (Ovcha kupel) and Bankia, mineral baths of Narechen, Burgas, Sliven, Stara Zagora and Haskovo that form a dense network of centres with undeveloped potential.

Despite the fact that all municipalities will be able to obtain financing for the development of tourism, the National Spatial Development Concept of Bulgaria (NSDC) indicates two types: those that need measures to promote the balanced and sustainable development of tourism, and others that need to overcome the harmful effects of tourism expansion.

The NSDC offers plans for tourism development applying to the following requirements:

• At the national level—a reasonable balance between the coast and interior

• At the regional level—more balanced development in different tourist areas, and also in areas of level 2

• At the local level—sdiversification of tourist offer in new "alternative" forms according to resources in the regions and municipalities.

A special policy for tourism development should be given to the development of prevention and treatment activities in territories with the presence of mineral waters, suitable microclimates and products for thalassotherapy (seawater, mud, salt and lime, algae). This allows a chance for the development of centres for spa and medical tourism in the municipalities of Sofia (Bankia, Gorna Banya, Lozenec), Burgas, Pomorie, Stara Zagora, Kyustendil, Sliven, Haskovo, Devin, Velingrad, Sandanski, Hissar, Varshets, Dobrinishte, Separeva Banya, Pavel Banya, etc. In most communities there are favourable conditions for combining prevention and curative actions as well as to diversify the tourism product to make these complexes a competitive tourism market (National Spatial Development Concept of Bulgaria 2013 - 2025). .

National Strategy for Sustainable Development of Tourism in Bulgaria 2009–2013

Health tourism (balneotherapy, SPA, wellness) advantages:

• Increased competitiveness of natural potential

• 142 resorts, 58 of which are balneo, 56 mountain air and 28 sea—the climate is healthy and has a curative effect in certain regions of the country

• Modern spa and spa hotels not only on the Black Sea coast but also inland—Devin, Velingrad, Hissar, Kyustendil, Sandanski, Bansko

• Established traditions in balneotherapy

Health tourism (balneotherapy, SPA, wellness) disadvantages:

• Health tourism in Bulgaria is not displayed as a priority of the state policy on tourism

• Bulgaria did not attend the European health card travel

• Insufficient number of qualified medical and travel personnel

• Legislation relevant to this type of tourism is insufficient, and the existing one needs to be clarified and updated

• Lack of strategy for health tourism, the price of mineral water for medical purposes is very high and it does not stimulate the development of this type of tourism

• Lack of extensive targeted advertising abroad for balneotherapy, spa and wellness products

• Lack of updated and adequate operational statistic information about users of the product (National Strategy for Sustainable Development of Tourism in Bulgaria 2009-2013).

Measures

• Development of a program and an action plan for the development of health tourism tourism in Bulgaria

• Creating opportunities for year-round use of multifunctional resort areas with a complex integration of their resources

• Protection and conservation of natural resources, environmental policy for "healthy living environment" as a guarantee for healing functions of resort-tourist areas

• Regulatory status, construction, equipment and furnishings, services provided and staff requirements of balneology, spa and wellness centres

• Specific requirements regulating the construction, equipment and service providers in the balneology, spa and wellness hotels

• Determine the optimum recreational capacity of the resorts

• Setting standards for the tourist service of balneology, spa and wellness tourism

• Settlement activities and competence of state bodies and local authorities relating to the most effective use of hot and cold mineral waters with proven health and economic characteristics and qualities for organizing various business activities, including for health tourism

• Ensuring the rational use of mineral therapeutic waters to strengthen and restore the health of citizens and to increase public welfare

• Establishing a system of measures for the promotion of hydrothermal resources of the country as a public natural resource and a factor for the social and economic development of the areas rich in mineral waters

• Promote and create conditions for the protection, conservation and improvement of the quality of the hydrothermal potential of the country to the complex socio-economic development of the area, particularly tourism

• Offering health programs as part of the overall tourist entertainment

• Improve the mechanisms of interaction between institutions having expertise in the field of health tourism

• Encouraging employers to invest in tourism in the formation of new professional skills, training and qualifications of personnel in health tourism

• Development and dissemination of information and promotional materials, organizing and participating in seminars, round tables, conferences, exhibitions and other tourism PR, advertising and marketing campaigns and more initiatives dedicated to the development of health tourism.

The new Tourism Act sets specific criteria to be met by the hotels, so they can be classified as a spa, wellness or balneology centres (National Strategy for Sustainable Development of Tourism in Bulgaria 2009-2013).

According to the Law on Tourism 2013, a medical spa centre is a separate or adjoining accommodation centre located in the city or resort area which applies health treatments based on natural healing resources (mineral waters, peloids, climate, etc.). The centre offers a variety of treatment programs that take place in specialized equipment rooms, halls and premises designed to make the diagnosis, treatment, rehabilitation and prevention possible. The centre is a hospital within the meaning of the law on hospitals.

A spa centre is a separate accommodation centre located in the city or resort area applying different procedures, programs and rituals based on natural health resources by applying eastern (breathing, movement, manual and meditative) and aqua methods of influence aimed at treating stress, relaxation, psycho-physical recovery, aesthetic and beauty modelling, consulted and implemented by non-medical personnel with specialized training in the field of spa and wellness. At a spa there can be found, at a minimum, indoor and/or outdoor swimming pools, a gym and a juice bar, and the beauty area has at least two rooms with baths and jacuzzis, Turkish baths, a steam room, a sauna, and a relax room (Tourism Law Prom, 2013).

A wellness centre is a separate accommodation centre located in an urban or resort environment in which the use of water and natural factors is not mandatory. The centre offers a variety of recreations and applies cosmetic and beauty treatments, programs and anti-stress rituals based on a combination of natural healing resources and holistic approaches to achieve physical, emotional, spiritual, intellectual, professional and social well-being, implemented by non-medical personnel with specialized training in the fields of spa and wellness.

A thalassotherapy centre is a separate accommodation centre located in the city or resort area on the beach, which offers programs and rituals involving the use of sea water and/or derivatives of natural products, and/or estuary mud through classic and non-traditional therapeutic methods of influence aimed at restoring mental and physical health, focusing on the beauty of the human body (Tourism Law Prom, 2013).

The main organizations involved in balneotourism and their activities

The Bulgarian Union of Balneology and SPA (BUBSPA) is a non-profit organization that defines itself as an organization operating in the public interest. The Bulgarian Union of Balneology and SPA is a member of the European Spas Association (ESPA) and the International Institute of Medical Hydrology and Climatology (ISMH).

The Memorandum of Cooperation was signed by the Bulgarian Union of Balneology and Spa Tourism (BUBSPA) and the Association of Physical Medicine and Rehabilitation. The memorandum combines common interests and efforts of both organizations for the development of spa tourism in Bulgaria. Another main objective of cooperation is to create preconditions and conditions for public and expert participation in solving problems related to the development of regulations regarding balneological, spa and wellness tourism in Bulgaria, as well as cooperation in the process of constructing a development strategy for mineral water resources in the country (http://tourist.bg/novini/bulgaria/2012-09-02/memorandum-za-razvitie-na-balneo-i-spa-turizma.html).

Both associations will put their joint efforts into the development of common European criteria and standards in the field of tourism and medical spas, turning Bulgaria into a health centre in Europe.

BUBSPA Vision offers:

• Bulgaria to be reborn as a year-round tourist destination based on wellness-balneologic, spa and wellness tourism

• Bulgaria to be transformed into a centre for the prevention and rehabilitation of the population in Europe.

BUBSPA Mission:

• Recognition of Bulgaria in the international market as a year-round tourist destination of balneology and spas

• Revival of traditions from Roman times

• Return of confidence in the quality of the product in Bulgarian and foreign citizens and partners

• The negation of Bulgaria as a "bogus spa" destination.

BUBSPA Values:

• The union is strength

• Together we can achieve more.

BUBSPA Main goals:

To represent the unified interests of balneology, spa and wellness tourism to their development as a priority sector.

To protect the rights of non-government associations and tourism companies (traders according to the commercial law) developing and representing balneology, spa and wellness tourism in Bulgaria.

To promote the conservation, development and effective use of Bulgarian resources of mineral water, mud, climate, etc.

To create conditions for the solidarity and cooperation of relevant public institutions, NGOs and enterprises for the development of balneology, spa and wellness tourism in Bulgaria.

To develop, finance and implement national and regional projects in balneology, spa and wellness tourism.

To liaise with central and local government authorities to improve the regulatory and legal framework in the field of balneology, spa and wellness tourism.

To lobby for a legislative initiative to adopt regulations in accordance with the modern world trends in the industry, protect the interests of the companies working in the field of balneology, spa and wellness tourism and the interests of tourists, the users of balneology, spa and wellness products.

To produce programs for the use of mineral water, healing mud, climate and other natural phenomena in the direction of prevention, treatment, rehabilitation and residential applications, as well as creating bio products for balneo-prevention and restoration of health.

To develop new marketing and advertising products in the field of balneology, spa and wellness tourism aimed at attracting users from the EU and worldwide.

Effectively implementation of strategies and practices of pre-accession funds and other international financial institutions for the development of balneology, spa and wellness tourism in Bulgaria.

The association states the following objectives:

Raising of funds through admission fees, membership fees, sponsorships, donations, etc. to the financial assurance of activities.

Prepare and implement international projects funded from the structural and cohesion funds and international donors.

Give opinions, suggestions and recommendations for the consideration and implementation of investment projects of national importance affecting the tourism industry, including in terms of the planning, urbanization and construction of resort areas and the creation of effective public-private partnerships.

Provide information (advertising and scientific), researches and implementation of new programs, methods and equipment in the field of balneology, spa and wellness tourism.

Cooperate with similar organizations in Europe and worldwide, as well as leading international and foreign educational institutions in the field of tourism.

Arrange congresses for Bulgarian and international participation, scientific and practical conferences, roundtables and other forums, and publish export achievements.

Organize training and qualification in tourism.

Promote legislation and experience of the European Union of Balneology, spa and wellness tourism and services associated with it.

Projects and proposals presented to the competent authorities for the improvement of the legal framework in the field of balneology, spa and wellness tourism.

Cooperate with municipalities, local governments and local authorities with regard to the improvement of existing facilities and development of balneology, spa and wellness tourism, and conservation of mineral waters and environment.

Promote the activities of the association through the Bulgarian and foreign media.

The Association performs the following activities:

Participate in the formulation of government policy and support the implementation of this policy of balneology, spa and wellness tourism.

Protect the interests of balneology, spa and wellness tourism in the formation and implementation of the tax, investment, transport, communication policy and competition and environmental protection.

Provide general representation of the interests of balneology, spa and wellness tourism in the institutions of the European Union and other international bodies, and the promotion of tourism opportunities in the country.

Initiate and/or participate in the development and discussion of strategies, opinions, proposals and draft amendments to the international and domestic regulations in the field of balneology, spa and wellness tourism.

Provide legal, tax procedure, spatial, methodological, statistical and other consulting services to its members.

Participate in the organization of international, national and regional events in the field of balneology, spa and wellness tourism, including events in national tourism advertising.

Consultations and expert evaluation of the characteristics and qualities of mineral waters and other natural resources, assist and participate in the research studies and promote their implementation in practice.

The Bulgarian Association for Balneotourism (BAB) unites interests and acts as an information bridge between all its members: municipalities with natural curative resources, rehabilitation hospitals, rehabilitation centres for recreation, private companies, businessmen, physicians-field spa tourism, architects, engineers and scientists (http://med-spa.kitt.bg/bg/bubspa/cat-10/).

The association's objectives are:

• Collaborate to develop a national strategy for balneotourism

• Perform assessment of training needs in the field of spa tourism and create a training module for staff in this field

• Advertise balneology products

• Organize sightseeing tours in the spa centres

• Motivate business ethics in business subjects and promote fair competition between them

• Assist in the alignment of the legal framework with that of the European Union
• Attract investment for the development of spa tourism
• Exchange scientific information
• Protect the environment, encouraging the conservation and rational use of natural resources of mineral water and mud.

The activity priorities are:

• Collection of data on accommodation and spa centres
• Provide members of the BAB with diverse information about their activities
• Assist in the reduction of tariff charges of mineral water
• Develop and provide standard contracts between the tour operators and balneology products
• Cooperate with public institutions, and national and regional NGOs
• Maintain good relationships with the allied association of the Republic of Serbia and Montenegro
• Participate on the National Tourism Council
• Participate in conferences and discussions on issues of tourism
• Develop a project to compile a program for the sustainable development of spa tourism, creating a model for replication.

Discussions

The lack of strategy for balneotourism is the reason for fragmented work, but such a strategy is now being constructed. The training of personnel takes place which allows hoteliers to participate in the process. It is found that hotels where wellness programs are offered need development and implementation of health programs and vice versa. It is necessary to balance both directions as they dictate the kind of equipment and the staff training.

Analyzing practices in Europe, it could be concluded that thermal waters attract tourists but balneotourism cannot rely on this factor alone because there are mineral waters all over Europe. However, balneotourism combined with cultural, ethno, ecological, religious, rural and hunting activities gains great appeal. Balneotourism can occur as a stimulus for the development of other types of tourism which complement each other. Thus, spa resorts may appear as centres around which tourism develops completely, and not just health tourism (http://brainworkshop.org/index.php?option=com_content&view=article&id=90:2011-03-30-14-21-30&catid=37:tourism&Itemid=59).

Balneotourism is inherently an interdisciplinary branch which means that emerging issues affect many areas of science such as medicine, economics, culture, archeology, ethnicity and religion. It encompasses the problems of education, architecture and urban planning, infrastructure, conservation of water and mud deposits, etc. To solve these problems, many institutions have to work together.

A debate is scheduled for April 13–19, 2013 entitled "Tourism is a Policy" between the leaders of the political parties that will participate in the upcoming elections.

National Board of Tourism (NBT) has proposed a ten-year master plan for the development of tourism in Bulgaria, according to which will result in an additional increase in the share of industry in GDP. Discussions will be organized between all political parties that will have the opportunity to demonstrate their commitment to the sector.

The Institute for Analysis and Assessment in Tourism has said that the draft-law has many flaws. The disadvantages are connected with an overlooked Brussels requirement to create a so-called "guarantee fund," that definitions of "tourist" and "tourist attraction" are not clear, and there are problems with the categorization of accommodation and many others.

The president vetoed the Law on Tourism, as it contains unconstitutional texts.

The tourism sector has never been more divided. Around 159 institutions exist now, which complicates the dialogue between the industry representatives and hinders the development of the sector. Surprisingly, there is not a separate minister for tourism, since this sector forms between 10% and 12% of Bulgaria's GDP.

At this stage of management the government has taken no steps to change the country's tourism, which could transform the country into a desired tourist area, targeting the middle classes of Europe.

The share of the tourism industry is due to the professionals in this sector, and not the efforts of government. For this important economic sector to be successful a long-term program must be created and performed (http://med-spa.kitt.bg/bg/bubspa).

Conclusion

Bulgaria is one of the most favourable countries for recovery and healing, although it gave its leadership to other countries. Bulgaria is the second in Europe (after Ireland) in the wealth and variety of mineral waters and climatic factors. The unique feature of this small country is the combination of sea, mountains and mineral waters, making it a place for relaxation and healing throughout the year. What Bulgaria offers as a complex product cannot be found elsewhere in Europe, comprising mineral waters, peloids and the climate enabling the successful implementation of prevention and rehabilitation of socially significant diseases.

The favourable ecological nature of Bulgaria provides real opportunities for longer seasons and the more efficient use of bioclimatic resources for recreation, treatment and prevention of a number of socially significant diseases. With this potential, Bulgaria can become one of the most popular destinations for people seeking health and recovery after fatigue or stress. In order to make this idea successful, tourism products need to be brought to a whole new level. The Bulgarian balneo-product needs to be included in the global tourist catalogues to aid the development of the new image.

To meet the increased demands of international tourism markets and terms of increased competition, the development of this type of tourism requires substantial investments to build a modern infrastructure, as well as initiatives and a wide range of therapeutic recreational tourism programs.

A law of balneotourism using the experience of many experts with different specialties should be formulated. Nature has provided a treasure that must be well managed, controlled and rationally used, and it is necessary to develop an integrated strategy for mineral resources in the country.

Balneotourism is one of the most promising sectors in the industry, but its responsibilities are blurred between four institutions: the Ministry of Health, the Ministry of Regional Development and Public Works, the Ministry of Environment and Water, and the Ministry of Economy, Energy and Tourism. These departments, with different regulations, govern the use and protection of mineral waters. There is no communication between the various agencies in this industry, particularly the ministries. To solve this problem it is necessary to establish good communication between ministries, associations, agencies, etc. that would allow spa tourism to progress to a higher level as a tourist segment.

The establishment of good relations between the Bulgarian Association of Balneology, the Bulgarian Association of Balneotourism, the National Association of Spa and Wellness Tourism, product tourism organizations and travel agencies will contribute to improving the quality of spa tourism.

The development of tourism clusters of balneotourism and networking between nuclei in balneotourism create enrichment of the tourist product.

References

Cras, Mineral Water (2008). SPA and balneotourism in Bulgaria, Cras Plus Ltd

Tourism Law Prom., SG. 30 of 26.03.2013, effective from 26.03.2013, the National Spatial Development Concept of Bulgaria 2013 - 2025

National Spatial Development Concept of Bulgaria 2013-2025

National Strategic Reference Framework 2007-2013.

National Strategy for Sustainable Development of Tourism in Bulgaria 2009-2013

Operational Programme "Regional Development" 2007 - 2013

Ordinance amending Ordinance № 30 of 2004 for establishing medical standards Physical Medicine and Rehabilitation, SG 85 from 2004

http://brainworkshop.org/index.php?option=com_content&view=article&id=90:2011-03-30-14-21-30&catid=37:tourism&Itemid=59 - Balneotourism-alternative to mass tourism

http://tourist.bg/novini/bulgaria/2012-09-02/memorandum-za-razvitie-na-balneo-i-spa-turizma.html - Memorandum for the development of balneology - and spa tourism

http://med-spa.kitt.bg/bg/bubspa/cat-10/ - Bulgarian Union of Balneology and Spa Tourism is a member of the International Institute of Medical Hydrology and Climatology /ISMH/.

http://med-spa.kitt.bg/bg/bubspa - Bulgaria is one of the most favorable countries for recovery and treatment

CHAPTER SIX

THE ABRUZZO LANDSCAPE CULTURAL HERITAGE: BETWEEN PROTECTION AND ENHANCEMENT

ALEJANDRA MEDA

Introduction

The territory of the Central-Southern Apennines, taken as a "system," displays a natural reality which is strongly marked by human influence in the building of a territory able to meet the needs of the single inhabitant and the whole community concerning the political, economic and social organization.

An in-depth analysis of the landscape shows the quality of the changes and their importance over the centuries, performed by populations who found a place to stay in a territory where the past can be today traced.

The mountainous inner landscape of the Abruzzo region, especially the Gran Sasso mountain range in the province of L'Aquila, represents the environmental conditions and changes produced by mankind; nature and human features combine with each other under different perspectives, leading to the creation of the territory (AA.VV.2001).

The value of history and architecture in the historical centres

The definition of "historical centre" has been widely analyzed in recent years with the aim to identify those values of a community which are to be protected, taking into account its features and traits regarding the urban space, and the morphological, typological architectural decisions developed during the period.

The Italian scholar Roberto Pane stated that "the distinction between poetry and architectural literature is remarkably justified through the acknowledgment of the fact that the performances are those factors which compose the environment of our old towns and contribute to define a special local trait and not the few monuments present."

Fig. 6.1. View of Castel del Monte

Hence, a historical centre is not only composed of artefacts of the utmost monumental value and able to draw everybody's attention, but also by the "minor" building sector as evidence to the different historical perspectives of the place.

The historical centre becomes valuable in terms of history and architecture when it is generally considered to be "monumental." (Montella 2009).

For the minor architectural field the concept is justified if the urban structure is taken as comprising all elements of buildings, roads, places, materials, colours and the communities, along with their own activities, each contributing to the creation of the identity of their old towns.

The preservation of the historical centres is given the significant role to recover the community traits, including the unique features of that type of building and that type of that place. As result, the old town becomes important if it is lived in any aspects. The memory of our historical identity and the art of the ancient architecture are the result of the values integrated in the ancient buildings and as such they are to be passed on through proper and accurate protection strategies.

Significant role is given to the recovery of the community traits, including the unique features of that type of building and that type of that place for the preservation of the old towns. As result the old town becomes important if it is lived in its many aspects. The memory of our historical identity and the art of the ancient architecture are the results of the values integrated in the ancient buildings and as such they are to be passed on through proper and accurate protection strategies.

Instead, the interventions started in the small historical centres often highlight a limited knowledge of the ancient architectural art, since they are mostly performed by workers who apply modern techniques and materials which are unrelated to the local culture.

It is quite easy to come across buildings which, even if they look slightly modified from the outside due to the new living needs, have been significantly changed inside, influencing their identity as evidence of the architectural tradition (AA.VV, 1980). Therefore, the approach to emptying the architectural structures was too over-simplistic, leading to the loss of those elements which were the focal points of the historical and artistic buildings.

In the case study of Castel del Monte, an historical centre in the Province of L'Aquila, a new process has been witnessed in the repossession of the buildings located in the historical town, and this is also due to the modernization of the services provided which also favour tourism during the winter season. As a consequence, the recovery of artefacts has not been logically followed by the transmission of historical and traditional values, resulting in the loss of "local architectural pieces."

Fig. 6.2. View of a vault ("sporto")

Therefore, it is necessary to choose a strategy able to direct the actions towards a substantial and detailed recovery, starting from the knowledge of the old town through a typological and architectural analysis, by continuously comparing the rules of the ancient architecture with the new technological techniques. The latter have to apply advanced and convenient methods to old building patterns which are still evident now (AA.VV, 1974).

The Historical Centre of Castel del Monte: the Study of the Territory

By analyzing the human influence which has been continuously changing the territory, it is possible to trace the history of a community which has tried to adjust the areas to meet the personal needs of the people in relation to the political and economic structure of the society.

The morphological patterns of a territory can be reasonably considered as the combination of the elements which belonged to and altered the territory, some of which still have an influence through human activity (aa.vv 2001).

Nowadays, technology speeds up the changes of the territory, and it is therefore necessary to set out a planning system which takes into account methods to recover specific conditions while exploiting resources.

Through the centuries the historical centre of Castel del Monte experienced suffering and dark events by changing its shape and adjusting to the political, social and economic scenarios.

The economy of the town was mainly based on sheep-farming and the relationship it had with Apulia region through transhumance activity. These features had an influence on the urban and architectural plan, integrated in an area which was not adjustable and where life had to strictly follow its rules. It was a difficult territory to live in due to the morphological features but also due to the inhabitants from different social classes; at the same time the territory was rich in grazing lands and in hard and resistant stones used in building villages, considered the symbols of our historical memory.

The sheep farming particularly "marked" the territory through tracks descending from the mountains to the Adriatic coast and influenced the dislocation of old towns.

The side of the Gran Sasso mountain range in the province of L'Aquila is the unique territory where the community of Castel del Monte lived and lives. The area is within the protected National Park "Gran Sasso and Monti della Laga," which was one of the territories under the control of the baronage Baronia di Carapelle, also encompassing towns such as Santo Stefano di Sessanio and Calascio. Castel del Monte shared the stylistic patterns of its architecture with these mountainous villages.

The villages listed above shared mutual features, such as economies based on sheep farming and breeding of numerous herds. The animals lived in stalls which were not allowed to be built in the village, and therefore the high population densities, adding to the integrity of the buildings next to each other for protection purposes and the steep hillsides of the territory, were the reasons for building tower houses.

This original aspect marks the territory, which is once again influenced by human traits and natural elements (Chiarizia, Gizzi 1987).

The historic centre: the development of the urban core

The first evidence of the historic town Castel del Monte dates back to the eleventh century, when the town was built around the peak of the mountain for protection purposes.

Ricetto (a small fortified area), the name still used to the present day, was composed of two gates (signs of which can still be found) and the internal street was, at that time, the courtyard of the original castle, whose tower is one of the main features of the Church Chiesa Matrice used as a bell tower.

A small population of quite a few inhabitants gave birth to the town which is today identified for its medieval features and which planned "to protect and to welcome families." Therefore, the origins of the town do not belong to a mere act of a squire, who might have built his own palace, since the Ricetto is without doubt the coarsest and widest area with walls which were difficult to climb, and was protected by cylindrical towers.

The analysis of the urban implant, starting from the "castle," emphasizes the growth of the old town along the main streets. The main streets, together with the steep and winding secondary roads, cross through archways and vaults surrounded by the walls, giving a remarkable degree of integrity.

After the eighteenth century building was allowed outside the town walls and, even if the simple building of the stalls for the animals were being constructed, this fact has to be implied as a very first step in overcoming the limits of the walls. According to an old by-law it was forbidden to have grazing animals inside the town, and stalls were therefore built in a designated area called the "pesola."

As social and political conditions changed, the need for accessibility and the vicinity of new communication roads rose.

Between 1890 and 1919 the focal point of the town shifted beyond its walls. The tangible marks for the ongoing modernization process are the construction of the aqueduct with the new fountain on the main square of the town in 1901 and the opening of the new arterial road towards the municipality of L'Aquila.

The town of Castel del Monte significantly developed during the twentieth century, but the period marked the decay of the town life within the walls. Economic changes and new necessities led the population to abandon the old houses and to build new ones in the areas outside.

In the 1960s and 1970s the rediscovery of the old town, considered now more as a combination of properties than a combination of traditional values, encouraged the restoration of the buildings, aiming for better living conditions rather than focussing on the urban structure and historical and architectural heritage.

The enhancement of old town centres has been an issue for about twenty years, and nowadays the town of Castel del Monte, being aware of its valuable historical prosperity, knows how to stop the architectural decay in order that protection will be the means of expression of the new culture (Clementi et al., 2001).

The Restoration of Castel del Monte

The decision undertaken by the town administration to start the restoration process of the historic centre of Castel del Monte comes from the real need to redevelop the town, and finds the drive to restore and enhance the old town in the province of L'Aquila thanks to two recent events.

The first event is related to the initiative undertaken by the National Association of Italian Municipalities (in Italian ANCI) to create a guide called "*I Borghi più Belli d'Italia*" ("The Most Beautiful Villages of Italy"), considering that a valuable urban, architectural and historical treasure is kept in the small towns which are often interconnected and able to form a significant network, attracting international tourism. This network is composed by the territories of Castel del Monte, Calascio and Santo Stefano di Sessanio. Since the town of Castel del Monte has decided to join the initiative, which acknowledges the features for the unique cultural legacy, it is committed to following a number of criteria set out by a commission appointed for the possible acknowledgement of the town as one of "the most beautiful in Italy." In order to "pass" the difficult "exam," the town administration bodies have prepared several plans for the restoration.

The second event, as mentioned before, has to be traced back to the implementation of the natural gas methane to directly heat homes and buildings and which has increasingly improved the quality of life, attracting a greater number of tourists.

The strategy implemented by the town bodies was to operate in two directions at the same time: entrusting planning tasks to draft a pilot project which involved the old town on one hand, and undertaking a serious awareness-raising campaign for those subjects potentially interested in recovery actions on the other.

Indeed, the action performed came from the fact that protection and enhancement actions related to the environmental, historical, cultural and architectural heritage, even for recreational purposes, represent remarkable opportunities to vary the economic activities of the mountainous territory. These interventions have a beneficial influence on the community, deriving from rural tourism which plays a leading role relative to protecting/requalifiying and improving/enhancing the environmental and cultural heritage (Falini, 1986).

The territory taken into account shows the patterns of an economic decay, and it is therefore necessary to recreate a dynamic social context to draw and manage the investments required for tourism development. Hence, for the highest participation of the community and the various field operators, as well as the political and administrative bodies of the territory, Castel del Monte is an integral component, and has led architects and the town administration to inform about the efforts performed.

The exhibition created by the architects was dedicated to the plan for the requalification of the historic centre; this was part of summer cultural events with the participation in the opening of the political bodies of the Regione Abruzzo, the authority of the National Park "Gran Sasso-Monti della Laga" and of the National Association of Italian Municipalities, besides mayors and operators of the old towns which have been the driving forces triggering the long-awaited reflection process.

The exhibition, which was remarkably useful in showing what kind of approach had to be chosen to start an active cooperation among the key actors, has created a deep change and the need for setting out a number of rules coming from the ancient architecture, being able to assess the proper methods to protect the values of the past. It is highly desirable that homeowners and operators, as persons involved in the redevelopment process, become conscious of the historical architectural patterns as well as of those actions and values which are so important to protecting the old town.

The objectives have taken shape in *Manuale di Castel del Monte* (a book written by the author of this chapter and the architect Pietro Caravagiio), which hopes to point out and guide the actions in order that the historical heritage of Castel del Monte will be neither lost nor forgotten.

In order to facilitate the task, the public administration has decided to link a hands-on-laboratory experience to this guide and to fill out the further "pages" of the manual with an ongoing update of contacts, debates and materials.

Identifying the fields on which to undertake actions, setting out the priorities, and defining, carrying out and managing the projects are the steps of a process which require the community and the public administration to come together and confront each other in a place already defined as "Laboratorio del Borgo," similar to other experiences created in several Italian realities.

Inside an old medieval inn of the historic centre called "Matrice," the main topic discussed is the Manuale, and through confrontation not only are valuable action proposals carried out, but the numerous problems the community has to tackle in order to meet specific needs are also addressed and the questions linked to the proper use of the materials are clarified.

The laboratory turns out to be the opportunity to involve homeowners in the restoration project, since they have the chance to carry out surveys through the help of technicians. Here, they can not only enrich the legacy of data already measured, but can also broaden their knowledge of their own building and consider the recovery plan with a modern attitude.

Therefore, the laboratory is the daily management place of each action undertaken in the architectural field, through the help of architects, commercial operators, workers, the administration office and the community of the old town as the irreplaceable key actor.

The Methods: From the Survey to the Project

Guidelines of the pilot project

The small towns bear witness to the history of the province of L'Aquila in the foothills area, which is strongly marked by the events experienced by the population, and represent the result of the skilful integration with the landscape, built with material of the same "colour." The stone, extracted by expert hands from the mountains, has survived over the centuries in the historic centres of the small urban areas.

The beauty of the artefacts, which is too often hidden through restorations of poor quality, lies in the simplicity of those who built them; further conditions to set the rules were climatic, economic and defensive reasons. The question to be debated is "how" to perform the restoration and not whether it is right to change, protect or reuse the buildings. It is beyond any doubt that abandoning an old town is always to be criticized and that the decay of urban centres inevitably leads to a condition of non-development (Casiello, 1990). In particular, the local authorities in charge of the territorial protection have to be truly focused on this scenario in order to set out a number of initiatives which are to be undertaken to control the abandonment of the population from the mountains; therefore, through work and human influence it is possible to protect and enhance the territory as well as the historical, artistic, cultural and social heritage.

Therefore, it is crucial to re-establish the relationship with the territory, and for this purpose the municipality of Castel del Monte decided to start a masterplan through cooperation with the aim of recovering the invaluable historical and artistic heritage of the territory.

The project is composed of two interrelated proposals:

- the urban restoration of a historical centre in order to improve its accessibility
- the architectural recovery of the historical centre.

The goal of the project regarding the urban restoration is to improve its accessibility, from east to west, for public electric motor-driven means of transport. In order to achieve this goal, it is necessary to start from the gateways at the entrance of the town on some slightly sloped roads by removing the few existing steps to make a viable 500 metre road.

Fig. 6.3. Flying over Castel del Monte

If the viability of some roads of the historical centre allows us to reach the innermost areas, the possibility of building parking spaces making some streets accessible is also significant, and it will be also important to build car parks near the bridge by requalifying abandoned areas.

The architectural recovery is focused on the analysis and study of the urban structure "surrounded and protected" by the walls, starting from the oldest area called Rione Ricetto which adjusted naturally to the morphology of the ground. A group of buildings give the roads and the small squares compactness and importance, becoming the real core of the town life.

Castel del Monte is composed of buildings, roads and small squares made from stone, a simple and strong material extracted and skilfully carved by shepherds.

The aim of recovering the historical centre of Castel del Monte implies a strong commitment to rediscovering and enhancing an old town which apparently seems to be abandoned.

Further, the development project will have a significant influence on the financial conditions of the town by focussing the effort on the public sector, leading to a re-launch of Castel del Monte in terms of tourism and culture.

The decision of the enhancement process in that specific area has been undertaken by the administration, based not only on the strategic importance of the town, but also on the willingness of the community and the owners of the buildings overlooking the roads to perform actions on artefacts, which are incentivized through the municipal bodies.

The enhancement of the primary urban structure, such as the restoration of the road and the choice of underground electricity cables, leads to the project of an open-air theatre as planned by the municipal bodies to use the natural spaces to organize cultural events during the summer period.

Recovering one area means firstly planning an integrated and interlinked process between the private and the public sector which will lead to an effective enhancement of the old town by improving accessibility.

To perform an internal intervention means to restore the supporting walls, the façades and all the features belonging to the old architectural pattern, or simply to remove those elements which are not compatible with the original architecture, such as aluminium frames, the plastic canopies and anything else that has nothing to do with the environment.

To recover also means to focus the attention on what was important for the skilful artisans of the past, who were able to draw materials from the surrounding nature without destroying and contributing to the construction of a landscape.

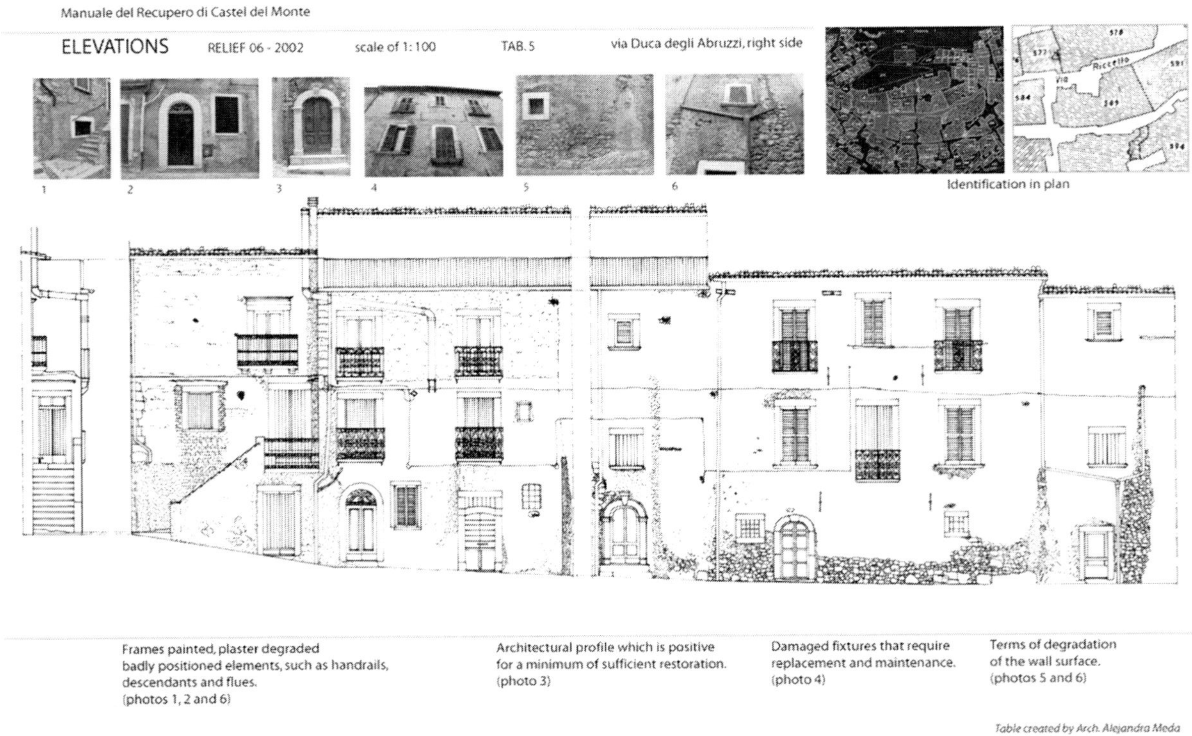

Fig. 6.4. Graphic sheet, photographic survey. Graphic reconstruction of the actual state (Alejandra Meda)

The rediscovery of simple and substantial antique methods capable of shaping the material is significantly important in carrying out t the restoration of the buildings (Civita 1987).

The evaluation: graphic sheets to identify negative architectural outlines

The research highlighted the lack of studies related to the urban structure and the components of the historical architecture; therefore, it was necessary to apply a methodological approach, starting from the classification of the typology pattern of the castle to estimate the understanding of static and developmental principles

Fig. 6.5. Portal with stones

Unlike the simple land registry records available, the complete absence of documents regarding the buildings due to dramatic events, such as the fire of the archive of Castel del Monte set by bandits in the nineteenth century, led to researches taking into account the beauty of the artefacts. The operation undertaken when drafting the masterplan and the manual could not be completed since the old town is spread throughout the area and the buildings are difficult to reach. The concept is finding further existing elements through a recovery step analyzed in the Laboratorio di Borgo. The data collection which was gathered through a direct evaluation and the documents reconstruction allowed for the classification of the architectural elements of the historical centre.

From the identification of negative architectural outlines to project information

Several factors have been set out for the protection of the buildings, such as the lack of accessible roads, the lifestyle of the inhabitants residing in the mountains, and the widespread use of stone as a construction material (Docci and Maestri, 1994).

The recovery project of historical artefacts requires an in-depth analysis aimed at examining their composition under many perspectives, from the structure to their use. As a result, information about traditional and local building and manufacturing techniques is acquired and allows the architect to firstly identify the original elements and secondly to make more "consistent" decisions during the recovery process. The concept of "consistency" in the field of restoration is not simple to be defined depending on the subjective viewpoints of the action, from antithetical positions through surface and radical replacement of the elements comprising the building structure to the strict and intense "standstill" of structures which are no longer able to perform those functions.

Consistent decisions are those undertaken from a "conservative" viewpoint aimed to protect the original typology of the building and its historical and cultural traits, and providing the substitution, if necessary, of the highly decayed elements (Franceschi and Germani, 2003).

The specialist is required to adjust to the different scenarios, even after an in-depth preparatory analysis, which results in continuous changes regarding the operation performed on the building artefact under recovery. The specialist often performs their activity in those historical centres where decay is mainly caused by the abandonment of whole buildings, which decline to an irretrievable point. These are situations where simple and ordinary maintenance works could have protected the building and architectural structures, bearing limited expenses, by preventing large-scale recovery actions under a conservative and economic perspective.

It would be beneficial to perform a conservative "minimal intervention," meaning actions performed in time which are remarkably inexpensive and prevent the danger of alterations and operations which in the past changed and, in some cases, undermined the invaluable cultural importance of the buildings.

The "minimal intervention" necessarily requires knowing which materials are employed and their effects on the building components, and their recovery or their replacement with compatible elements (Marucci, 1991).

Results

Purposes and methodology of the manual

The projects regarding the recovery process in the public and private fields often require not only urban plans but also the willingness and ability of the local body to learn the actual state and to organize the group of technicians by directing public resources to the recovery process based on the integration and co-operation of public and private actions.

The building recovery is comprised of complex actions which require an interlinked series of procedures, methodologies and tools so that the already carried out operation and the operation to be performed comply with the principles of consistency and continuity.

Among the specific tools implemented in the case study are three elements: the typological analysis, the manuals about the place, and a regulation framework based on the real needs of the contemporary inhabitant.

In particular, the manuals are the tools which depict the cultural situation and the technological evolution of the place, providing ideas for an innovative and well-aware project (Giovannetti, 2000; Montagni, 2002; Agostinelli et al., 2002).

The manual on the recovery process of Castel del Monte (the very first manual of the Abruzzo Region), drafted by the group in charge of the project, is identified as a guidance tool outlining a series of behavioural regulations for the building recovery and for the operations ongoing in the historical centre which are subject to conservative procedures aimed to enhance the cultural traits of the architectural heritage.

The duty of the local body, supported by cultural organizations and professionals, is to “guide” and “direct” the private operator. The starting point of the manual is to raise awareness in learning and protecting historical buildings by providing a guidance service for the private sector.

In summary, the aim is to distribute the knowledge of recovery. It is crucial to inform on the series of duties which have been given a more operational role. This has led to the formulation of the identification criteria of the elements and description of the materials and the recurring flaws. As a consequence, it shows the parameters and operational techniques which are consistent with the architectural components for maintenance, reshaping and possible replacement of those components, as well as planning criteria to be followed to improve housing features, and to carry out reinforcing measures in compliance with the historical nature and the existing building philosophy (Amoroso and Camaiti, 2006).

The recovery procedure is in compliance with a regulation framework which is complex and not clear for the group of technicians. Yet, being aware of how to gain financial funding in order to develop old towns, financial benefits as well as clearly knowing the entire process have significantly encouraged the recovery actions. For this purpose, the manual is a means of mediation (Carbonara, 2004).

The main objective of the manual is to revive the interest for the old factory identified as a precious collection of experiences, materials and small artefacts that are only recalled by a few people.

The real recovery knowledge necessarily goes through a cognitive process which allows for the assignation of a historical and cultural value to architectural features.

As it happens on an archaeological site, the manual aims at rediscovering old wooden attics, vaults made with stones and bricks, doors and windows, and detailed hardware objects by reinstating their important roles from the past.

At the same time, the objective of the manual is not to provide a list of solutions; it indicates an approach to solving the problems related to the recovery process, through knowledge and in-depth analysis, in order to acknowledge, value and guide the planning decisions.

In terms of effectiveness, the manual for is important not only for a local territorial aspect with similar features, but can be implemented where the same recovery and protection operations are taking place in old towns.

The establishment of the Laboratorio del Borgo

Based on the experience of the European countries and following the urge to tackle the topic related to the housing quality, the first laboratories were created in the 1980s, launching a new method to deal with problems experienced by the community.

If one compares the experience with the present time, the scenario has changed remarkably because through the analysis of artefacts it is necessary to consider criteria comprising the urban tools and the housing quality standards.

The recovery project, as a tool able to fulfil the housing needs, has to originate from the knowledge and the confrontation of the community, organizational facilities, private operators, artisans, and small companies which operate in that specific territory.

For this purpose, within the masterplan experience of the Abruzzo region, a new meeting and confrontation place was created in 2004, the so-called Laboratorio di Borgo, located in the historical centre, aiming to become a crucial tool for performing effective recovery actions, fulfilling the real needs of the community.

Hence, this strategy gives the historical centre the opportunity to protect and to enhance itself, becoming an important facility which meets the real and changed needs of the community, as it was in the past.

Conclusions

Operational tools and methods implemented and applied to plan the recovery process of the historical centre of Castel del Monte as a case study had to deal with the common problem linked to the low protection and development degree of the historical, cultural and landscape heritage. The recovery process is an essential social and economic resource and it encourages the sustainable development of the community surrounded by an invaluable landscape.

The articulated methodology, supported by in-depth analysis and accurate proposals, shall lead to shared actions. Assumptions and treatments are to be integrated on a multi-dimensional scale, taking into account the ability to assess several development possibilities in terms of products and services provided to meet the needs of the community.

In the recent years the concept of the cultural heritage has increasingly become an important trait in the life of the community by rediscovering and protecting the territory, and by enhancing cultural and historical assets kept within the territory. This concept is really experienced among the communities living in the mountains, where the territory is being abandoned but, at the same time, is the holder of cultural assets implied as environmental and human systems with high quality and historical memory (Fazio, 1977).

The tangible and intangible cultural heritage expresses its ability to encourage activities and prosperity marked through the analysis outcome focused on the protection and the enhancement of the cultural heritage, aiming to functionally recover abandoned architectural buildings which are used as housing or production facilities and especially valued by foreign buyers.

The measures proposed within the masterplan regarding the recovery of Castel del Monte, in compliance with the intrinsic qualities of the artefacts, establish the proper guidelines concerning heritage protection and development management originating from company benefits.

The increasing request for cultural and recreational services encourages the direct and generated value creation, since the local cultural heritage and the historical and artistic geography play important roles for the well-being of the community as well as for the image beyond the town boundaries through new developments.

The objective of an accurate protection and enhancement process is to draw funding from the outside, promote local products, and encourage tourism linked to the cultural features of the town, such as products from the wine and food sector and handcrafted objects.

Furthermore, the significance of the role played by the "culture" in the current economic development of a town or a territory cannot be denied. Professor Xavier Greffe suggests a further step in his document addressed to the French Government: "the cultural attractiveness of the territory shifts the emphasis on the economic value of a culturally creative nation."

References

Agostinelli, M.; Diotallevi, P.; Scoccianti, M. (2002). Manuale del Recupero Urbano della Città di Jesi. DEI, Roma

Amoroso, G.; Camaiti, M. (2006). Scienza dei materiali e restauro. Alinea Editrice.

AA.VV. (2001). Censimento relativo alle emergenze a carattere monumentale ed ambientale nei comuni ricadenti in tutto o in parte all'interno di Parchi naturali regionali o nazionali. Dipartimento della Protezione Civile, Roma

—. (1989). L'immagine della città storica. Elemond, Milano

—. (1974). Risanamento e speculazione dei centri storici. F.Angeli Editore, Milano

—. (1980). Risorse, struttura, forma dell'insediamento minore. Facoltà di Architettura "G. d'Annunzio", Pescara

Carbonara, G. (2004). Trattato di Restauro Architettonico, vol I-II-III-IV. UTET

Casiello, S. (1990). Restauro. Criteri, metodi, esperienze. Electa, Napoli

Civita, M. (1987). Attualità e dialettica del restauro. Chieti

Chiarizia, G.; Gizzi, S. (1987). I centri minori della Provincia di L'Aquila. Regione Abruzzo

Clementi, A.; Redi, F.; Avolio, Cialone G. (2001). La memoria delle pietre.

Docci, M.; Maestri, D. (1994). Manuale di rilevamento architettonico e urbano. Laterza

Falini, P. (1986). Il recupero rinnovato. Edizioni Kappa

Fazio, M. (1977). Il destino dei centri storici. Italia Nostra

Forlani, M.C. Tecnologie locali e costruzioni della casa in Abruzzo. Università "Gabriele d'Annunzio", Chieti

Franceschi, S.; Germani, L. (2003). Manuale operativo per il Restauro Architettonico. DEI, Roma

Giovannetti, F. (2000). Manuale del Recupero di Città di Castello. DEI, Roma

Marucci, G. (1991). I centri storici, Esperienze di recupero e modelli di sviluppo. Atti del Seminario. Camerino

Montagni, C. (2002). Trattato di Restauro Architettonico. Materiali per il restauro e la manutenzione. UTET

Montella, M. (2009). Valore e valorizzazione del patrimonio culturale storico. Mondadori Electa, Milano

CHAPTER SEVEN

INTERNAL MIGRANTS IN KOSOVO AFTER 2000: REASONS AND MOTIVATIONS

MIMOZA DUSHI

Introduction

Migration is a selective phenomenon of mainly individuals, such as men with high levels of education. They have greater opportunities for employment, and hence can more easily take the decision for moving compared to females. However, the situation created in Kosovo after the war initiated migration in groups, where whole families moved together. Taking part in those migrations were people with different socio-economic and demographic backgrounds, and therefore their reasons and motivations vary, and which are the topic for discussion in this chapter.

The transformation process that occurred in Kosovo after the war in 1999 has brought many changes to all aspects of life, including the level and intensity of internal migration, especially toward the capital city Prishtina. The post-war reality found many settlements destroyed and many houses burned, mainly in rural areas. The population did not have the basic opportunities for life, so they begin to seek shelter in other settlements in Kosovo. Prishtina, compared to other cities, had a more ethnically mixed population and many citizens who had left Kosovo for their own countries of origin left behind empty houses which served as shelter for the population who had survived the war. In the beginning they were temporarily housed, but seeing the advantages of living in the city they began to solve the ownership problems and gradually became permanent residents of Prishtina. This was the beginning of internal migration to Prishtina after the war in 1999, continuing with other migrations almost with the same intensity, but mainly for economic reasons.

Internal migration in Kosovo, especially to Prishtina

Until the 1950s the spatial mobility of population in Kosovo was very limited. The largest percentage of the population was rural, which mean that they dealt with farming and livestock. Those economic branches required high numbers of manpower, which dictated living in extended families with many family members and with one common household property. The property of the family in which the agricultural activity was developed was very close to the house and its members did not need to go long distances for the work. These conditions dictated a limited way of living. Furthermore, all biological and socio-economic functions such as production, consumption, reproductive, education and socialization were realized within families, and hence the need for movement was limited even more.

This situation was almost homogeneous over the entire territory of Kosovo. Transformation started after 1960s and 1970s, when industrialization urged the development of urbanization and modernization and influenced changes in the educational levels of the population (Dushi 2009). This new situation brought differences between regions in terms of development, which intensified the movement of populations mainly from rural to urban areas. The opportunities to increase profits outside of households and for income not from agriculture have increased among the population, bringing about the phenomenon of splitting the extended family into small families. In some cases they are divided forever, thus taking part in permanent migration, while in others, due to the inability to provide permanent residences for family members, they participate in daily migration for work. This means that daily migrants are the main actors of urban lifestyle innovation in villages (Islami 2008), which results in a decrease of interest for agriculture among the population in rural areas, thus increasing the interest for settling permanently in urban areas.

Prishtina, as the capital, took on the administrative functions after the Second World War in 1947. At this time Prishtina counted 19,631 inhabitants (15% of the total population of Kosovo). During the 1960s industrialization developed the economic life which allowed many people from different parts of Kosovo to migrate. Within these years the population of Prishtina doubled to 38,593 inhabitants or 18% of the total population of Kosovo. In accordance with this number, the demands of the population were growing. In the city health institutions and a certain number of primary and secondary schools were opened. In continuation with this, in 1970 the University of Prishtina was established as an independent institution, which soon becomes one of the main factors attracting the largest number of population. Migration played the central role in shaping the social and demographic development of the region. The intellectual and economical structure of the population has changed and opportunities for employment in industry and services have increased. Under these circumstances, Prishtina became attractive for many people and noted the first wave of mass migrants. In the census of 1971, Prishtina had 69,514 inhabitants, while in 1981 had 108,083 inhabitants (Islami 2008). According to statistical evaluations, which cannot be considered accurate as they are based on previous records, in 1991 Prishtina had 199,654 inhabitants (SZS 1993), and in early 1999 had about 230,000. This increase of population was followed by the relevant infrastructure where a certain number of collective housings was constructed. However, despite these changes in population structures, the migrants did not manage to change their ways of life. In general, the elements of rural structures were still observed. Immigrants carried the characteristics of traditional cultures and behaviours for some years, and many of them, despite the economic changes, failed to change their lifestyles, cultures of dress, food, housing, etc. for some time (Islami 2008).

The 1990s were fatal for Prishtina. Following the reduction of Kosovo's autonomy by the Serbian government in 1989, a repressive regime was imposed throughout Kosovo and Albanians were largely purged from state industries and institutions. The University of Prishtina was seen by the Serbian regime as a hotbed of Albanian nationalism (Schmit 2012), so lectures were sacked and students expelled. This created a very hard economic and social situation for the Albanian population, which in response formed parallel structures founded by private contributors to provide free services. Even though there were some efforts to survive, dissatisfaction among the population was very high and many Albanian's migrated abroad, especially the young people. Therefore, the population of Prishtina remained static for much of the 1990s, only beginning to change following the war in 1999.

A post-war period brought many economic, social, political and cultural changes in Kosovo. The freedom of movement, at least within the territory of Kosovo, was the most obvious. Internal migration was mainly shaped by rural-urban migration with Prishtina being the main destination. In this time the second wave of mass migration to the city took place and was the biggest in its history, leading to its over-population. According to estimations within the first years after the war around 80,000 immigrants settled there (Islami 2008), mainly from the rural areas of Kosovo. During the war the greatest damages were in the rural areas, where many houses were burned or destroyed. The populations had no basic living conditions and were forced to migrate and move into the vacant houses left by the

Serbs, having emigrated to their countries of origin. Seeing the advantages of urban life and knowing of the impossibility of returning to their settlements, those immigrants quickly began to integrate into urban life. This situation was surely accelerated by the execution of administrative functions and services left in the hands of Albanians and the arrival of a large number of international organizations that opened new placement sites. These things slowly returned Prishtina's desirability, which continues to attracts new residents even today, currently home to approximately 400,000 people (Osmani 2011).

However, this mass migration has resulted in the harsh degradation of the city's physiognomy. Its inability to accommodate the influx of migrants, increasing many times faster than it can house them, has created wild urbanization (Islami 2008). The city itself has increased in area, mainly by creating new residential suburbs. In addition, illegal constructions have emerged in the city while the green spaces have declined and the traffic has become denser during rush hours. These characteristics make life much harder for the residents of Prishtina and the competent body has not yet managed to maintain an urban order. With this situation it is estimated that immigrants are yet to enjoy the urban life despite the improvement of living standards, and their ways of life have not yet changed, sticking to their rural ways. They try to combine different social-cultural norms and values in their lives, and this can be seen during weekends when the city becomes silent as the immigrants visit their relatives in their home towns (Flam&King 2005).

Methodology of Study

Moving from one place to another is an unstoppable process, through which people express dissatisfaction with one district and improve their living conditions in another. Such movements take place for different economic and non-economic reasons and undoubtedly cause accompanying effects. Economic reasons are mostly related to employment, greater profits, improving working conditions and improving quality of life. Meanwhile, non-economic reasons are related to demographic, geographical, social, physiological and political factors. These make migrations quite complex, because if they happen in great volume they can cause demographic and socio-economic disorders and disrupt the population structure according to age and sex, educational and economic structure, etc. At the country level, in the short term, this can cause changes in economic development, and in the long term can cause changes in the reproductive rate of population.

The movement of the population in Kosovo, from rural to urban settlements, has always been present. Initially, people migrated for economic reasons, due to increasing the industrialization at the country level and the increasing demand for qualified workforce, and for educative reasons. However, those migrations that started as temporary, after finishing education and finding a job, became permanent. This form of migration was present until the late 1990s, after which the form of migration, and its reasons and volume, have changed. At this time, due to the luck of basic living conditions, people were pushed to migrate towards urban centres, and this is one of the main reasons for internal migrations right after the war. Those are characterized as mass migrations into urban settlements that caused a reduction of the workforce for farming and livestock. Therefore, the agricultural land altered into lea and the number of livestock fell every day. In the end, this situation influenced the decrease in development of the primary sector of economy at the country level.

To analyze this process and to find the main reasons of these migrants, an empirical research with people who immigrated in Prishtina after 1999 was realized in 2012. Two hundred respondents took part in the research, all participants of the second wave of mass migration to Prishtina. The questionnaire asked a lot of questions about migration, including its causes and consequences. From the analysis of these data we found:

- the form of migration (individual and group) and reasons
- the demographic and socio-economic characteristics of migrants
- the consequences of the migrants' leaving.

Reasons of mass migrations towards Prishtina

Looking first at the time of arrival in Prishtina, according to the research 25.0% of total immigrants came in 1999 and 21.0% in 2000 (Fig. 7.1). This means that the majority of the population migrated right after the war. As a preliminary result, this corresponds with the fact that a considerable number of migrants changed their place of residence because of the destruction of their homes and properties during the war. As mentioned, the Serbian people moved to their country of origin and left behind many empty houses, which served as shelters for Albanians after the war. This is confirmed by answers that respondents gave to the question "why did you decide to migrate?" where 13.0% responded due to their houses being destroyed (burned) during the war, and 41.5% due to better living conditions. In these cases the characteristics of migration is movement in a group, meaning moving the whole family together. Therefore, the results of this question are fully compatible with the following question where 84.5% of respondents proved that they have moved with their family members, spouses and children or parents with siblings.

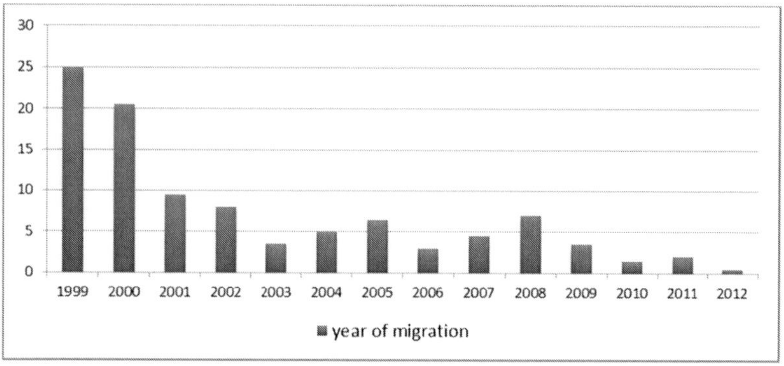

Fig. 7.1. Percentage of immigrants in the city of Prishtina (N=200)

Analyzing the reasons that pushed people to migrate to the city of Prishtina, from the respondents included in the research it is found that 34.0% moved for employment, and 41.5% for better living conditions, which means for economic reasons. The employment rate among them is quite high, and 78.0% were working in the time of the survey. From those who came for economic reasons (34.0%), 17.5% of this group found a job before moving into Prishtina. A second group of 10.5% found a job within their first month, and the smallest group, 6%, within the first year. These results show that migrants have reached their goals in a short period. In addition, there has been a group of migrants employed in their place of origin, but after migration improved their positions in the occupation. This is derived from the answers given to the question "where do you work now and where did you work before?" with 18.0% of migrants working in public administration, and 8.0% in financing, education, trade and construction and other occupations. The variability of employment compared to the time before movement shows improvement, because 51.0% of respondents claimed to have worked in their places of origin, mainly in agriculture, trade, construction and public administration.

With these changes in their lives, more than 70% of migrants have managed to fulfil their needs in life. This means that their lives have were improved after migration. The remaining who have failed to improve their lives and survive in the urban settlements need economic assistance from other family members working abroad. In this way they are living with remittances, and according to them 60% of their value are used for daily consumption.

Another group of migrants in a smaller percentage mention social reasons such as education and family reunion as push factors. It is an undeniable fact that some of the migrants came from some rural places where the network of primary and secondary schools were poor and the chance for education of their children quite limited. Therefore, they assessed moving to Prishtina as representing an increased educational opportunity for their children. Moreover, the University of Prishtina was, until few years ago, the only university in Kosovo, and therefore some families moved to Prishtina to increase the opportunities for their children and their relatives for continuing education and preparing for future occupations. Taking care for relatives shows that there were chain migrations. Initially, some family members moved due to the insufficient economic conditions. Then, after obtaining a residence and improving their living standard, they expressed a tendency for joining other close and distant family members. This is driven by information from relatives who migrated earlier. In these cases young men are more mobile being driven by opportunities for education, employment and exploring a big city through a desire to be free from daily routine and jobs in their small rural settlements, or to be released by social norms. The elderly are the one who do not prefer to move, staying close to their real estate and riches accumulated during their lifetimes. However, the luck of their close relatives, sometimes combined with some attractive elements in urban areas, also drove them to be part of the movement and join their families.

Emigration according to demographic and socio-economic characteristics of migrants

According to data about demographic components of migrants included in the research, the majority were male (59.9%). This result is not surprising because mostly heads of households took part in the research. In the Albanian society in Kosovo, the patriarchal system of society and families dominates, providing the settlement for the entire family and with total authority over the household. Therefore, the role that they have in the family gives them greater responsibilities. This makes them more mobile, compared to females in the house. They are highly educated and easily find a job. These are normally responsibilities that are taken in middle age, when they marry and create their new family. This is demonstrated in this research as well, because the average age of respondents was 43.3 years.
86.6% of them were married and with children. The total fertility rate among immigrant families was 3.2, and the average household size was

5.5 members, which is within the average rate at the country level.[1] We cannot reflect more on family size because it is still too early to talk about changing attitudes toward the number of descendants, and more time needs to pass before the migrants can change their attitudes. Furthermore, the woman have to be involved in economic processes in order to change the way of life, families should be completely separated from lifestyle and the social norms of their places of origin, and the positions of children within families should be improved.

Women, especially in rural areas, are dependent persons in families; therefore, in most cases they are not initiators for movement towards the urban areas. They can take the initiative in cases only for family reunifications (Caro 2011). Men are more mobile, but based on their drives are divided into groups. Those aged 45–54 mostly migrate for employment, and based on the empirical research another reason was that their homes were destroyed during the war. Characteristics of this category are group migration, meaning that most people of this age are married and have children. So, they are the leaders of the families and have to take care for all family members. Men aged 25–34 move mostly for education. They are preparing for their professions, and therefore migrants in this category are mostly individuals.

Socio-economic variables are measured through education, employment and place of origin. In total, 40% resulted with high school and 35 % graduated. It is interesting to see that migrants are faced with demands of life that were not present in their places of origin, such as the demand for skilled manpower. Seeing employment as a basic condition that must be fulfilled to ensure the basics for living and familiarity with the new life environment, a considerable number of them continue education after immigration, often alongside their children, just to increase their chances for employment. Therefore, the employment rate among immigrants it is quite high, with 78.0% stating that they were employed at the time of the survey. In the end, 70.5% of migrants come from rural areas and the rest form urban settlements within the entire territory of Kosovo. Among the immigrants from villages, the dominant reasons for moving were their homes being destroyed in the war and family reunification. Education and employment as reasons for migration were not related to the type of residence, and were present in both urban and rural migrants, almost at the same value.

[1] According to the population census in Kosovo in 2011 the average family size is 5.88 members (ASK 2012).

The consequences of migration

In general, every population movement, depending on volume, leaves behind demographic and socio-economic consequences. In the Albanian society in Kosovo, internal migration in recent years has happened in a very large volume, leaving behind huge consequences in the place of origin and creating new problems in the destination. In the place of origin, the short-term consequences are mainly related to the lack of manpower and neglect of agriculture and livestock, while the long-term effects may appear later with a reduction of the biological reproduction of the population. After some time, the changing attitude towards the number of offspring will be transferred from urban to rural areas through the migratory population. In this way they will deliver less, which will change the population growth rate at the country level (Dushi 2009). On the other side, the flux of migrants into places of destination can cause various problems. Existing infrastructure and social founds are insufficient to support a huge flux of migrants. To fulfil their needs immigrants start building houses and shops without permission, thus disrupting the physiognomy if the city. Furthermore, the traffic capacity of the city is unable to cope with the flux of cars, and therefore becomes very busy. From the social aspect, migrants brought a rural mentality with them, traditional behaviour and culture thus creating a peasant environment. Undoubtedly, these things create dissatisfaction among the urban population.

More or less the same picture of the consequences from migration comes from the empirical research conducted with immigrants in Prishtina. Respondents cite as a primary problem their real estate in their places of origin. According to the results, 78.0% of migrants included in the research moved the whole family. Of them, 60.0% have left houses behind, which are now uninhabited, and 75.0% have left agricultural land; in other words, 82.0 % left behind both houses and land. This real estate in some cases is used by other parts of family members or villagers, while the rest claim that they remained unexploited. All these damages are not the only consequences in emigration places as there is also stagnation in the investments. Immigrants, after abandoning their places, do not invest in economic aspects, such as building new houses, infrastructure, trade or small business. Moreover, the fact that 56.5% of respondents show no interest in the future of their places of origin is disturbing, as they are planning to spend all of their daily activities in the places they are living.

In addition, the consequences could be of a social nature. Level of education and migration have strong correlations. Mostly populations with high levels of education take part in migration because the labour markets in urban areas require skilled manpower, which leads social disorder in the intellectual structure of the population (Gashi&Haxhikadrija 2012). Those with low levels of education remain in the places of origin, whereby the chances for development of these rural areas are further reduced. The population has continued to be involved in the primary sector of economy that is agriculture and livestock. It should be mentioned that those areas are now faced with the problem of teaching staff shortages in their schools and lack of health personnel in clinics. This worsens the situation of the young generation and pushes the remaining population to migrate towards urban settlements.

Migrations, whether internal or external, small or large, also destabilize the situation in the destination. The mass migrations in Kosovo were from rural to urban areas, and this population has brought its traditional culture of living, housing, dress, food, etc. and created a great mass in the urban areas (Islami 2008). Firstly, according to the research, 50.5% of respondents, even after migration, preferred to continue living in extended families, because of the economic and social security. However, this concept runs contrary with urban culture and its way of life. The livelihoods of a large number of members cause frustration to the family themselves and to the neighbourhoods. With this high density of population the public order is disrupted, the public institutions are over charged and the quality of life is diminished. Secondly, to fulfil their residential needs this population started building and overbuilding without planning permission, thus ruining the aesthetics of the city. Thirdly, lack of economic development at the country level and insufficient opportunities for employment in Prishtina have caused high unemployment among the population. This has created many other social problems. Younger people fall into illegal work, delinquency, stealing and robbery, thus making the capital an insecure place for life. All these problems drove us to the conclusion that we need to pass at least one generation until these immigrants become familiar with urban life and culture.

Conclusion

Migrations are a permanent phenomenon of human evolution. A certain type of migration, initiated by different causes, characterizes each period of development. They are mainly conditioned by different levels of economic and social development between places. The higher the difference of development between places, the more mobile the population will be in spatial terms. In this case, the reasons pushing people to migrate are mainly economic, such as employment, higher wages, higher standards of living, and improvement of housing conditions.

In all stages of migration within Kosovo, migrations were mainly pushed by economic reasons, even though they were sometimes pushed by political reasons as well. Industrial development of the 1960s and educational development in the 1970s increased the spatial mobility of the population, especially in terms of larger urban centres where the industry and educational institutions were concentrated. At this time the first wave of internal mass migration to Prishtina was recorded. A similar situation continued until the 1990s and the annexation of Kosovo, when the University of Prishtina was closed and many Albanian employers were purged from state industries and institutions. The hard economic situation pushed many residents to participate in external migration. Then political changes that occurred in the country at the beginning of the twenty-first century brought the largest flux of migrants to Prishtina. This was a second wave of internal mass migration, which for a short time changed the demographic structures of the population.

To analyze changes in population structures, an empirical research was carried out in 2012 with 200 immigrants of the second wave of mass migration to Prishtina. From the preliminary results we can conclude that these migrations were group migrations, concerning the whole family. This means that their demographic and socio-economic backgrounds were different, and their reasons were therefore not the same. Firstly, most of the migrants came right after the war, which means that their homes in rural areas were destroyed. Secondly, migrants come for economic reasons, such as greater employment opportunities, improved occupations, possibilities of more money and improving the quality of life. Among the social reasons are education and family reunification, and the greater opportunities for the education of children have forced many families to leave their places of residence.

Certainly, those migrations have left behind many consequences and created problems in the new residences. The former are economic reasons following the short time of mass migration, and the demographic consequences have not begun to be felt to a large amount. For now, the most visible is disorder in population structures, age and sex and their impact on economic activities at the country level. The population in rural areas was mostly engaged in agriculture and livestock, and after migration into urban areas land remained unworked. The problems that are created in urban places are wild construction and unplanned residential buildings, increase in the level of unemployment and the maintenance of rural ways of life and traditional cultures. These things have undoubtedly brought tediousness to the local population in Prishtina.

References

ASK-Agjensia e Statistikave të Kosovës. (2012). Regjistrimi i popullasisë, ekonomivë familjare dhe banesave në Kosovë 2011, Rezultatet përfundimatre. Të dhënat kryesore. Prishtinë.

Caro, E. (2011). From Village to the City. Rosenberg Publishers, Amsterdam.

Dushi, M. (2009). Femra dhe evolucioni i lindshmërisë në Kosovë. ASHAK, Prishtinë.

Flam, H.; King, D. (2005). Emotions and Social Movements. Routledge, New York.

Gashi, A.; Haxhikadrija, A. (2012). Social Impact of Emigration and Rural-Urban Migration in Central and Easter Europe. Final Country Reports Kosovo on behalf of European Commission.

Islami, H. (2008). Studime demografike, 100 vjet të zhvillimit demografik të Kosovës. ASHAK, Prishtinë.

Osmani, J. (2011). Qytetet. Kosova, vështrim monografik, pp. 109-130. ASHAK. Prishtinë.

Schmit, O.J. (2012). Kosova, histori e shkurtër e një treve qendrore ballkanike. Koha. Prishtinë.

SZS -Savezna Republika Jugosllavija, Savezni Zavod za Statistiku (1993). Popis Stanovništvo '91. Nacionalna Pripadnost, Detaljna Klasifikacija 3, Beograd.

CHAPTER EIGHT

THE FACTORS DETERMINING FOREST PRODUCTIVITY IN THE WESTERN PART OF THE TAURUS MOUNTAINS (SW ANATOLIA)

IBRAHIM ATALAY AND RECEP EFE

Introduction

In the widest meaning, vegetation and/or vegetation productivity mostly depend on climate. Productivity increases from the arid and semiarid areas to hot-humid, mild-humid and cold-humid climatic regions (Efe 1996; Efe 1998a, 1998b; Efe 2010). The other factors which affect forest productivity are the cation exchange capacity (CEC) of soil and parent material, the amount of clay and organic materials. The CEC of parent materials is mostly related to the weathering degree and the release of plant nutriments from parent material. Productive forests are widespread on well and deeply weathered parent materials in the humid tropical regions. Poor and/or low biomass productivity is common on the less weathered and siliceous parent materials (Atalay 2008; Smith & Smith 2003).

The forest productivity of the western part of the Taurus Mountains (southern Turkey) is related to climatic factors—mainly the amount of precipitation and temperature, and the weathering process of parent materials. The productive forests which on the first and second site indexes occur on karstic lands, unconsolidated sandstone and colluvial deposits, and well and deeply weathered serpentine-peridotite bedrocks. These regions also receive abundant rainfall (Efe et al. 2007; 2011; 2012; 2013; Sönmez et al. 2012).

Cracks and fissures within the limestone and soft colluvial deposits generate the easy taproot development of trees through deep parts. One of the low productive forests is widespread on the horizontal layered clay and marl Neogene deposits due to the fact that their parent materials mostly prevent the development of taproot system (Atalay & Efe 2010).

The goals of this study are to get an overview and determine the relationship between the forest productivity and climate, soil and parent materials in the western part of the Taurus Mountains (Fig. 8.1).

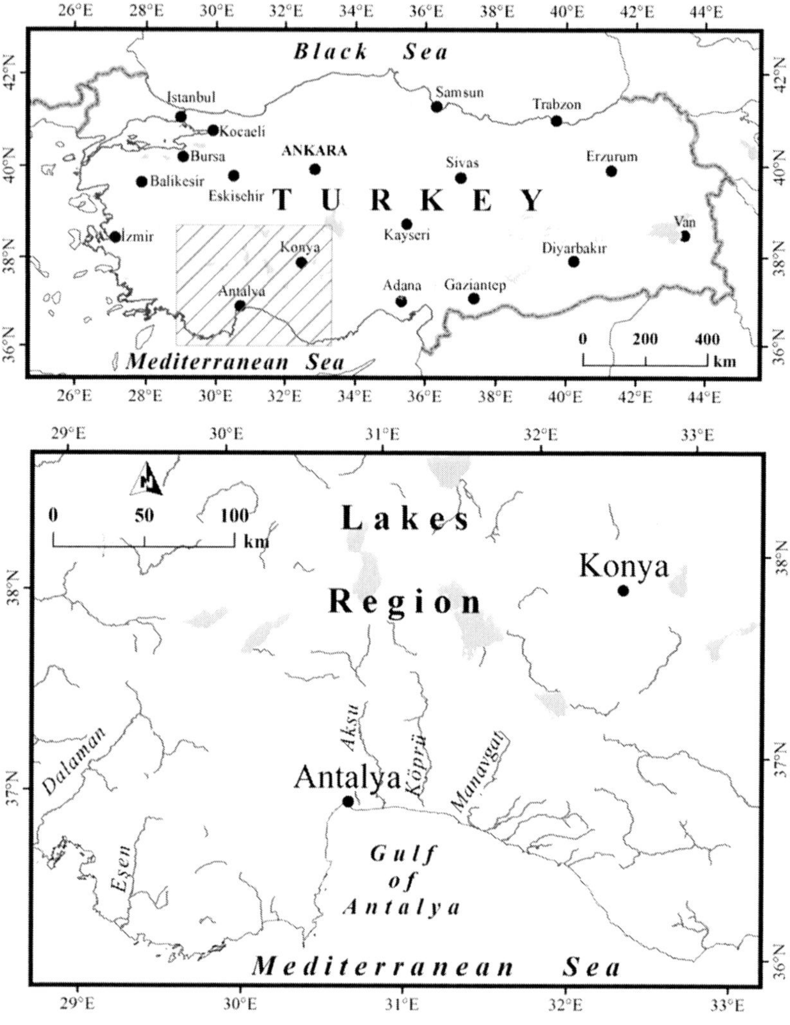

Fig. 8.1. Location map of study area

Material and Methods

The study area is located on the Taurus mountains in the south western part of Turkey. In order to determine forest productivity a lithologic map, geological stratification and geological cross-section were generated. Soil samples collected on the different parent materials were analyzed at the laboratory of West Mediterranean Forest Research Directory to determine the physical and chemical properties of the soil and parent materials. The site index of forests is determined in the field observations. Climatic data were obtained from the Meteorological station of the study area, and topographic maps are also used. Some data were collected from the previous studies (Atalay 1987a; 1987b; Atalay et al. 1998; Atalay et al. 2008; Tetik & Yeşilkaya 1997). Forest vegetation is classified according to altitudinal belts.

Geographical Setting

(a) Geology and lithology: Quaternary, Tertiary and Mesozoic terrains are common from the lowland to the upland of Taurus Mountains (see Fig. 8.2 below). Quaternary gravelly fan deposits and alluvions cover the Mediterranean coastal belts. Coarse colluvial deposits are widespread on the edge of steep slopes. Neogene marly, clayey limestone, sandy, and gravelly layers alternating with each other occur on the southern lowlands of the Taurus Mountains. Tertiary limestone with thin stratified layers with a thickness of 5 to 15 cm are common on the upland areas.

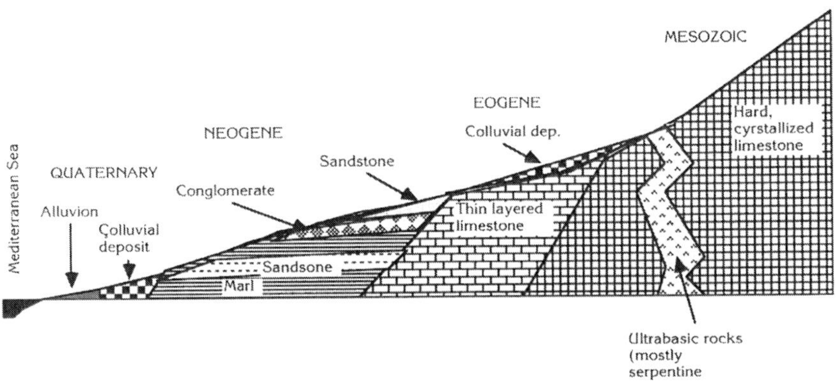

Fig. 8.2. General geological cross-section representing the study area

Clayey and marly compact limestone mostly occurs in the tectonic depressions. Toward the upper part of Taurus Mountains Mesozoic crystallized comprehensive limestone are seen.

Ultrabasic rocks mainly composed of serpentine-peridotite occur in some parts of the Taurus mountains (Fig. 8.2).

(b) Topography: Taurus orogenic mountains, one of the main branches of the Alpine-Himalayan orogenic belts, abruptly rise along the coast of the Mediterranean Sea in Southern Turkey. The upper levels or summits of the mountains are higher than 2,000 m (Tahtali Mountains—2,366 m, W of Antalya Gulf; Bey Mountains—3,070 m and 2,744 m; Alacam Mountain—2,328 m; Ak M.—3,016 m; Sarp M.—2,004 m, N of Antalya Gulf; Kuyucak M.—2,505 m and Ak M.—2,505 m). Except for Gulf of Antalya, all parts of the study area have very rugged and dissected topographies. This situation is related to the uplifting of all parts of the Taurus Mountains that occurred at the end of the Neogene and the Early Quaternary periods. Indeed, the rivers established on the Taurus Mountains deeply cut their valleys in accordance with the uplifting movements. In other words, uplifting movements led to the rejuvenation of the fluvial process in the Taurus Mountains. Therefore, the Taurus Mountains have been deeply cut by rivers, i.e. Esen, Aksu and Manavgat, all of which flow into the Mediterranean Sea. On the other hand, the uplifting movements have also caused the shifting of the karstification process from the surface to the deeper parts in the limestone areas. Therefore, the cave systems at different altitudes, ground rivers, deep lapies on the surface, karstic depressions mostly dolines have formed as the result of the karstification process. In the karstic land, "U" shaped deep valleys cutting in a north to south direction in the Taurus Mountains have formed mainly as the result of chemical erosion of surface waters and rivers. On the other hand, faulting movements are responsible for formation of depressions such as Elmali, Korkuteli, Bucak-Celtikci and fault scarps in the Taurus Mountains. The uplifting and faulting movements accelerating fluvial and karstic erosion have created the rugged topography of the Taurus Mountains. The relative altitude between the bottom of the valley and the upper part encircling the highlands area is more than 1,000 m (Figs. 8.7 and 8.9).

(c) Climate: In the study area two types of Mediterranean climate prevail. A Mediterranean climate is characterized by mild and rainy winters and hot and rainless summers on the lowlands between 0 m and 1200 m elevation on the southern edge or the Mediterranean Region coastal belt. Here, mean annual temperature is 18–20°C, the mean July temperature rising to 25°C, and falling to 10°C in January on the coastal area. The mean January temperature, which is about 10°C on the coastal belt of the Mediterranean Sea, falls to 5°C at an elevation of 1000–1200 m in the Mountains (Fig. 8.3).

An Oro-Mediterranean climate prevails in the highlands with an elevation of more than 1,200 m in the Taurus Mountains. The mean annual temperature is about 12°C, during the winter it is snowy and the temperature is generally below freezing.

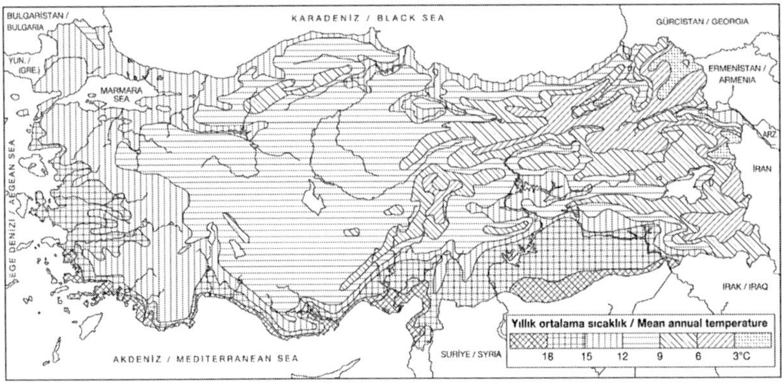

Fig. 8.3. Mean annual temperature distribution in Turkey (Atalay 2012)

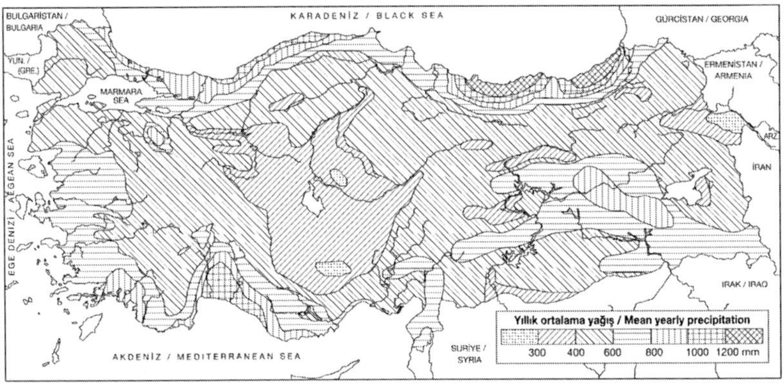

Fig. 8.4. Mean annual precipitation distribution of Turkey (Atalay 2012)

The amount of precipitation ranges from 500 mm to 2,200 mm and mostly occurs in the winter period between October and April. The summer is generally rainless, but the relative humidity is over 60%. Tectonic and deep karstic depressions correspond to the low precipitation areas due to the rain shadow effect. The precipitation, which is more than

1,000 mm on the coastal belt, decreases to as low as 470 mm at Buk-Lutfi Yildirim Forest Research Station, at an elevation of 460 m and 20 km far from NW of Antalya. The NW facing slopes of the Taurus Mountains NE of Antalya Gulf comprise the rainiest area with a mean yearly precipitation of 2,200 mm (Fig. 8.4).

Humid air comes from the Mediterranean Sea during the summer season. Therefore, the relative humidity is higher during the summer season than the winter, and the south-facing slopes of the Taurus Mountains are covered by fog, and here orographic rain occurs.

(d) Soil and parent material: The climatic soil type of the study area is reddish Mediterranean soil (Alfisol, suborder Xeralf) common on all parent materials on the flat lands in the eastern part of the Antalya Gulf; its depth changes ranges from a few cm to more than 1 m. Soil depth increases along the weak zone of colluvial materials found on the coastal belt. Reddish Mediterranean soil is dominant in/on thin stratified limestone of Tertiary and Mesozoic cracked limestone. There is no soil cover on the sloppy karstic land due to atmospheric water easily infiltrating along the cracks. Soil formation is seen along the cracks, the weak zone and among the surfaces where water is held (Atalay 1997; 2011a; 2011b; Eler 1988).

Physically, reddish Mediterranean soil generally has a clay texture, because the calcium carbonate in limestone is completely carried away as calcium bicarbonate with water (Atalay 1997; 2011a; see also Table 8.1 below). Top soil contains very low calcium carbonate content where mean annual rainfall is over 600 mm. In other words, carbonates in the reddish Mediterranean soil are carried away from the soil profile in the rainy areas of the Taurus Mountains. Organic material is low due to the high temperature in the lowland areas, but is high in the uplands. Soil is medium alkaline in karstic land, but in the highland around 1,800–2,000 m weak acid reaction is common. The cation exchange capacity (CEC) of reddish Mediterranean soil is around 50 me/100 g soil (Table 8.1)

On the other hand, yellowish Mediterranean soils are found in the weak zone with cracks more than 30–40 cm wide in the strongly cemented conglomerates, especially in the vicinity of the Köprülü Canyon locality. Here, widened cracks also contain thick soil with a sandy loam texture. Its pH is generally a weak alkaline reaction and CEC capacity changes between 21 and 26 me/100 g due to low organic content and clay materials. Deep crack or holes support the growth of humid maquis elements composed of *Nerium oleander*, *Mrytus communis*, *Phillyrea latifolia*; *Arbutus unedo* and *A. andrachne* (Fig. 8.5, 8.12 and 8.14).

Clayey and pale brown and yellowish submature soil is common on the clayey limestone and marly deposit of Neogene, because soil formation takes a very long time on the clayey materials because the decaying process is slower than other parent materials (Atalay 2012).

Pale reddish soil which is more than 50 cm in thickness is common on serpentine which is found on the gently undulating and flat lands. Soil texture changes according to the weathering process of serpentine.

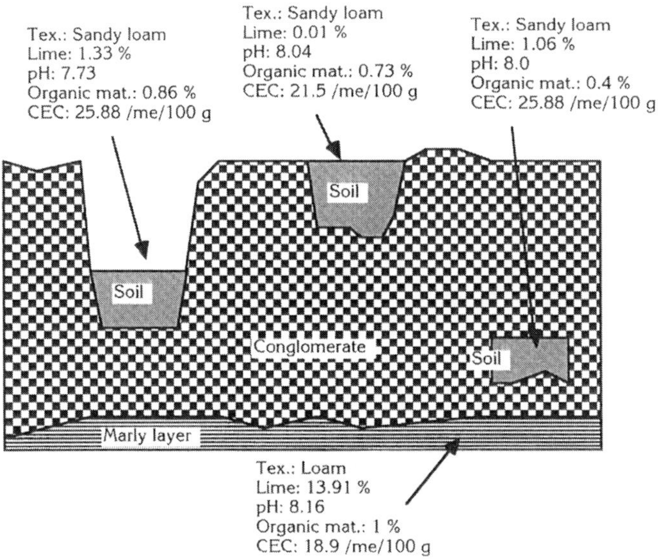

Fig. 8.5. Soil formation developed weak zone and wide cracks in the strongly cemented Neogene conglomerate in the Koprulu Canyon locality, NE of Antalya.

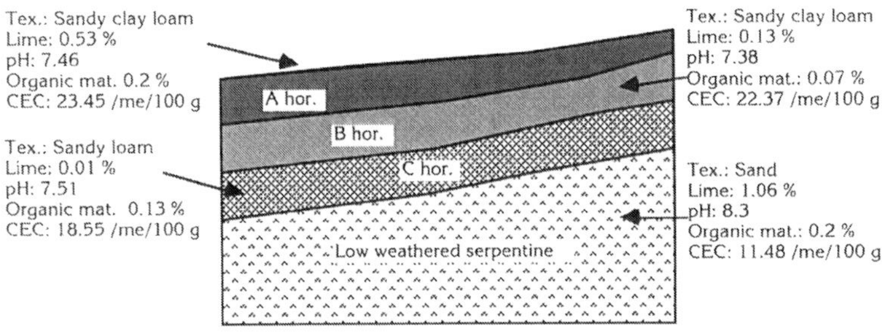

Fig. 8.6. Some physical and chemical data of soil that developed on the serpentine rock.

In the topsoil, loam and clay loam textures change to sandy loam and loam sand textures in the subsoil. Sand texture is common in the C horizon due to the low weathering process. pH is generally neutral with a weak alkaline reaction, and free calcium carbonates are carried away due to abundant precipitation. CEC capacity is 25–37 in the topsoil, and decreases to as low as 11 me/100 g soil toward the subsoil and low weathered serpentine bedrock.

The lowest CEC is recorded on low weathered and unweathered serpentine as compared to other soil and parent materials (see Table 8.1 and Figs. 8.6, 8.15, 8.16 and 8.17 below). For instance, productive and good quality Pinus nigra forests mostly grow on well-weathered serpentine in the Karsantı locality (N of Adana, Dirgine locality [NW of Anatolia], and Eskele locality [N of Mediterranean Region]), while poor stands of Pinus nigra forests are found on the unweathered serpentine rocks.

There is no soil cover on gravelly dejection fans and colluvial deposits because the rain water easily infiltrates the wide pores. However, toward the deeper part gravelly deposits in submature soils and ex situ soil are found in the pockets and wide cracks.

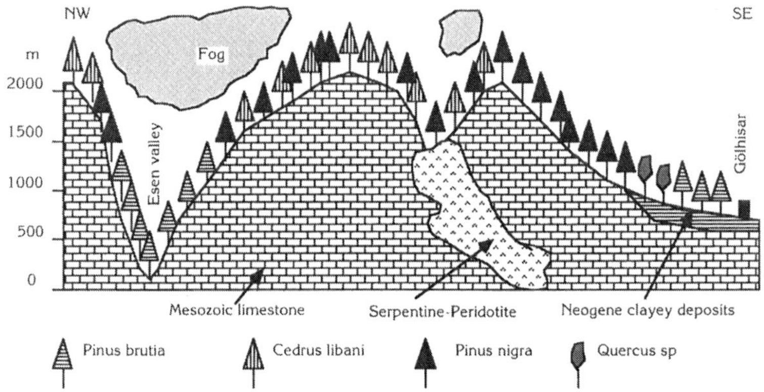

Fig. 8.7. Changing of vegetation composition from Esen valley (receiving a humid air mass) to the inner part of the Taurus Mountains. *Pinus nigra* forest is dominant in the inner part of the Taurus Mountains and well-established *Cedrus libani* forest occurs on the humid area receiving fog.

(e) Vegetation: Three altitudinal vegetation belts are found based on the altitude in the study area.

Eu/thermo belt. This belt is the main spreading area of *Pinus brutia* forest which is the climax vegetation of Mediterranean climate. Red pine is very resistant to summer drought and its direct solar radiation requirement is very high compared to other trees. The main vertical spreading area of *Pinus brutia* is between 0 m and 1200 m, but it climbs up to 1,500 m on the south facing slopes of the Taurus Mountains (see Figs. 8.7, 8.9, 8.16 and 8.18 below). Maquis vegetation and naturally occurign lower levels of *Pinus brutia* forests are common where *Pinus brutia* forests have been completely or partially destroyed (Atalay et al. 2008; Boydak et al. 2006).

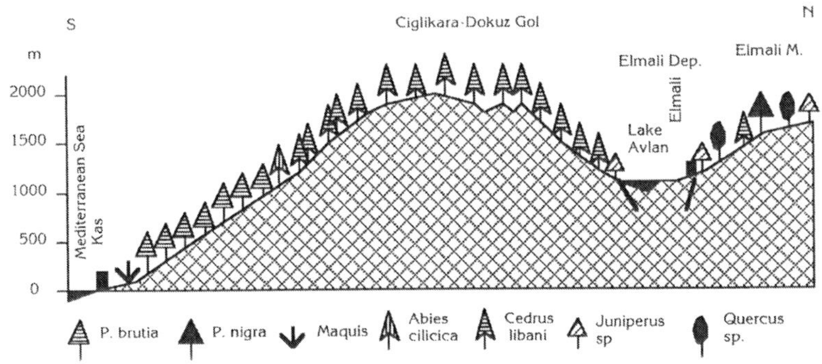

Fig. 8.8. The altitudinal vegetation belt and distribution of vegetation between the Kas and Elmali depressions in the western part of the Taurus Mountains. The productive cedar forests are common on the north-facing slopes of Ciglikara-Dokuz Gol locality

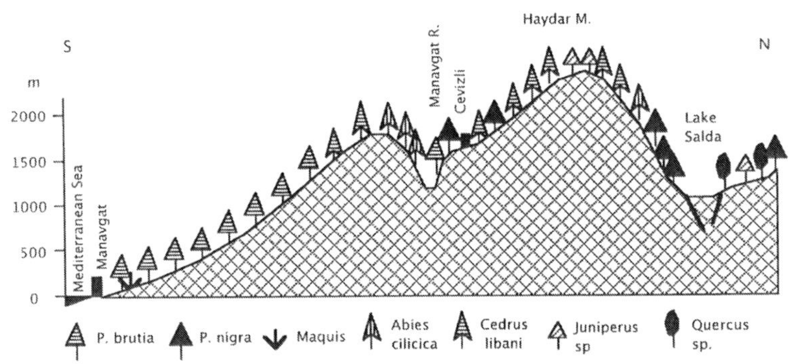

Fig. 8.9. The distribution of vegetation depending on altitude is an aspect factor in the eastern part of the Taurus Mountains

Oromediterranean belt. This belt comprises the native areas of montane forests composed of *Pinus nigra*, *Cedrus libani* and *Abies*

cilicica. It occurs between 800–1,200 m and 2,000 m in the upper part of the Taurus Mountains. *Cedrus libani* grows very well on the south-facing slopes receiving humid air from Mediterranean Sea. *Abies cilicica* (Taurus fir) prefers humid and diffuse radiation area on the slopes facing north. *Pinus nigra* is widespread in a somewhat continental inner part of the Taurus Mountains. The Juniperus community consists of *Juniperus excelsa* and *J. foetidissima* growing in the areas where Oromediterranean forests have been partly destroyed or clear cut (see Figs. 8.7–8.9 above).

Subalpine grass belt. This occurs over the natural timberline which begins at an elevation of 2,000–2,100 m on the south-facing slopes of the Taurus Mountains, and exceeds 2,300 m in the inner parts due to continentality. This belt is very rocky in the karstic lands.

Productivity of Forests

Forest productivity is determined by several factors such as climate, parent material and soil, altitude and aspect factors according to the ecological location of the area.

Climatic factors: As a general rule, the productivity of forests decreases from the coastal belt to the upper parts depending on decrease in temperature. The good stands of *Pinus brutia*, *Cedrus libani* and *Pinus nigra* forests occur where mean annual rainfall is over 1,000 mm in the north west slopes of mountains in the north east of the Antalya Gulf. The productivity of forests decreases due to low temperatures after 1,700–1,800 m. On the other hand, the most productive *Pinus brutia* forests occur only between 400 and 800 m on the south-facing slopes of the Taurus Mountains due to low respiration during the night under somewhat cool conditions (Atalay et al. 1998; 2008; Atalay 2002; Boydak 2008, Çepel et al. 1977). However, there is close relationship between productivity and the amount of precipitation. The site index is low in the rain shadow area and backward depressions of the Mediterranean, such as the Elmali depression and the Buk-Lutfi Yildirim Forest Research area located in a karstic depression, while productive forests are widespread where the annual precipitation is high and orographic rain occurs in the summer.

The areas receiving humid air from the Mediterranean Sea are the occurrence areas of the productive forests. Well established cedar forests occur on the upper part of the Esen River Basin and the north-facing slopes of the mountains in the south of the Elmali depression (Fig. 8.7–8.9).

Soil and parent material factors: Soil is an important factor for the germination of seeds and young plant phases or the early succession of trees. With the development of the root, soil factor is not prominent for the growth of trees because it takes common nutriments from the parent materials. If parent materials are well weathered and plant nutriments released enough the growth of the trees are in good condition, the productivity of trees is high.

Parent material is one of the main decisive factors for the productivity of vegetation in the study area. As such, horizontal marly layers and marly deposits mostly prevent taproot development. As a general rule, the roots of *Pinus brutia* develop along the horizontal marly layers, and in this situation taking nutriments and water is inadequate for its optimum growth. For that reason, *Pinus brutia* forests are in poor and medium site indexes (Fig. 8.20 and 8.21). However, in the weak cemented conglomerate and sandstone deposits, taproot development is common due to its high porosity. That is why the site index of *Pinus brutia* on the conglomerate deposits is higher than that of marly deposits (Fig. 8.16 and 8.18, Kantarcı 1982, Ozturk et al. 2002).

Karstic lands have a special importance in terms of soil forming and capillarity process, reflection of sun radiation, root development, and plant nutriments. Productive forests are found on the karstic lands in general. Indeed, best growing areas of the forests mostly occur in the incline-layered karstic lands, because the roots of trees easily develop along the cracks and among the layers (Fig. 8.11 and 8.12). The seeds that fall into cracks containing soil easily germinate and regenerate due to partly shadowed micro environments, and their roots follow the seepage of water along the cracks; after one vegetation period the root stretches for more than a metre along the cracks. For this reason, the best regeneration takes place on rocky karstic land (Fig. 8.13). The cracked structure of limestone almost prevents the capillarity process and runoff, and so too the majority of the precipitation absorbed in the karstic lands. The pale or whitish colour of limestone creates cool environments for plant growth due to its high albedo rate.

The CEC of soils developed in the karstic land is higher than that of other parent materials in the Taurus Mountains. As a general rule, the ECE of red Mediterranean soil in the karstic lands is nearly twice as high than the conglomerate, and three times as high than the less weathered serpentine (Fig. 8.10).

As to ultrabasic serpentine, good stands are common on the thick soil cover on the slightly undulating and flat lands. However, the poorest stands or productive forests occur on the exposed serpentine where soil has been completely eroded in the sloping area. Here, *Pinus brutia* stands are in the chaparral due to hard root development and low CEC with 12 me/100 g parent materials (Fig. 8.10, 8.25 and 8.26).

In brief, first site indexes and productive forests more than 20 m high mostly occur on karstic land, especially on cracked and layered limestone. Medium-site index forest of 15–20 m in height is observed on the well weathered serpentine and low cemented colluvial and sandy deposits. One of the poor stands appears on the strongly cemented conglomerate. However, the poorest or the lowest productive forest is common on unweathered serpentine (Fig.8.10).

According to the parent material-soil depth quality relations of natural *Pinus brutia* forest in the Antalya region, the first and second site indexes of *Pinus brutia* forests are common on the limestone. Flysch and schist parent materials are in the second order, and the remaining parent materials composed of serpentine, marl, sandstone and conglomerate are in third order (Tetik & Yesilkaya 1997).

GEOLOGY	LITHOLOGY	SOIL	FOREST PRODUCTIVITY
QUATERNARY	Alluvion, dejection fan, travertine	Reddish Mediterranean soil	Medium-poor (Site index II-IV)
NEOGENE	Conglomerate, sandstone	Submature soil, Entisol,	Poor and Medium (Site index II-III)
	Marl, clayey limestone	Inceptisol, entisol	Poor and medium (Site index III-IV)
EOCENE	Thin stratifed limestone	Rich reddish Mediterranean soil along layers	Productive (Site index I, II)
MESOZOIC	Comprehansive cyrstallized blueish limestone	Reddish Mediterranean soil along cracks	Medium, productive (Site index II-IV)
LOWER MESOZOIC	Ultrabasic rocks (serpentine)	Submature Red. Soil Inceptisol, Entisol	Productive, medium, very poor (II-V)

Fig. 8.10. Forest productivity and related ecological properties

Conclusions

- Thick soil cover belonging to reddish Mediterranean soil is common on the flat and slightly undulating areas on which productive forests are common on the middle part of the southern slope of the Mediterranean belt.

• Productive forests are widespread on the karstic lands composed of very cracked and thin-layered limestone because in these areas soil is found along the cracks, and root development is good along the weak zones, especially within the cracks.

• Horizontal marly strata also contain poor forest stands due to mostly preventing taproot developments. A moderate site index is rarely seen in weathered Neogene marly and sandy deposits only.

• Weak-cemented colluvial and sandy deposits create good conditions for the growth of both maquis and forest vegetation. For this reason, medium site index stand forest is common.

• Poor stands are observed on strongly cemented Neogene conglomerate. Deep sandy soil which develops along the wide cracks and holes contains humid maquis vegetation.

• The poorest stands occur on serpentine parent material on which there is no soil cover.

Fig. 8.11. Reddish Mediterranean soil developed on and in cracks of crystallized limestone

Fig. 8.12. Reddish Mediterranean soil formed within the cracks and cedar roots developing along the cracks

Fig. 8.13. Reddish Mediterranean soil with CEC 50 me/100 g formed among thin-layered limestone. The good stands of cedar forests are found on this limestone

Fig. 8.14. Native cedar regeneration on the karstic lands. Cedar seeds falling into cracks containing soil easily germinate and the roots of young cedar plants develop easily along the cracks of limestone

Fig. 8.15. Productive cedar forest with first site index and ram cedar of 38 m in height, 2 m in diameter, and 677 years old at an elevation of 1,550 m in the Ciglikara locality, SW of Elmali depression

Fig. 8.16. Medium productive *Pinus brutia* forest with second site index on the conglomerate deposits in the Manavgat basin

Fig. 8.17. Exposed *Pinus nigra* lateral and taproots systems on the old colluvial deposit as a result of erosion

Fig. 8.18. A productive *Pinus brutia* forest on the colluvial deposit containing soil and a good native regeneration of *Pinus brutia*

Fig. 8.19. Neogene sandstone and marly layer intercalations that mostly prevent root development

Fig. 8.20. Low site index *Pinus brutia* forest on the slightly inclined marly deposit and limestone

Fig. 8.21. Medium site *index Pinus brutia* forest on horizontal marly sandy deposit of Neogene

Fig. 8.22. A deep and wide holes create a suitable condition for the growth of humid maquis vegetation

Fig. 8.23. Soil developed along the wide crack of conglomerate in the Koprulu Canyon locality

Fig. 8.24. Soil developed under the weak zones of conglomerate deposits of Neogene in the Koprulu Canyon locality

Fig. 8.25. Productive *Pinus brutia* forest on the deeply weathered serpentine, in the vicinity of Kemer, west of Antalya

Fig. 8.26. The horizontal root system developed on less weathered serpentine

Fig. 8.27. Very poor site index of *Pinus brutia* cluster on very low weathered serpentine

Acknowledgement

We would like to thank Cem Selçuk Karaca (forest engineer) from the General Directorate of Forestation (AGM) Antalya for his support and guidance during field trips in the Taurus Mountains.

References

Atalay, I. (1987a). Introduction to Geomorphology of Turkey (2nd Edition). E.U. Edebiyat Fak. Yay. No.9, İzmir.

—. (1987b). General ecological properties of the natural occurrence areas of cedar (*Cedrus libani* A. Rich) and regioning of seed transfer of cedar in Turkey. Ministry of Forestry Publ., 663/61, Ankara.

—. (1988). Vegetation levels of the Taurus Mountains of Mediterranean region in Turkey. Aegean Geographical Jour, 4:88-122

—. (1994). Vegetation Geography of Turkey. Ege Univ. Press, Izmir.

—. (2001). Ecology of Forests in Turkey. Silva Balcanica, 1/2001, 25-34.

—. (2002). Ecoregions of Turkey. Orman Bak. Yay. No: 163.

—. (2006). Effects of the Anatolian Mountainous Areas on the Biodiversity: A Case Study from the Northern Anatolian Mountains and Taurus Mountains. In: High Mountain Remote Sensing Cartography International Symposium, (Eds. V. Kaufmann, W. Sulser). Grazer Schriften der Geographie und Raum Forschung, Band 41:17-26.

—. (1997). Red Mediterranean Soil in some karstic regions of Taurus Mountains, Turkey. Catena 28:247-260

—. (2008). Ecosystem Ecology and Geography). Meta Press. Izmir.

—. (2011). Toprak Oluşumu, Sınıflandırılması ve Cografyası (Soil Formation, Classification and Geography, 4th ed.). Meta Pub., Izmir.

—. (2011). Türkiye'de Karstlaşma ve Karst Ekolojisi. Fiziki Coğrafya Araştırmaları Sistematik ve Bölgesel: sayfa. 203-222. Türk Coğrafya Kurumu Yay. No 6., Istanbul

—. (2011). Soil formation, Classification and Geography, 4th edition. Meta Press., Izmir.

—. (2012). Applied Climatology, 2nd ed. Meta Press, Izmir.

Atalay, I. and Efe, R. (2010). Ecology of Anatolian Black Pine (Pinus nigra Arnold subsp. pallasiana (lamb.) Holmboe) and its zooning into regions in terms of seed transfer. Forest Seeds and Tree Breeding Research Direc. Publ. No. 37, Ankara

—. (2010). Structural and distributional evaluation of forest ecosystem in Turkey. Journal Environment Biology. 31:61-70.

—. (2012). Ecology of Scots Pine (*Pinus sylvestris* L. var. *sylvestris*) Forests and their zoning into negions in terms of seed transfer. Forest Seeds and Tree Breeding Research Direct. Publ. No. 45, Meta Press, Izmir.

Atalay, I.; Efe, R. and Soykan, A. (2008). Mediterranean Ecosystem of Turkey: Ecology of Taurus Mountains. In Natural Environment and Culture in the Mediterranean Region: pp.1-38. In (Eds.) R. Efe, G. Cravins, M. Ozturk and I. Atalay. Cambridge Scholar Publising.U.K.

Atalay, I. and Mortan, K. (2011). Turkiye Bolgesel Cografyası (6th Ed.). Inkilap Publ. Istanbul.

Atalay, I.; Sezer, L.I. and Cukur, H. (1998). The Ecological Properties of Red Pine (Pinus brutia Ten.) and Their Regioning In Terms Of Seed Transfer. Research Directory of Improvement of Forest Trees and Seed, Pub Nu. 8, Ankara.

Boydak, M.; Dirik, H.; and Çalıkoğlu, M. (2006). Biology and Silviculture of Turkish Red Pine (*Pinus brutia* Ten.). OGEM-VAK Publ. Ankara

Boydak, M.; and Çalıkoğlu, M. (2008). Biology and Silviculture of Lebanon Cedar (Cedrus libani A. Rich.). OGEM-VAK publ., Ankara.

Çepel, N.; Dündar, M. and Günel, A. (1977). Türkiye'nin önemli yetişme bölgelerinde saf sarıcam ormanlarının gelişimi ile bazı edafik ve fizyografik etkenler arasındaki ilişkiler. TUBİTAK Tarım ve Ormancılık Araş. Grubu TOAG Seri 65, Tubitak yay.no. 354.

Efe, R. (1996). Yunt Dağı ve çevresinin dogal bitki örtusunun ekolojik şartları. Turk Cografya Der., no. 29, pp. 71-114.

—. (1998a). Ermenek Çayı Havzası; Doğal Ortam Özellikleri. FA.Ü. yayınları no 1. İstanbul. pp. 210, ISBN 975-303-001-0

—. (1998b). Yukarı Gediz Havzasında iklimin dogal bitki ortusu dagılışına etkisi. Turk Cografya Der., 33: 79-99.

—. (2010) Biyocoğrafya- Biogeography (2.Basım). MKM Yayıncılık, ISBN 978-605-5911-21-8

Efe, R.; Sönmez, S.; Cürebal, İ. (2007). Ecological Properties of Mancinik Mountain. International Symposium on Geography. GeoMed2007" Proceedings. I. Atalay, R. Efe (Eds.) 5-8 June, 200- Kemer/Antalya-Turkey. Inkilap Publ. Istanbul, p. 119-234. ISBN: 978-975-00219-7-8

Efe, R.; Soykan, A., Sönmez, S., Cürebal, İ. (2011). Burhaniye-Doğal Kaynak Değerleri. 384 s. Sistem Ofset, ISBN: 978-605-62253-1-4, Ankara

Efe, R.; Soykan, A.; Cürebal, İ.; Atasoy, E.; Sönmez, S. (2012). Ecological conditisons and Distribution of Vegetation on Southern Slopes of Kaz Mountain, Turkey. Journal of Balkan Ecology. Vol 15, No.4: 373-382.

Efe, R.; Soykan, A.; Cürebal, İ.; Sönmez, S. (2013). Balıkesir'in Ağaçları ve Çalıları. Akmat Basım. ISBN 978-605-62253-5-2

Eler, U. (1988). Türkiye'de boylu ardıc (*Juniperus excelsa* Bieb.) ormanlarında hasılat araştırmaları. Ormancılık Araş. Enst. Tek. Bul. Ser. No. 192.

Kantarcı, D. (1982). Akdeniz Bolgesinde dogal agac ve calı turlerinin yayılısı ile bolgesel yetisme ortamı ozellikleri arasındaki iliskiler. İ.U. Orman Fak. Yay., no.330. Istanbul.

Ozturk, M.; Celik, A.; Yarcı, C.; Aksoy, A and Feoli, E. (2002). An overview of plant diversity, land use and degradation in Mediterranean Region of Turkey. Environtal Management. 13, pp.442-449.

Smith, R.L. and Smith T.M. (2003). Elements of Ecology (5th ed.), Benjamin Cummings.

Sönmez, S.; Efe, R.; Cürebal, İ.; Soykan, A. (2012). Madra Dağı ve Yakın Çevresinin Floristik Zenginlikleri. Kazdağı ve Madra Dağı Belediyeler Birliği, Madra Dağı Ulusal Çalıştayı. 12-14 Ekim 2012, Ayvalık-Balıkesir.

Tetik, M. and Yeşilkaya, Y. (1997). Antalya Yöresi Doğal Kızılçam Ormanlarında Anakaya-Toprak Derinliği-Bonitet İlişkileri (Parent Material-Soil Depth-Site Quality Relationships of Natural Pinus Brutia Forests in Antalya Region.

CHAPTER NINE

THE MEDITERRANEAN MOUNTAIN: THE OTHER DEVELOPMENT DIMENSION. THE CASE OF THE ABRUZZO REGION

MARINA FUSCHI AND PAOLA MUCCIANTE[1]

Introduction

The Mediterranean mountain: its importance as territory and its role in the history.

> "Mountains are all around in the Mediterranean. They come right down to the sea, taking up more than their share of space, piling up one behind another, forming the inescapable frame and backdrop of every landscape. They hinder transport, turn coast roads into corniches and leave little room for serene landscapes of cities, cornfields, vineyards or olive-groves, since altitude always gets the better of human activity…. The Mediterranean plains, for lack of space, are mostly confined to a few coastal strips, a few pockets of arable land… Life was simply not the same in the hills as in the plains. The plains aimed for progress, the hills for survival" (see Fig. 9.1 below).

These are the words uttered by Fernand Braudel in his work "Memory and the Mediterranean" (1998). He emphasizes the importance of the physical feature, beyond the cultural aspect, played by the mountain territory within the Mediterranean area. This territorial framework, outlined by the end of the 1960s, is an element of surprise for the unique capacity of the villages to adjust to the current mountain background. Although they had developed under an economic and socio-behavioural angle, especially with regards to a deep urban-centred lifestyle, they do show features of marginalization, mostly regarding the territorial dynamics

[1] Although the study has been prepared as a single entity, Marina Fuschi wrote sections 1-2-3-5; Paola Mucciante section 4 and cartographic processing.

within the widespread regional frameworks, corroborating a historical paradigmatic comparison between the core and the periphery.

In the first place, the importance of the mountain in the Mediterranean area has its roots in the role played by the countries on the coast and their altitudinal morphology, taking into account that it is an orogeny area aroused by the contact between the African and Eurasian plates (Atalay et al. 2014; Efe et al. 2008; Soykan et al. 2008) . Generally, the surface of the mountains is less than 1.8 million sq. km., equal to 21% of the total basin surface with 66 million inhabitants, equal to 16% of the total population. In detail, the mountain territory influences 16% of the countries on the southern and eastern coast and 39% of those on the northern coast; however, there are deep discrepancies within the regions in this first classification and there are marked contrasts on a national scale as well. In the first case, the difference in area between Africa (barely 9%) and Asia (up to 59%) is sharp, while the EU countries situated on the northern coast have a mountain surface of 37% and the Balkan area shows a mountain surface of 54%.[2] From a territorial perspective, the mountainous nature which spreads throughout the Afro-Asian area cannot be found in the south-eastern area, where countries such as Libya and Egypt have hardly any mountain elevations (1% and 6%, respectively), while the European area shows a very sharp and continuous mountain surface, where six countries out of nine have mountainous surfaces of well over 50% (Greece, Italy, Albania, Bosnia and Herzegovina, Slovenia, Serbia and Montenegro[3]). Overall, the countries with the most mountainous territories are Lebanon (83%) and Turkey (70%) in the Asian area, Bosnia and Herzegovina (73%), Albania (72%) and Slovenia (67%) in the European area, while among the African countries only Morocco shows a certain degree of mountainous surface (46%) (Table 9.1).

Demographically speaking, the territory framework shows opposite data, with 66% of the population living in countries on the southern and eastern coast and 34% of the inhabitants in the countries situated on the northern coast. This outlines the depopulation process in the mountain area of the North and an increase in the Southern area, despite emigration (Dubost 2003). According to a sub-regional research, the most inhabited mountainous area is the Asian zone (32% of the total population) followed by the Balkan area (27%), whereas as on the national scale Bosnia and Herzegovina is the country with the highest percentage of its population living in the mountains. On the one hand the aforementioned results do not

[2] Cyprus is also included in this percentage.
[3] They have been included in the classification.

show the high fragmentary nature of the regional frameworks (Table 9.1 above), and on the other the results lead to discussions about the role played by the mountainous nature of the Mediterranean area. The Mediterranean area, along with its social, economic and territorial features, is deeply marked by a high degree of coastal development: indeed, 34% of the population live on the coastal regions compared to 13% on the Basin surface (Table 9.2).

The mountain-coast evolution has reshaped the framework of the settlements and overturned old internal gravitations, especially in the years following World War II. The evolution was encouraged through the breakdown of the rural economy and the crisis of traditional lifestyles on the one hand, and through a higher degree of productivity and income resulting from the flat coastal line on the other, and this stands for available building areas, simplified transportation, industrial infrastructure[4] and agglomeration profitability. This results in the high urban development of the coast (through migration from the countryside and natural increase), where the number of coastal towns has nearly doubled in less than fifty years[5] (Moriconi-Ebrard 2001), while tourism draws the attention by consolidating the coast as a focus point at the expense of the mountain territory, and the Mediterranean is a popular tourist destination worldwide, with over 200 million arrivals yearly (data of the World Tourism Organization).

The uniformity of the mountain-coast evolution, briefly described as a core-periphery relationship studied by Gunnar Myrdal, requires a consideration about the role played by the mountain area in the past, which also led to building the coastal and urban areas or, in general, the flat centrality, and about the connections which still exist between the mountainous nature and the strong economic areas to create a balanced and sustainable territorial development.

Under a historical perspective, the mountain area has always been considered a shelter from enemies from the sea and the unhealthy conditions of the coastal areas suffering from malaria. However, it has also represented a place where the population had to adjust itself to often harsh conditions, living a lifestyle mostly marked by subsistence farming and

---

[4] For this reason the coastal industrialization needs to be mentioned as it emphasized, especially in the 1960s–70s, the function of the ports through "hard" localizations, such as Etang de Berre of Marseille in France, Sfaxand Gabès in Tunisia, Iskenderun in Turkey, and Taranto and Brindisi in Italy.

[5] The number of towns with over 10,000 inhabitants has increased from 777 to 1,449 (Moriconi-Ebrard 2001).

often highlighted by the eternal "fight" between agriculture and sheep farming (Fuschi 1994).

Table 9.1. The Mediterranean: surface and population of the mountain areas for countries and macro-regions

Mediterranean Countries and Sub-regions	Surface of the mountain areas			Population of Mountain Areas (2000)		
	Km^2 x 1000	Percentage (%) of the total area of the Mediterranean mountains	Percentage (%) of the total area by country	In million (Peopl.)	Percentage (%) of the total population of the Mediterranean mountains	In percentag e (%) of the total population by country
Countries (Eu Members)	535	31	37	15.9	24	10
France	107	6	20	3.4	5	6
Greece	71	4	58		2	15
Italy	148	9	50	6.5	10	12
Spain	209	12	42	4.6	7	12
Southeast Europe	146	9	54	6.7	10	27
Albania	20	1	72	1.3	2	39
Bosnia-Herzegovina	38	2	73	1.9	3	50
Croatia	17	1	31	0.3	0	6
Slovenia	13	1	67		1	24
Rep. Fed. Yugoslavia	55	3	54	2.8	4	26
Cyprus	3	..	37		..	6
Sub-Total of Northern Mediterranean Countries	681	40	39	22.6	34	12
Middle East	578	33	59	27.8	42	32
Israel	4	..	21	1.0	1	18
Lebanon	8	..	83	1.1	2	37
Syria	21	1	11	2.6	4	16
Turkiye	545	31	70	23.1	35	37
Northern Africa	478	27	9		24	11
Algeria	192	11	8	7.9	12	26

Marocco	188	11	46	6.9	10	24
Tunisia	19	1	12	0.6	1	7
Egypt	60	3	6	..	..	
Libya	19	1	1			2
Sub-Total of South Mediterranean Countries	1.05 6	60	16	43.2	66	19
Mediterranean Countries		100	21	65.8	100	16

Source:Plan Bleu, Icalpe, 2003, p. 3

However, the most recent historiography studies (Maroni 2004) tend to re-evaluate the image of the mountain area and consider it not only as repository of ancient relationships based on an integrated and a deep ecological connection between the human being and their environment, but also, in going beyond the stereotype of its being static, regarding its innate seasonal or temporary migration and role as passage or crossing pathway (this latter aspect contributed to organizing the urban settlement feature) (Landini et al., 2011).

The mountain area in the Mediterranean environment has always been a repository of natural resources such as water and forests, and a holder of an invaluable and precious biodiversity (if we only consider the endogenous aspects of three particular continental areas of Europe, Asia and Africa). In the past the mountain area developed rich territorial areas and "ainsi la Méditerranée des villes n'a pu se dévelloper qu'enpuisant sur les ressources de la montagne- so the Mediterranean cities were able to develop themselves only by drawing on the resources of the mountain" (Dubost 2003). Nowadays, the mountain-coast connection unveils the vulnerability of the population-resources relationship, starting from the water demand (this resource is to be confronted with the summer drought and the high-pressured settlement during the summer holidays). The mountain-coast relationship is presented as a strategic feature to be re-evaluated in terms of cooperation (and not in terms of opposition) in order to achieve a consistent and sustainable territorial plan.

The "plurality" of the mountain territory: the diversity of the evolutionary model on a Mediterranean scale and the other territorial leading role on a regional scale

The mountain-coast relationship is the most accurate key to better understand the Mediterranean environment and mend its vulnerability, starting from an unbalanced structure which highlights, at the high and low ends of the territorial scale, the expenses due to the urban settlement pressure, artificiality and soil consumption along with the territory abandonment and erosion, depopulation and social, economic and cultural decay.

Table 9.2 Population in the countries and in the coastal regions of the Mediterranean area (2000)

Countries	Total amount (thousands of inhabitants)	Population on coastal regions (thousands of inhabitants)	% of population on coastal regions	Population density (inhabitant/sq. km.)	Population density on coastal regions (inhabitant/sq. km.)	Population density on coastal regions / population density (Index)
Spain	39.815	15.560	39,1	79	163	2,1
France	59.412	6.265	10,5	109	133	1,2
Italy	57.456	32.837	57,2	191	198	1,0
Greece	10.558	9.482	89,8	80	95	1,2
Malta	389	389	100,0	1.231	1.231	1,0
Monaco	34	34	100,0	17.000	17.000	1,0
Slovenia	1.965	102	5,2	97	23	0,2
Croatia	4.473	1.529	34,2	79	59	0,7
Bosnia-Herzegovina	3.972	452	11,4	78	76	1,0
Serbia-Montenegro	10.856	409	3,8	106	64	0,6
Albania	3.114	1.269	40,8	108	140	1,3
Morocco	28.505	4.492	15,8	62	107	1,7
Algeria	30.332	12.271	40,5	13	261	20,1
Tunisia	9.615	6.762	70,3	62	148	2,4
Libya	6.038	5.179	85,8	3	44	14,7
Egypt	66.007	22.929	34,7	66	200	3,0
Palestine	3.150	1.119	35,5	511	3.066	6,0
Israel	5.851	5.045	86,2	282	1.301	4,6
Lebanon	3.206	2.906	90,6	313	594	1,9
Syria	15.936	1.533	9,6	86	365	4,2
Turkey	65.627	13.691	20,9	84	112	1,3
Cyprus	785	785	100,0	85	85	1,0
	427.096	145.040	34,0	49	149	3,0

Source: United Nations Population Division, 2000

However, taking into account the uniformity of the mountain-coast model, which distinguishes the Mediterranean region, dissimilarities regarding the different geographic scales are to be found, from the small scale, referring to the mountains of the north and the south, to a bigger scale, such as the macro-regional which shows a development comprising several steps, for instance the mountain landscape in Albania and France, or a regional scale such as the example of Italy as to the diversity and variety of possible scenarios which mark out the Alps and the Apennines (Fuschi 2008).

The opposition between the northern and southern mountain areas tends to differently adjust to the harsh morphological conditions. The southern area is marked by pastoral nomadism, showing productive and political features (Fabietti 2001), while the northern area is marked by transhumance, whose economy gained a gradual trading and financial importance through the strong relationship with commercial activities, especially those related to the wool trade (Piccioni 1993). The development of historical events shows a distinguished dualism between the mountains of the north and the south, which stands for a higher birth rate in the Southern area due to Islamic traditions and a rural "shared" lifestyle, even when considered as a common rural migration. Furthermore, the reforms of the land made after World War II were less sharp and worsened the demographic pressure on resources, especially the soil and the water as evidenced by a subsistence economy suffering from erosion and hydro-geological issues. Instead, the north is the area which has especially suffered from abandonment by the population, creating high ageing rates among the inhabitants and a breakdown in the social and economic life, weakening the environment.

However, beyond this first brief interpretation concerning the Mediterranean scale, the plurality of the mountain landscape can be found at several scales. For instance, the north western mountain area has shown some progress in the economic, social and infrastructural fields by finding enhancement and territorial balance through the tourism, food and agriculture industries, historical and cultural heritage and environment protection. It is not the same for the Balkan mountain area where the inland has poor and unchanging economic and social conditions, being considered the weakest and most marginal part of the territory compared to the coastal and ground level areas, which take advantages from their locations to guide the post-collectivist economic transition. Once again, there are a number of scenarios behind this framework. The western, "advanced" mountain area is marked by economic wealth as well as emigration and marginalization, in addition to development of local and political resources, decadence of farming activities and the protection of woodland heritage, tourism settlement and decline in the landscape identity.

For this purpose, the Italian mountain area reflects a "plurality" of features rather than uniformity. Therefore, the Alps and the Apennines, which are completely different from each other considering their altitudes, climates, locations (the Alps are aligned to the parallels, while the Apennines are aligned to the meridians) and latitudes, gave birth to two different "regional" realities, according to which the Alps:

> *«because of their greater morphological roughness and their extension that, especially in the past, have been isolated and more autonomous, they had to find greater homogeneity in the type of organization and evolutionary dynamics»* while the Apennines still have *«structure to "stain of leopard", where every stain is rich and evolved, or poor and backward, not so much in reason for its own intrinsic potentialities, but to the more or less great remoteness and marginality of the strong or weak plain areas on which it depends».* (Bernardi 1994, 10)

However, it is important to highlight that the Alps were able to carry out models regarding endogenous territorial planning based on the development of local resources and the newly rediscovered feeling of awareness by the local communities, starting from the strong relationship with the rich ground level areas through actual settlement typologies, such as winter tourism, where the demand from the urban area temporarily increased the population in the mountains, followed by the ownership of a second home.

Hence, the diversity of the mountain area is being understood as an open system where several scenarios are deeply influenced through the relationships created with the bordering areas over the years, especially with the ground level and coastal areas which became the main areas of the region at different periods, by overturning old gravitations at the very beginning that were solely damaging the inland. Nowadays, the step-based development and wealth of the postmodern society tend to make those areas more attractive with alternative strategies, even though there are deep differences among the regions. The self-focused central role based on the actual potentiality of the mountain areas and independent theoretical and management decisions mostly results from important intervention plans carried out by the state and the regions.[6]

---

[6] There have been numerous legislative measures taken to encourage the development of the mountain areas through European Union policies in Italy and other countries, especially in Europe. In Italy, the first action referred to Law no. 991, dated 1952, about assistance, followed by Law no. 1102, dated 1971, of the "Comunità Montane" (Mountain municipalities). Despite several limitations, it suggested a plan which could be adjusted for the territorial entities (as subject to

The Apennine mountain of Abruzzo: a model of a Mediterranean mountain.

An interpretative analysis. The Abruzzo Region is a Mediterranean mountain region *par excellence*, which shows the typical marks described before and resembles the relationship between mountain and hill-coastal areas. The main role played by the mountain areas, which cover 68% of the territory (see methodological note and Fig. 9.2)[7] (equal to about

---

territorial reorganization in 2013). Further regulations have been carried out, including Law no. 142, dated 1990, about "Ordinamento delle autonomie locali" (Organization of local authorities), focussing on the mountain municipalities, and Law no. 97, dated 1994, regarding "Nuove disposizioni per le zone montane" (new regulations for the mountain areas), while the global scenarios after the Conference in Rio de Janeiro in 1992 and the "Anno internazionale delle montagne" (International year of mountains) proclaimed by the United Nations General Assembly in 2002 encouraged a greater awareness towards mountains as a resource and towards a need for sustainable development. In addition to the transnational measures, such as the Alpine and Carpathian Convention, the EU made interventions launched in 1975 and encouraged from the second half of the 1980s, through the reform of structural funds and the policy regarding the economic and social cohesion, action based on the so-called regions programmes (including most of the mountain areas), followed by the community initiative programme (CCI), such as Interreg and Leader.

[7] The choice of mountain territorial units shows an internal problematic nature linked with the definition of "mountain" itself, as the authors of works regarding geography broadly acknowledge (Febvre 1980; Blache&Blanchard 1933; Almagià 1934; Veyret 1962, 1964; De Vecchis 1996). However, the criteria regarding the physical, morphological, climatic and/or human features of a mountain "life being," taken into account as landmarks, do not solve the difficulty experienced in defining the limits of the mountainous nature, agreeing with Veyret that, especially regarding its lower limit, "tocca all'uomo apportare quella precision che manca alla natura- it's up to man to bring precision that is lacking in nature" (1962, p. 35). The choice made here to define the mountains of Abruzzo is based on a combination of some parameters, while recognizing that the continuity of geographical space relativizes the accuracy of any criterion. The parameters taken into account are: altitude (where the lower limit is 700 metre a.s.l) and the areal contiguity that led to gather all municipalities in the province of L'Aquila—whose mountain location is, however, determined by the continentality and plasticity of the territory (for this context, Sulmona and Avezzano are identified as two basin mountain cities)—and the municipalities of the coastal provinces of Teramo, Pescara and Chieti, which have been further screened through the criteria regarding the difference between the valley bottom and the peaks (energy of the relief) and the features of the Rural Region (Regione Agraria) which they belong to. For the first criteria, the "energy" with over 1,000 meters was taken into account; for the

730,000 hectares), is characterised, in relation to its orographic-morphological profile, by: an average altitude of 700 metres and by the two highest mountains in the Apennines, Gran Sasso and Majella (approximately 3,000 metres a.s.l.); an insufficient population level (according to the analysis in 2011, about 380,000 inhabitants live in the areas, equal to 30% of the total regional surface, and with a density of 53 inhabitants/sq. km[8]); a broad involvement of 176 municipalities,[9] about 58% of the total surface and 93 smaller municipalities where less than one thousand inhabitants live, as a result of the ongoing migrations from the mountain areas of Abruzzo region[10]; a weak demographic structure which can be assessed on the incessant ageing of the population and a low birth rate, as evidenced through the ageing index, figures which have more than tripled from the 1970s by 190% (in 2011), and the dependency ratio of 53% with a containment if compared to the results from 1971 as a consequence of the low number of young people (see Table 9.3 below); and a weak economic and productive system marked by slightly less than 27% of the total regional employees, showing a deficit value of 17.420 when compared with workers (see Table 9.4 below).

Conversely, the development of the historical events gives the mountain areas in the Abruzzo region an important role within the balanced and well-managed relationship between human and the environment. Humankind is able to adjust to the territory and its harsh conditions in terms of resources, and finds in activities such as transhumance and wool production the tools to establish a flourishing trade and proto-industrial economy to give cities such as L'Aquila and Sulmona a well-defined role in the middle-aged urban structure:

> *«the mercantile vocation of farming had made the mountains of Abruzzo and its two capital, L'Aquila and Sulmona, one of the main centers of production that has characterized the entire Apennine chain and has had an outlet in the manufacturing towns of Tuscany and the Po Valley»* (Mercurio 2000, 812).

---

second criteria, the municipalities belonging to mountain Rural Region were considered (R. R. Medio Sangro mountain; R. R. The East Slope of Majella mountain; R. R. The North Slope of Gran Sasso mountain).

[8] The data are referred to demographic and economic indicators, well known in literature, evaluated in diacronic terms and compared with the latest census data (2011) as well; the latter provides data only for demographic variables.

[9] On a total regional amount of 305 municipalities.

[10] Among these smaller municipalities, Montelapiano, Carapelle, Calvisio and Montebello sul Sangro are really suffering from this deficit with less than 100 inhabitants, and about fifty of them have a population under 500 inhabitants.

On the other hand, the guide Via degli Abruzzi, one of the main commercial, diplomatic, cultural and military itineraries of the fourteenth century (Gasparinetti 1967), displays the Apennines area as a passage and crossing pathway by highlighting the strategic position and the role played by the Municipalities in the mountain areas. This explained the increase of the number of inhabitants in the mountains compared to those living in the hill and coastal areas until the second half of the nineteenth century (Sori 2004) (Fig. 9.3).

However, the scenario has undergone a transformation over the years, and more than 70% of the population now resides in the hill and coastal areas (equal to over 900,000 inhabitants) with a density of 261 inhabitants/sq. km, showing an annual average growth of 0.53% for the census period 1861–2011.

This change occurred during the last decades of the nineteenth century when the hill and coastal territory experienced a development phase through the improvement of the rail and road connections,[11] the recovery of the agricultural activity on the hills and along the seashore which had been reclaimed, and then later made the decisions to undertake measures for industrial development[12] and the rise of middle-class tourism.[13] Simultaneously, the regional urban focal point was shifted to the coastal area where Pescara was the main city (the only city which had more than 100 thousand inhabitants from the 1970s), while the other municipalities situated on the hills, such as Chieti and Teramo, partially absorbed internal migrations, which, after the Second World War, strengthened the attraction towards the seashore, anticipating an ongoing coastal settlement especially on the central and northern parts, related to the most widespread economic development of tourism and industry.[14] The coastal area of the Abruzzi region, comprising 19 municipalities, would especially draw both demographic and economic resources by attracting the main extra farming activities (25% in manufacturing; 37% in the service industry, and over

[11] The rail connection Ancona-Pescara-Foggia was built in 1863. while the works for the highways A14 (E45), A24 (E80) and A25 (E80) were performed after World War II.

[12] This refers to the industrialization process that arose through subsidies by the Cassa per il Mezzogiorno (Fund for the South).

[13] The first beach resort was opened in Francavilla al Mare in 1873.

[14] This is the result of a post-war political framework discovering the "pathway to the coast" within the policy regarding industrial and touristic investments in complete accordance to the increasing opportunities available through the development of owning transport vehicles.

40% in advanced tertiary in 2001[15]), and achieving a 33% rate of the total amount, growing by 400% during the census period 1861–2011.

Table 9.3. Demographic indexes

Aging index (P65+/P0-14*100)

	1971	1981	1991	2001	2011
Mountain municipalities	61,55	86,05	123,11	172,33	192,23
Non-mountain municipalities	46,86	61,43	94,20	136,36	157,75
Total	51,81	69,02	103,14	146,87	167,13

Dependency ratio (P0-14+P65+)/P15-64*100)

	1971	1981	1991	2001	2011
Mountain municipalities	56,34	56,93	54,32	55,26	53,45
Non-mountain municipalities	56,10	54,14	48,02	51,14	52,93
Total	56,19	55,05	49,99	52,40	53,08

Replacement index of the active population (P60-64/P15-19*100)

	1971	1981	1991	2001	2011
Mountain municipalities	67,38	72,01	n.a.	104,21	143,40
Non-mountain municipalities	92,36	106,99	n.a.	106,99	130,20
Total	83,03	93,25	n.a	106,13	133,99

Source: own elaboration on ISTAT data

Table 9.4. Economic profile of the mountain and non-mountain municipalities

	Municipalities n.	LU	UL% val	Workers a.v.
Mountain municipalities	176	30120	28.71%	126008
Non-mountain municipalities	129	74808	71.29%	297950
Total	305	104928	100%	423958
	Workers % v.	**Employees**	**Employees % v.**	**Surplus/deficit Employee a.v.**
Mountain municipalities	29.72%	108798	26.44%	-17210
Non-mountain municipalities	70.28%	302654	73.56%	4704
Total	100%	411452	100%	-12506

Source: own elaboration on ISTAT data

The close relationship between mountain and non-mountain areas legitimates the centrality in the regional territorial planning on one side, while on the other it increases the discrepancy with the mountain area that is more and more "emptied" (Censis 2002), trapped in an involution process (according to Myrdal), that, based on demographic vulnerability and controlled income level and consumption, the existence of the most

[15] This is the most recent economic census data available.

basic services has brought into question, leading to a further depopulation. The mountain and non-mountain environment has been favoured by the improved network and the increased productive potentiality on the coastal and hill area (see Fig. 9.3 below), the latter aimed at creating a territorial connection with the coastal area based on its endogenous potential with a farming tradition resulting in opportunities for the counties to develop besides decisions belonging to an industrial policy focused on the valley floors, without ignoring productive devolution and the most current urban settlement steps which lead to discovering and improving the typical features of the territory, starting from the agro-industrial ones.

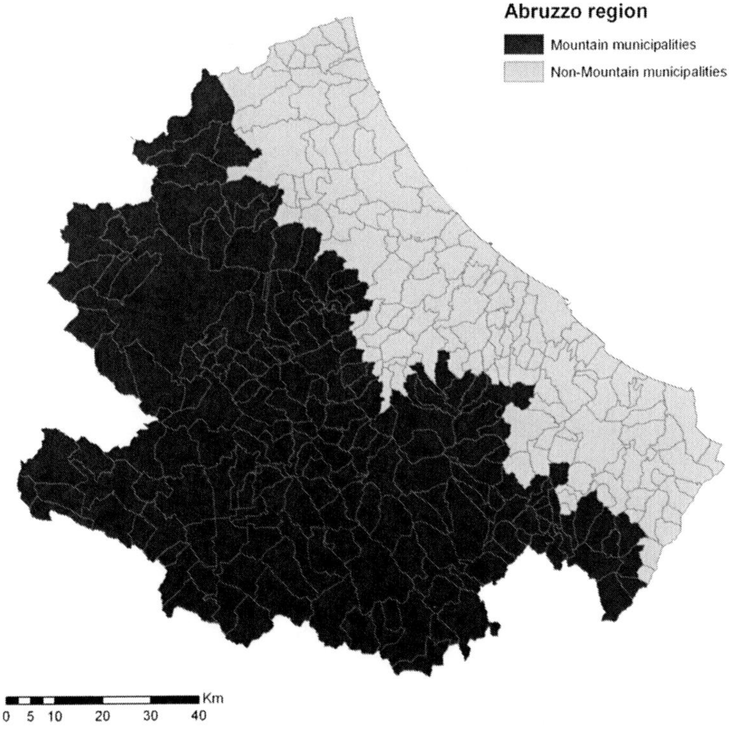

Fig. 9.1. The mountainous and the non-mountainous Abruzzo Region.

However, the mountain areas in the Abruzzo region are outlined by "plurality" showing pure urban origins, and the municipalities such as L'Aquila, Sulmona, Avezzano and Castel di Sangro are the main historical centres of the territorial planning and features,[16] although industrial, rural and tourism scenarios are found in other areas. Among these are: the Carsoli area bordering Lazio, consisting of the industrial district Piana del Cavaliere and the flourishing economies of Carsoli and Oricola (with an employee surplus index of, respectively, 211% and 46%[17];); the Eastern foothills of the Majella with Fara San Martino, famous as a "pasta production centre" (index of 87%) and Pretoro, known for winter holidays (12%); the Marsican territory, where agricultural vocation displays a high degree of productivity and income (an opposite scenario can be found on the other side of the rural mountains, which features the destruction of the agricultural system and self-assistance forms), giving the territory a stable typicality and disseminating the results (indeed there are other municipalities besides Avezzano, the latter with a better defined urban-industrialized profile, such as Scurcola Marsicana, Aielli and Ortucchio which have an employee surplus index of 16%, 12% and 3% respectively); the Altopiano delle Cinque Miglia area, including Roccaraso, known as the "capital of winter tourism" (19%), shows specific physical, orographic and infrastructural features, and has been fostered by the urban demand in Southern Italy from the beginning of the twentieth century; Parco Nazionale d'Abruzzo, Lazio e Molise, where the municipality of Pescasseroli (4%) is situated, famous for its nature tourism and already established as a summer holiday resort; the Upper Valley of the Pescara, with the historical electrochemical pole of Bussi sul Tirino, despite being in a reindustrialization phase (6%), and the town of Tocco da Casauria (26%) with a vocation related to the agro-food products (wine, liquor) and more widely to the agri-"cultural" ones (see Fig. 9.3 below).[18]

---

[16] In this regard, between the 1960s and the early 1970s the Abruzzo mountains were also affected by the incentive policy of Cassa per il Mezzogiorno with the creation of three nuclei of industrial development, in relation to the cities of L'Aquila, Avezzano and Sulmona.

[17] The employee surplus/deficit index measures the ability or inability of economic attractiveness from the territory, expressed in terms of actual localization, and is quantified by the following relationship:

> 100*(employees in X – non-agricultural workers in X)/non-agricultural workers in X (this is because the agriculture Census data doesn't provide the number of employees)

It is evident that all considered activities (industry and services) recommend the usefulness of proceeding, with a subsequent analysis of detail, to the disaggregation of data for sectors and subsectors.

[18] The decision to proceed with a "dualistic" representation shows extreme results with regard to the area of economic hardship, while regarding the territorial

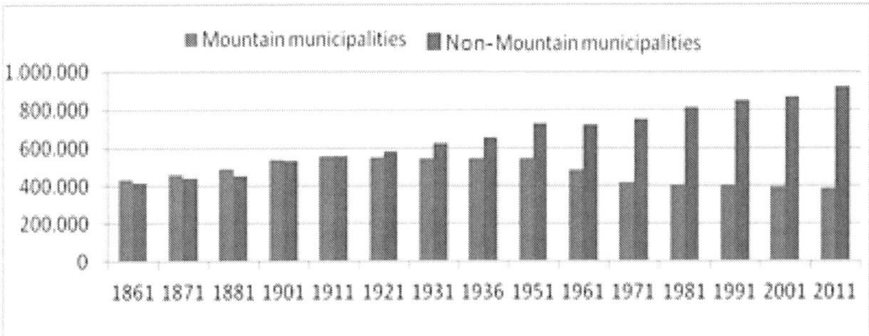

Fig. 9.2. Population time-series in the Abruzzo region
Source: own elaboration on ISTAT data

Therefore, in contrast with the monolithic concept of the mountain areas showing marginalization and underdevelopment features there is a more complex and deeper territorial reality which is composed by important cities from a historical viewpoint (first of all the administrative centre L'Aquila). This is supported by a certain degree of infrastructure (highways A24 and A25) and is partially fostered through specific development strategies, such as the environmental protection (Cardinale and Fuschi 1998; Fuschi 2001), supported by the Regional Authority through the National Framework Law no. 394, dated 1991. These results demonstrate the most distinguishing reality for visibility and extent.[19]

The extreme concentration of the urban feature of the mountain areas in the Abruzzo region, along with the excessive administrative destruction of the remaining settlements, may explain the dimension of the marginalization and partially justify the low influence played by the initiatives which were planned and implemented in recent years,[20] even if

excellence, with reference to the small towns, the result has been barely significant due to the small size of absolute data.

[19] Through the aforementioned framework law two further National Parks (Gran Sasso-Monti della Laga and Majella), a regional park (Sirente-Velino) and 37 natural reservoirs, refuges and oases for plant and wild life protection have been established in addition to the Parco Nazionale d'Abruzzo (comprising Lazio and Molise). The surface now comprises 35% of the total regional territory.

[20] To be mentioned are the APE project (Appennini Park of Europe) with several pivotal projects; the drafting of "Convenzione degli Appennini" with the aim at networking with the National Ecological Network and with the National Strategic Network dated 2007–2013 to better implement the EU funds available; the "Invest Abruzzo" initiative to encourage real estate offers for valuable historical and architectural buildings to be used for tourism purposes by buyers; several territorial marketing initiatives supported by public and private institutions to improve

the regional authorities are those responsible "to share the objectives and to build the technical and administrative system, by making the achievement of these projects faster and less complex" (Comitato Tecnico Interministeriale per la Montagna 2007).

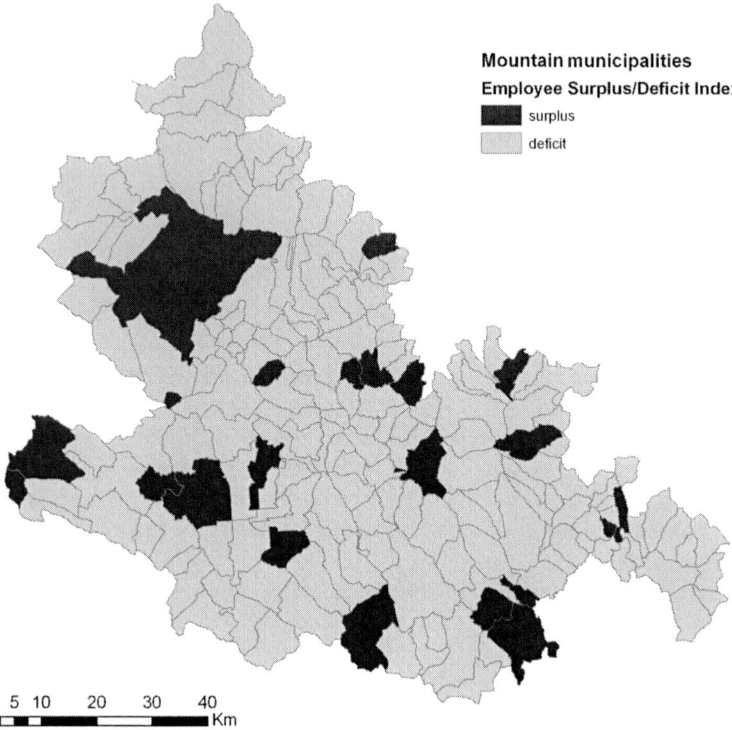

Fig. 9.3. Employee Surplus/Deficit Index in the Abruzzo mountain

For this reason, taking into account the high number of initiatives which often remain unaccomplished, it is to be highlighted that a network planning for the sector integration and the territory has been lacking. This feature is a reality as administrative reorganization processes are required which cannot be postponed anymore during this period of financial crisis taking its toll on the economy of the Western world since 2008.

---

tourism in several municipalities in the mountain areas, by participating in nation-wide projects, for instance the guide "I Borghipiù belli di Italia" ("The most beautiful villages in Italy") or other initiatives created by local municipalities, such as Borghi della Majella ("The villages of the Majella") (Fuschi & Ferrari 2009).

Furthermore, the local vulnerability due to the catastrophic earthquake which hit the mountain areas of the Abruzzo region on April 6, 2009 it is to be taken into account. L'Aquila is now one of the most seismically active areas in Italy, concerning frequency and intensity (Adamoli 1984).

The Apennine mountains of Abruzzo, an analysis of the latest economic figures. A methodological proposal.

The need to extend the analysis to the most recent figures (the official 2011 census data were not yet available for economic variables) led us to use those published by ISTAT (National Institute of Statistics) on the Local Units[21] of industry and services and their employees[22] for the year 2008.

The information derived from the Statistical Register of Local Units (Unità Locali) of active enterprises (ASIA-Local Units) was acquired through a direct survey addressed to local units of larger companies (IULGI) and administrative sources that provide data on companies and individual local units with reference to the local labour systems (LLS)23 of the Abruzzo region.

The decline of regional production areas highlighted by the comparison of the data with those for 2001 (see Table 9.5 above), especially with regard to mountain systems, emphasized a heavy critical component for employees. This aspect is even more serious if referred to a period that marks only the beginning of a crisis that has become structural in recent years (further aggravated by the Abruzzo earthquake in 2009).

There has been talk, above, of the "plurality" of mountain territories, and for this we measured such diversity through techniques of multivariate analysis. Thus, the study area is the Apennine Mountains of Abruzzo and the territorial investigation unit is the "municipality." The figures for 2008 (Source ASIA—Local Units) were available at a disaggregated level only for municipalities with a population greater than 5,000 inhabitants, while

---

[21] This term does not exist in English as it is specific to the Italian Chamber of Commerce, but for convenience is translated into "local units." It refers to a physical place where the juridical-economic unit (company, institution) carries out one or more economic activities.

[22] Persons employed in a juridical-economic unit as self-employed or employees (full-time, part-time or contract work and training), even if temporarily absent from work.

[23] The local labour systems (LLS) are aggregations of municipalities derived from a survey conducted by ISTAT and IRPET in collaboration with the University of Newcastle, UK. The criteria used for the definition of LLS are self-containment, contiguity, and space-time relationship.

the data on the remaining municipalities were aggregated for each local labour system. To extrapolate the data at a municipal level, we made a rough estimation as shown below.

Table 9.5. Regional production areas highlighted by the comparison of the data with those for 2001. Source: ISTAT/ASIA-Local Units

LLS	n. of Municipalities	n. of LU 2008	n. of EM 2008	n. of LU 2001	n. of EM 2001	% vatiation of UL	% vatiation of EM
Avezzano	29	7460	25448	7629	29003	-2,22%	-12,26%
Castel Di Sangro	18	2023	4729	2185	5469	-7,41%	-13,53%
Celano	4	1098	3063	1174	3518	-6,47%	-12,93%
L'aquila	29	6713	23968	7696	33838	-12,77%	-29,17%
Pescina	8	875	1747	1014	2674	-13,71%	-34,66%
Sulmona	24	3745	11799	4096	14262	-8,57%	-17,27%
Basciano	7	1032	3613	1137	4033	-9,23%	-10,41%
Castilenti	6	697	2732	684	3069	1,90%	-10,98%
Giulianova	12	10397	37632	9510	36313	9,33%	3,63%
Montorio Al Vomano	5	853	2653	902	3077	-5,43%	-13,77%
Pineto	3	3539	11218	3377	11938	4,80%	-6,03%
Teramo	8	7311	24925	6632	28701	10,24%	-13,16%
Penne	8	2289	7420	2211	8280	3,53%	-10,38%
Pescara	26	33910	112272	30452	116574	11,36%	-3,69%
Popoli	28	2710	9300	2828	10542	-4,17%	-11,78%
Atessa	49	9323	42122	8903	40097	4,72%	5,05%
Guardiagrele	9	1701	5108	1698	5766	0,18%	-11,41%
Ortona	10	3050	10195	3035	11064	0,49%	-7,85%
Vasto	26	7062	25823	6619	28204	6,69%	-8,44%

Knowing the number of Local Units (LU) and Employees (EM) in each municipality (of year 2001), we suppose that:

$$LU_{ki}^{08*} = vu_k * LU_{ki}^{01} \; ; EM_{ki}^{08*} = va_k * EM_{ki}^{01}$$

Where:

k varies from 1 to the number of LLS in Abruzzo
i varies from 1 to the number of mountain municipalities in each LLS

LU_{ki}^{08*} = number of local units estimated for 2008 in the i-th municipality of the k-th LLS
EM_{ki}^{08*} = number of employees estimated for 2008 in the i-th municipality of the k-th LLS
LU_{ki}^{01} = number of local units in 2001 in the i-th municipality of the k-th LLS;
EM_{ki}^{01}= number of employees in 2001 in the i-th municipality of the k-th LLS.

$vu_k = \frac{TotUL_k^{08}}{TotUL_k^{01}}$ is the constant of proportionality of total LU in all municipalities with fewer than 5,000 inhabitants in the k-th local system compared to the same in 2001.

$ve_k = \frac{TotADD_k^{08}}{TotADD_k^{01}}$ is the constant of proportionality of total employees in all municipalities with fewer than 5,000 inhabitants in the k-th local system compared to the same in 2001.

Essentially, the basic assumption was to exploit the functional relationship between quantities characterized by a non-zero numeric constant. This choice can be justified by the peculiarities of the spatial character of the labour system as well as by a certain rigidity of components of the productive structure. On the other hand, the percentage of estimated data with respect to known data is of 41.90% for LU and of 35,07% for the EM.

The objective of the analysis was to identify homogeneous clusters with relation to their structural demographic characteristics and to their productive system. To do this we applied the Cluster Analysis technique of Hierarchical Agglomerative type, choosing as a criterion of similarity the Euclidean distance and Ward's method as a hierarchical aggregation algorithm.

The variables used are: resident population, population density, aging index, dependency ratio, replacement index of the active population, LU, EM, population growth rate, percentage change of employees (compared to 2001), percentage change in local units (compared to 2001).

The hierarchical procedure applied allowed us to construct a hierarchy or tree-like structure that shows the relationship among entities identifying four clusters categorized as follows:

- the municipalities of "discomfort" that represent approximately 12% of the total. The group is thus defined as characterized by strong aging and heavy depopulation

- the municipalities of "mediocrity,",about 57%, although in critical condition less dramatic
- the "hard-working" municipalities are 28%, reflecting a lively demographic-productive condition in spite of the small size; among these should be mentioned Celano and Guardiagrele in particular, the first inserted in Fucino has recently accelerated its specialization in key agro-industrial and the second, in the foothills of the Majella, is at the top of an industrial district identified by Istat, in addition to claiming a historic "urban" role confirmed by public functions and other services in addition to being the "capital" of the Majella National Park
- the municipalities of “excellence,” L’Aquila, Avezzano e Sulmona, at only 3% confirm their role as "urban poles" of primary importance in the area of the province.

The analysis confirms a large area of marginality in the middle range, with extensive involvement of the Alto Sangro area; a confirmation of the “hard-working” areas corresponding to well-defined areas (the “area marsicana,” Piana del Cavaliere, l’Aquila basin, Subequana basin, Piana delle Cinque Miglia, the eastern Majella foothills).

Conclusions

The mountain territories in the Mediterranean area marked by vulnerable population rates (even though there are some differences between the northern and southern areas), economically weak situations (even though the areas show a high potential which is not being utilized to achieve high standards of productivity, acknowledgement and trade) and environmental fragility (reflecting a feeling of pressure in the south and abandonment in the north). Hence, the mountain territories have to regain their importance within the territorial and economic development policy. For this purpose, there is a wide range of considerations and proposals made by the European and Mediterranean partnership policy; however, the lack of a methodical policy for the mountain territories in the northern and the southern areas has achieved an amount of expenses which is no longer affordable.

However, shifting from the perspective according to which the mountain areas are considered marginal territories which need help and are subject to by-laws and other laws, according to which the mountain areas are deemed as one territory where the mountain-coast relationship is to be thought of in terms of alliance, a long-term sustainable development and improvement for the whole Mediterranean region can be guaranteed.

The Abruzzo Mountain, in particular, needs to mend that relationship now accrued on oppositive terms. The starting point is to focus on a relationship based on mutual needs which has to firstly be acknowledged by institutions and civil society, beginning from its strengths and recognizing its weaknesses.

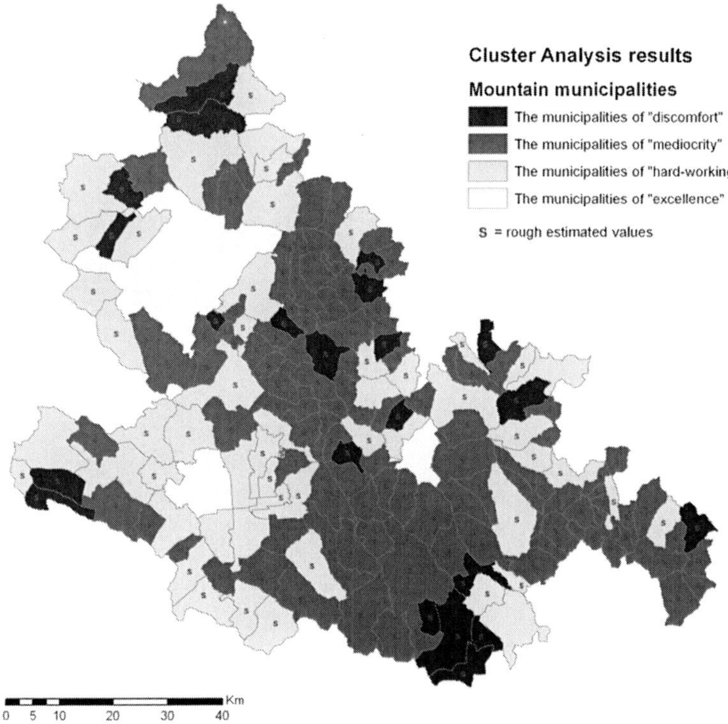

Fig. 9.4. Cluster Analysis results.Source: Authors' elaboration

As for the mountain areas in the Abruzzo region, a new territorial model has to be drafted and implemented by focusing on:

- reorganizing the self-centred urban system into a polycentric system by encouraging relational networking and removing old narrow-minded positions about recessive and territorial drifting. For this purpose, the cities of L'Aquila, Sulmona, Avezzano and Castel di Sangro should consolidate their relationships with the provincial territory by strengthening the infrastructure of highways and state roads;

- consolidating the territorial potential through aggregation policies between municipalities to overcome the issues regarding the demographic and the functional undersizing, involving the municipalities in projects and financial circuits and giving them an active and independent management role, preventing imported models which do not have any territorial influence;
- planning a new industrial policy, according to which the different specializations (electronics, chemistry and mechanics) should adopt the logic of the industry chain by leaving aside the structure for "nuclei/consortia in their own right," and which are in fact already obsolete. (Landini e Massimi 2010, 334);
- being involved into a broader policy which focuses on the sectoral and territorial integration, by starting to reallocate public services (education, health care, basic infrastructures) and implementing mutual development projects capable of networking the different potential factors from the agricultural field (composed of top-of-the-range sectors), from the cultural field (focussing on the main historical and architectural features) and from the environment (the Abruzzo region as the Region of the Parks).

These actions are necessary to consolidate the strengths of the territory, limiting the abandonment from the mountain areas, and to rebalance the organizational structure of the region based on a heavily weighted scenario on the coastal area and gravitating around the Pescara-Chieti area (Fuschi 2006), in a perspective of integrated regionalization that alone can project Abruzzo into a new and complex scenario imposed by globalization. One should not forget that the "alliance" between the mountain territory and the coastal area must first of all be a political project capable of adopting a polycentric mindset and thinking in terms of a project which overcomes the local dimension, albeit within a complex and a strategic multisectoral approach which reflects the peculiarities of the region and supported by "participation," as Magnaghi says: "***Reconstruction of the Community is the essential element of self-sustainable development***" (2010, 91).

References

Adamoli, L. (1984). Le cause dei terremoti in Abruzzo. In Terremoti e centri storici, ed. A. GhisettiGiavarina, pp. 129-143.Lanciano: Carabba.

Almagià, R. (1934). Montagna. In Enciclopedia Italiana, fondata da G. Treccani, p. 711. Roma: Istituto della Enciclopedia Italiana,vol. XXIII.

Aldenderfer, M.; Blashfield, R. (1984). Cluster Analysis, London, Sage Publ..

Atalay, I; Efe, R.; Öztürk, M. (2014). Effects of Topography and Climate on the Ecology of Taurus Mountains in the Mediterranean Region of Turkey.

Elsevier, Procedia - Social and Behavioral Sciences. Volume 120, Pages 142–156

Bernardi, R. (1994).Montagna e montagne.In L'evoluzione della montagna italiana fra tradizione e modernità, eds. R. Bernardi,S. Salgaro, C. Smiraglia, pp. 9-16. Bologna: Pàtron.

Blache, J.; Blanchard, R. (1933). L'homme et la montagne. Gallimard, Parigi.

Braudel, F. (1998). Memorie del Mediterraneo. Bompiani, Milano.

Buzzetti, L. (1998). Urban function in mountain regions. In Urban and Regional Development in Italy and in Poland, ed. A. Besana, pp. 17-34. VIII Seminario Italo-Polacco, Trento: Ed. Colibrì.

Cardinale, B.; Fuschi, M. (1998). Environmental protection in Abruzzo. In Urban and Regional Development in Italy and in Poland, ed. A. Besana, pp. 207-221. VIII Seminario Italo-Polacco, Trento: Ed. Colibrì.

Censis (2002). Il valore della montagna. Angeli, Milano.

Comitato Tecnico Interministeriale per la Montagna (2007).Relazione sullo stato della montagnaitaliana. Roma.

De Vecchis, G. (1996). Da problema a risorsa: sostenibilità della montagna italiana. Ed. Kappa, Roma.

Dubost, M. (2003). Quel avenir pour les montagnes mediterraneennes? Seminaire sur le smontagnes mediterraneennes (1-2 Avril, 2003, Sophia Antipolis). I calpe, paper.

Efe, R.; Atalay, I., Soykan, A. (2008). Mediterranean Ecosystems of Turkey: Ecology of Taurus Mountains. In: Efe, Cravins, Ozturk, Atalay (Eds). Environment and Culture in the Mediterranean Region. Part I, Chapter One p. 3-37, Cambridge Scholars Publishing. ISBN (13): 9781847186584, 2008, Newcastle, UK.

Fabietti, U. (2001). Sceicchi, beduini e santi. Potere, identità tribale e religione nel mondo arabo-musulmano.Angeli, Milano.

Febvre, L. (1980). La terra e l'evoluzione umana. Introduzione geografica alla storia. Einaudi, Torino.

Fuschi, M. (1994). Il carsismo nella montagna abruzzese: fra determinismo e sviluppo. In L'evoluzione della montagna italiana fra tradizione e modernità,eds. R. Bernardi, S. Salgaro, C. Smiraglia, pp. 331-341. Bologna: Pàtron.

—. (1999). Cambiamento globale e aree protette: una riflessione geografica. Boll. Società Geografica Italiana, 765-773.

—. (2000). Proposta di classificazione delle aree intermediedella montagna abruzzese. In La montagna appenninica italiana: conoscere per gestire, ed. R. Bernardi, pp. 107-124. Bologna: Pàtron.

—. (2001). Environmental protection and social protection: the Sirente-Velino Regional Park.In Global Change and Protected Areas, eds. G. Visconti, M. Beniston, E.D. Iannorelli,D. Barba, pp. 475-487. Dordrecht: Kluwer AcademicPublishers.

—. (2006). Per una regione medioadriatica. Città, territorio, economia. Angeli, Milano.

—. (2008). Il Mediterraneo. Geografia della complessità. Angeli, Milano.

Fuschi, M.; Ferrari, F. (2009). La montagna abruzzese: da icona dell'abbandono a immagine del recupero e della valorizzazione.In Territori contesi. Campi del sapere, identità locali, istituzioni, progettualità paesaggistica, ed. P.Persi, Atti IV Convegno Internazionale Beni Culturali, Pollenza 11-13 luglio 2008, pp. 408-414.Ciocca Grafiche.

Gasparinetti, P. (1967). La "via degli Abruzzi" e l'attività commerciale di Aquila e Sulmona nei secoli XIII-XV. Palombi, Roma.

Landini, P.; Massimi, G. (2005). I sistemi geo-economici abruzzesi. Una lettura integrata areale-reticolare. Cresa, L'Aquila.

Landini, P.; Massimi, G. (2010). Sismicità, insediamento, economia. Analisi geografica nelle aree del terremoto d'Abruzzo (2009). Boll. Società Geografica Italiana, 2, 287-339.

Landini, P.; Massimi, G.; Fuschi, M. (2011). Regionalismo e città diffusa. La "saldatura" medio-adriatica. In Tante Italie Una Italia. Dinamiche territoriali e identitarie, eds. C. Muscarà, G. Scaramellini, I. Talia, Vol. III, Terza Italia. Il peso del territorio, pp. 113-132. Milano: Angeli.

Magnaghi, A. (2010). Il progetto locale. Verso la coscienza di luogo.Bollati Boringhieri, Torino.

Mercurio, F. (2000). Reti di comunicazione e formazione delle gerarchie territoriali.In L'Abruzzo,eds. M. Costantini, C. Felice, Coll. "Storia d'Italia. Le regioni dall'Unità a oggi", pp. 807-876. Torino: Einaudi.

Piccioni, L. (1993). La grande pastorizia transumante abruzzese tra mito e realtà. In Abruzzo e Molise. Ambienti e civiltà nella storia del territorio, eds.M. Costantini, C. Felice, pp. 195-229. Mantova: Cheiron. Ed. Centro Federico Odorici.

Sori, E. (2004). Storiografia e storia della montagna appenninica: l'evoluzione demografica.In Economia nel tempo. Persistenze e cambiamenti negli Appennini in età moderna, eds.A.G. Calafati, E. Sori, pp. 21-38. Milano: Angeli.

Soykan, A.; Efe, R.; Atasoy, E. (2008). Functions of Mountains: A Case study in the Mountains of Turkey. International Conference. Education, Science, Economics and Technologies in the Global World. 12th-14th September 2008, Burgas, Bulgaria, Proceedings.

Veyret, P.G. (1962). Essai de définition de la montagne. Revue de géographie alpine, 5-35.

—. (1964). Petites et moyennes villes des Alpes. Revue de géographie alpine, 6-124.

Zani, S. (2000). Analisi dei dati statistici II: osservazioni multidimensionali. Giuffrè: Milano.

CHAPTER TEN

AN URBAN-GEOGRAPHIC STUDY OF URBAN SETTLEMENTS IN BOSNIA AND HERZEGOVINA AS AN ELEMENT OF AREA PLANNING

RAHMAN NURKOVIĆ AND
NUSRET DREŠKOVIĆ

Introduction

Urban settlements in Bosnia and Herzegovina, in accordance with socio-economic and other factors of urbanisation development, have been gradually changing over the time. Most of our urban settlements are still in an urbanisation stage characterised by strong polarisation, i.e. a fast growth of cities as a consequence of the concentration of jobs and population migration from villages. Such a form of urbanisation with, naturally, certain specificities, is characteristic of all countries in an industrialization stage. A rapid rural population growth and the resulting rural exodus have certain consequences which arouse attention and require certain interventions of society in Bosnia and Herzegovina. These are, on the one hand, problems in cities that appear due to overemphasized and often uncontrolled concentration (large population density, unemployment, "wild" construction, pollution etc.), and problems occurring in rural environments (ageing populations, appearance of old people's households, extensification of production etc. [Černe 1999]).

In larger urban settlements a higher level of socio-economic development and urbanisation has been reached. Urban population growth ranges mostly within dynamics of total population growth. The focus of urbanisation, in the sense of population concentration, jobs and housing settlements, is shifted to the fringes and suburban zones. In addition to pronounced differences in urbanisation forms in Bosnia and Herzegovina, primarily industry and mining, urban settlements developed the production of different industrial products and services, not only within its borders, but also in a broader gravitation area. Over the twentieth century, urbanization developed permanently under conditions of rapid population growth. Interest has predominately been oriented to housing construction, which gave a fundamental characteristic to the spatial and urban development of urban settlements. Communication and other characteristics had one of the most important roles in Bosnia and Herzegovina in the past. The objective of this chapter is to analyse intensity, forms and other characteristics of urbanisation in urban settlements of Bosnia and Herzegovina in the period 1879–2010 in the conditions of the aforesaid littoralisation.

Material and Methods

Urbanization development of the urban settlements in Bosnia and Herzegovina, with its own specificities in moderately small areas, requires using special methods. According to the methodology of urban geography scientific methods have been used, such as methods of analysis and synthesis, mapping methods, statistical methods, descriptive methods, methods of field research as well as methods of text analysis. Therefore, studying the general models and urban geography methodology in combination with the urban-geographic approach has been used. Agencies of the entity institutes for statistics, together with the Agency for Statistics of Bosnia and Herzegovina, have been conducting a survey on urban population in accordance with methodological rules and principles of the International Labour Organisation (ILO) and the requirements of Eurostat for five years, by which the international comparability of data in the area of statistics of urban settlements has been provided. As a basic method of collecting primary data sources, an in-depth interview method has been used where the main instrument was a reminder for an interview. About 25 urban settlements in Bosnia and Herzegovina have been surveyed. Research was also completed through an analysis of secondary source contents, interpretation and description of adequate data bases of the Agency for Statistics of Bosnia and Herzegovina.

Participation and growth of urban population as indicators of urban population level and dynamics

Urban population and its share of total population are among more relevant indicators of the urbanisation level in Bosnia and Herzegovina. The pronounced differences between total population growth, which is low, and the relatively high urban population growth correspond to the adequate stage of demographic transition with a low rate of natural growth, and the so-called industrial stage of urbanisation with a strong concentration in cities. In earlier periods Bosnia and Herzegovina also recorded a rapid growth of urban population as well as other neighbouring countries (Žuljić 1976).

In many developing countries the rates of growth of total and urban populations as well as their differences are considerably bigger, while in developed countries rates of growth of total and urban population are low, and the differences between them are insignificant. These data may be used as an orientation for the evaluation of development of our country and its parts. High rates of urban population growth are certainly indicators of population concentration in the cities. Such is the case with Bosnia and Herzegovina, and the development of the total population confirms this. Observed from a contemporary perspective, Bosnia and Herzegovina has a small number of more significant urban centres. Sarajevo, according to the first official Austro-Hungarian census, had 21,377 inhabitants in 1879. Despite the frequent wars, natural disasters and similar difficulties, the population in Sarajevo had been increasing rapidly, and, according to estimations, about 750,000 inhabitants lived in Sarajevo in 2010. In addition, Tuzla, Mostar, Banja Luka, Goražde, Doboj, Prijedor, Brčko, Bihać and Zenica are significant in the regional sense (Table 10.1 and Fig. 10.1).

Table 10.1. Review of population number in the more significant cities of Bosnia and Herzegovina, 1879–2010

Grad	1879.	1885.	1910.	1971.	1991.	2010/ procjena
Banja Luka	9.560	11.357	13.566	158.736	195.692	225.000
Bihać	3.097	3.506	3.943	58.185	70.732	63.000
Brčko	2.901	4.281	5.998	74.771	87.627	100.000
Goražde	846	1.226	1.460	34.685	37.573	17.000
Doboj	1.351	1.749	2.768	88.985	102.549	80.000
Mostar	10.848	12.665	14.370	89.580	126.628	112.000
Prijedor	4.681	4.746	4.999	97.894	112.543	95.000
Sarajevo	21.377	26.268	38.083	359.452	527.049	750.000
Tuzla	5.119	7.189	10.227	107.293	131.618	174.000
Zenica	2.101	3.073	4.226	60.910	145.517	118.000

The mentioned directions of urbanisation are mainly followed by the main lines of communication, which shows the significance of traffic accessibility in their appearance. The same conditions are also present in other regions of Bosnia and Herzegovina. The dynamics of urban settlement development and expansion of the urbanised zones in Bosnia and Herzegovina show the polarisation tendency in the urbanisation process. Distinguishing the urban settlements in Bosnia and Herzegovina is a permanently present problem. With the development of various activities, primarily of industry and mining, urban settlements in particular had been developing production of different industrial products and services, not only within their borders but also over a broader area. However, in the practice of differentiation of settlements in Bosnia and Herzegovina, a small number of indicators were applied. Most commonly this was a settlement size, then people's lifestyles, which was mostly expressed by the ratio of agricultural and non-agricultural populations, and other relative indicators. In 2010, diversity in distinguishing urban settlements in Bosnia and Herzegovina was expressed when the Institute of Statistics of Bosnia and Herzegovina separated three categories of settlements: urban, mixed and rural.

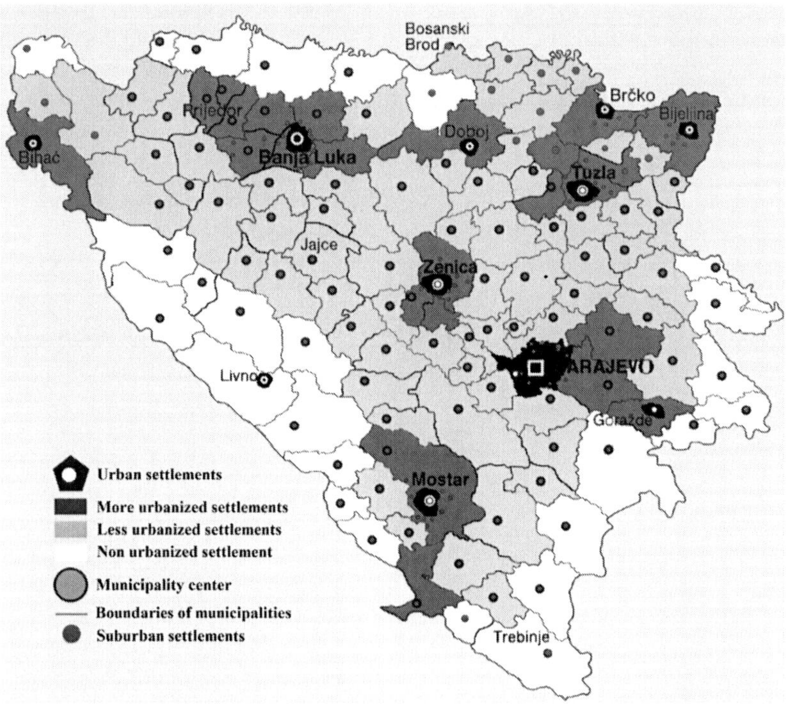

Fig. 10.1. Urbanisation level of urban settlements in Bosnia and Herzegovina, 2010.
Authors: Nurković, R. (2010). According to: Agency for Statistics of Bosnia and Herzegovina

For these categories a model of settlement size and participation of non-agricultural population was applied.

It may be concluded that this model satisfied the needs of differentiating the settlements in 1981, but in the differentiation of settlements some shortcomings have appeared. Therefore, it was necessary to adjust a model of differentiating the urban settlements with the existing development level (Lorber 2009).

If the separated cities of Bosnia and Herzegovina are analysed on the whole as a unique urban system, edifying results are obtained. In order of urban settlements size, which indicates the hierarchic features of urban systems, a certain irregularity was noticed that was also present in the urban development of Bosnia and Herzegovina in earlier periods. The first thing to notice in order of urban settlement size in Bosnia and Herzegovina is the overemphasised size of cities of Sarajevo, Banja Luka, Tuzla and Mostar, actually the largest urban settlements in relation to others. The differences in urban population share, respectively the urbanisation level, are the consequences of unequal dynamics of urban settlement development in all areas of Bosnia and Herzegovina. These differences are noticeable from data present in the first decade of the twenty-first century. For this purpose, it is possible to analyse the dynamics of total and urban population growth in the period 1981–1991. As the data on population numbers in 1981 and 1991 are not comparable due to different criteria of separating the cities, it is possible to compare the urban populations of separated urban settlements according to the mentioned model in 1981, as well as the population of cities in 1991.

Comparing the growth of the total and urban populations in Bosnia and Herzegovina in the period 1981–1991, it should be emphasized, first of all, that total population growth in Bosnia and Herzegovina was 38.2%. The differences between the growths of the total and urban populations were big. Sarajevo, Tuzla, Zenica and Banja Luka are the hubs in the urban network of Bosnia and Herzegovina, more important than others because they are the centres of larger numbers of connections. Such hubs have good accessibility. Settlements in which there are the largest numbers of central functions have the highest accessibility, as it is a precondition of a better spatial organisation. The least accessible hubs are in South and East Bosnia and in West Herzegovina, and are Foča, Goražde, Trebinje and Livno (Nurković 2003; Nurkovic 2011).

Data on the number and dynamics of urban settlement growth in Bosnia and Herzegovina are available, and may be judged according to census data from 1961, 1971, 1981 and 1991, as well as according to estimations for 2011. Until 1991, Bosnia and Herzegovina had 109 municipal centres, respectively urban settlements. In 2011, after the war events and a new territorial arrangement, Bosnia and Herzegovina had 141 municipal centres, which is more by around 30% than twenty years ago. With the changes in structure of settlements according to size and with an increase in the number and sizes of cities, urban systems of the regions have changed along with the country's urban system on the whole. In these changes, a positive trend is also noticed, respectively the creation of regular urban systems. The degree of primeness in Bosnia and Herzegovina was more than two. Data on population numbers in the cities of Bosnia and Herzegovina refer to certain differences in courses of urbanisation of the regions as well as to differences in the development of single agglomerations. Namely, it is obvious that in single regions a process of pronounced concentration is present, first of all, in regional centres, and that in some other regions polycentric urban development, with a metropolization tendency, is expressed (Vresk 1992).

Socio-economic transformation of urban and rural settlements as an urbanisation indicator

Socio-economic changes are the first form of transformation of rural and urban settlements in Bosnia and Herzegovina. Observed sociologically, it implies changes of population position on the social scale. When it comes to rural settlements it implies, first of all, restructuring of agricultural population into non-agricultural occupations and change of lifestyle. Socio-economic changes are also followed by other changes such as physiognomic and functional changes of settlements. In this manner, the urbanisation of rural settlements from the environment is performed. The main agent of the mentioned changes is employment of population in non-economic activities. As there are no jobs in rural settlements, or only a few, social restructuring has been enabled by daily migrations to work in a city. In order to evaluate the urbanisation level of Sarajevo, for example, settlements were differentiated according to socio-economic changes in four categories: urban settlements, more urbanised, less urbanised and rural settlements. Differentiation was performed by means of the model with two variables and specific parameters (Table 10.2)

Table 10.2. Model of differentiating the settlements according to level of socio-economic transformation in Bosnia and Herzegovina

Transformation degree	% agricultural population	% workers of active population
Higher degree	10.0 and less	75.0 and more
Lower degree	10.1–20.0	50.0 and more

With the mentioned model, more and less urbanised settlements and rural settlements were separated. All the settlements that do not meet the mentioned criteria in a model were included in the last group. The status of town was given to the settlements with more than 2,000 inhabitants, with less than 10% of agricultural workers, and with more than 50% workers in the place of living, if it has less than 10,000 inhabitants. Differentiation of settlements by means of the mentioned models shows that a larger number of settlements from around Sarajevo experienced some degree of transformation. Among the urban settlements, Osijek, Stup, Semizovac, Rajlovac and Hrasnica are distinguished for their importance and size. Among urbanised settlements of the surroundings, more urbanised settlements distinguish themselves by number. Along with the less urbanised settlements, they have a circular zone whose thickness varies considerably.

Transport connections caused a radial expansion of socio-economic transformation of settlements conditioned by travelling opportunities, particularly expressed towards Hadžići, Ilijaš, Kiseljak, Vogošća and Pale. It can be ascertained that transport connections in the mentioned directions enabled development of axes of urbanisation and development. Furthermore, satellite towns also contributed to the expansion of suburbanisation zones, which are important hubs connecting with Sarajevo, and also have a certain function of work and their gravitation zones of daily migrations. Therefore, intensity of transformation of settlements is somewhat more expressed around the satellite towns than the intensity of migrations in Sarajevo. This is particularly expressed around Ilidža, Vogošća and towards Hadžići. On the basis of the mentioned data on socio-economic transformation of settlements, it may be ascertained that the suburbanization of Sarajevo has become intensive. It is performed differentially, and is still dependant on daily migrations of the employed people in Sarajevo. Satellite towns certainly have an important role in this process. The migration of the population is a significant indicator of the importance and attractive power of each town and the values of their surroundings. At a certain level of each agglomeration, stronger migrations are, first of all, directed towards the town, and then redirected to the environment (Vrišer 1988; Pacione 2001; Lorber 2009).

Migration to city surroundings may be performed from rural settlements from the environment or other areas, and also from the parent city. In the directions of migration there are certain regularities that need to be taken into consideration prior to evaluation of suburbanistion level. In the earlier stage of development, migration to city surroundings was predominantly from distant and more isolated rural settlements. Sometimes it has a character of migration in stages towards the city. In later stages of development, the population of the parent city also moved to the surroundings. This can be defined by spontaneous or enhanced planned housing construction. For the evaluation of urbanisation, knowledge of the structure of immigrated population is also important. The poor population can move from the city to the surroundings due to a lack of more favourable apartments. The migration of a population of higher social status to the surroundings is a sign of its large value and usually appears in later stages of agglomeration development (Hallsworth 1994).

The intensified population migration of each country regularly follows industrialization and urbanisation. In Bosnia and Herzegovina, more intensive migrations, particularly to cities, started in the 1950s. In 1991, the participation of migrating population to cities was 62.4%. The motives and migrations to particular settlements are different. Migrations to satellite towns and to closer settlements were caused by opportunities presented by housing construction. Lower utility and cheaper construction land enabled the easier construction of houses. A policy of housing construction with a tendency of decentralisation contributed to the population migrating to satellite towns, particularly during the 1970s. In satellite towns, housing settlements used to be constructed in a planned manner, and at the same time the possibilities for the construction of housing units were poor. Due to prohibitions or the expensiveness of construction land, migration to the surroundings was enhanced. Migrations to rural settlements were mostly from neighbouring settlements, and were caused by land sales, marriages and other reasons.

On the grounds of level of socio-economic transformation and functional dependence on a city, it is possible to distinguish the surroundings of Sarajevo as separated and differentiated on the grounds of socio-economic features and their functional connection with Sarajevo. It is a level of employment shown by share of people employed out of own estates in total active population, and the share of daily migrants employed in Sarajevo. The share of daily migrants has been calculated in two forms: active and employed population. An additional characteristic in separating the surroundings is the level of socio-economic transformation of settlements. Separating and differentiating the surroundings was performed at the level of new municipalities (Table 10.2, 10.3).

According to the mentioned model, the surroundings of Sarajevo included municipalities in which the share of the employed active population was over 50%, and the share of daily migrants to Sarajevo was over 25%, respectively 30%. According to this, 6 municipalities were included in the surroundings and the eastern and southern sectors of the

Sarajevo agglomeration. The majority of municipalities have a higher level of transformation and connection with Sarajevo.

Table 10.3. Model of separating and differentiating the surroundings of Sarajevo

Degree of transformation	Employed people	Active	Employed people
Higher degree	>50	>40	>50
Lower degree	>50	>25	>30

Conclusion

The urbanized population of Bosnia and Herzegovina has had rather high growth rates in the whole period, reaching 49.5% of urbanisation in 2011 (inhabitants of settlements with more than 2,000 inhabitants are taken to be urbanised), showing that Bosnia and Herzegovina belongs to those countries with a medium level of urbanisation. Accelerated urbanisation in the period 1991–2010 may be attributed to the considerably larger role of smaller settlements that become secondary centres in the municipalities, which is in accordance with the proclaimed principles of polycentric development also supported by the first generation of spatial plans of municipality.

The interdependence of urbanisation forms and spatial mobility of the population in Bosnia and Herzegovina is clearly expressed. The permanent migrations of the population to centres of work still have a big importance in urban development, and this is fostering a further concentration and development of existing centres. A positive feature in such development is noticed in the slower growth of larger cities (Sarajevo, Banja Luka, Tuzla and Mostar) with certain tendencies of decentralization, which is expressed clearly with Sarajevo. The satellite towns of Sarajevo have a faster growth, as well as some smaller centres.

Constant migrations show the tendency of decline, and daily migrations are getting stronger, fostering the development of urbanised zones. A rapid population growth in urban settlements and the rural exodus, as a result, have certain consequences that arouse attention and require certain interventions of society. It is thought that economic forces have a dominant influence on the urban and economic development of cities in Bosnia and Herzgovina. In addition to material production, cities are also bearers of service activities. Except for industry, these are trade, education, health care, administration, banking, insurance etc. The populations of the surrounding places satisfy many existential needs in the cities, but the population of cities are also supplied with many food products from the surroundings. Therefore, there are complementary functional relations between a city and the neighbouring area. The intensiveness of the functional connection between a city and the surrounding area, as a rule, weakens with increase in distance from the city.

References

Černe, A. (1999). Prostorska identiteta – koncept različnosti. Anthropos, Časopis za psihologijo in filozofijo ter za sodelovanje humanističnih ved, letnik 31, štev. 4-6, pp. 296-300, Ljubljana, Slovenia.

Hallsworth, A.G. (1994). Decentralization of Rretailing in Britain: The Breaking of the Third Wave, Professional Geographer 46 (3), 296-307.

Lorber, L. (2009). Transition in Slovenian rural areas. Revija za geografijo – Journal for Geography, 4-1, Ljubljana, Slovenia.

Nurković, R. (2003). Influence of industry on urbanistic development and infrastructure of central settlements of the Tuzla valley, University in Belgrade, Faculty of Geography- Institute of area planning, Almanac, pages 355-364, Belgrade, Serbia.

—. (2011). Proceedings REAL CORP 2011 Tagungsband, Essen. http://www.corp.at ISBN: 978-3-9503110-0-6 (CD-ROM); ISBN: 978-3-9503110-1-3 (Print).

Pacione, M. (2001). Urban Geography- a global perspective, Routledge, London, UK.

Statististics annual /Chronicle of the Federation of Bosnia and Herzegovina, Editors: Manfred SCHRENK, Vasily V. POPOVICH, Peter ZEILE. page 24, Sarajevo, (2000).

Vresk, M. (1992). Bases of urban geography, pp 76-84, Zagreb, Croatia.

Vrišer, I. (1988). Centralna naselja v Sloveniji leta 1987, Geografski zbornik, XXVIII, str. 129-151. Ljubljana, Slovenia.

CHAPTER ELEVEN

SELECTIVE DEVELOPMENT OF TOURISM IN BOSNIA AND HERZEGOVINA

RAHMAN NURKOVIĆ AND ŠEMSUDIN DŽEKO

Introduction

With regards to the importance of selective tourism in Bosnia and Herzegovina in the past ten years, it is obvious that contemporary technical measures strongly affect the development of selective tourism in urban and rural areas. In addition, some new characteristics of selective tourism development according to diverse natural and social attractions, level of tourist valorisation and different reasons for travelling in Bosnia and Herzegovina have been emphasized in this chapter. The basic purpose of this research is to determine which concepts have been applied for the development of selective tourism in Bosnia and Herzegovina. Various issues and problems that appeared in the development of selective tourism have also been researched (Cigale 2004).

Bosnia and Herzegovina is an attractive country whose tourist development is based on natural and anthropogenic resources. Though a small country with less than 1% of the world's population, it has a significant role in tourism. A review of tourist resources shows that the majority of tourist areas and attractions have been insufficiently valorized so far. The attractiveness of the high mountains of Bjelašnica, Igman, Vlašić and others is reflected in the tourist facilities built. The percentage of tourists visiting these localities is around 30%, while the percentage of tourists visiting historical heritage in the cities of Sarajevo, Banjaluka, Mostar and Tuzla is approximately 60%. Based on research and geographic elements of tourism development in Bosnia and Herzegovina it can be concluded that there is a successful connection with other economic activities in a new regional development. This, first of all, refers to natural geographic features and anthropogenic contents of selective tourism development in Bosnia and Herzegovina. In the actual conditions of the market economy in Bosnia and Herzegovina, selective tourism may ensure an even faster economic development. Today, it effects, for the most part, the regional development and represents a challenge of a big spatial, ecologic and demographic change, but it certainly confirms that tourism has so far developed insufficiently in Bosnia and Herzegovina in relation to other branches of industry (Kurtović 2005).

Methods of work and data sources

For collecting primary data sources an in-depth interview method was used, in which a main instrument was a reminder for an interview. About 200 respondents were interviewed with regard to the importance of selective tourism in Bosnia and Herzegovina over the past ten years. The interview consisted of 20 questions on selective tourism development for each respondent, knowledge of the concept of tourism development, and motives, advantages and obstacles in the quality of cooperation of competent institutions in Bosnia and Herzegovina. Questions were both of a closed and open type. The obtained data were processed by means of the SPSS programme.

The research has been completed by analysing the contents of secondary resources, and the interpretation and description of adequate databases of the Agency for Statistics of Bosnia and Herzegovina. In the research, the interview method and tourist valorisation of tourist resources conducted by experts for tourism in Bosnia and Herzegovina have been used, which contributed to the relevance of data that were statistically processed afterwards. Family households were also included in the survey research and a big interest in inclusion in a process of offering tourist services was recorded.

Natural and cultural characteristics of selective tourism development

Starting from the natural and anthropogenic characteristics of Bosnia and Herzegovina, as well as from the fact that a quality tourist development requires abandonment of the concept of massiveness, or better said the alleviation of negative effects that were brought by development of a conventional form of tourism, and also in favour of selective, sustainable and elite tourism, it is indispensable to seriously give attention to the quality aspects of selective tourism that would bring not only a new quality and spatial redistribution of the offer to Bosnia and Herzegovina, but also an extended tourist season, sustainable tourism development and awakening of the area that is not affirmed in the tourist sense despite having real potential. It is believed that the enriched natural and anthropogenic tourist values in Bosnia and Herzegovina would properly valorize the tourist market. New selective forms of tourism would provide different behaviours of participants from the behaviours of tourists in mass trends, a different relation to environment, a smaller number of participants, and different principles on which a concept of sustainable development is based (ecologic, socio-cultural and economic) (Lorber 2005).

It is believed that the values of the area are generally known, but not sufficiently, even to people living in Bosnia and Herzegovina. Knowing that the area is multiconfessional, ethnically very diverse, the possibilities are unforeseen. Considering the fact that the heterogeneity of the tourist demand is more and more expressed, Bosnia and Herzegovina represents a valuable area that has real conditions for development of less aggressive forms of tourism, which will contribute to the sustainable development of transit tourism. Particular areas of Bosnia and Herzegovina that may be recognizable in the future through the development of ethno, eco, rural, sports, hunting and other types of tourism will therefore be explored (Fig. 11.1).

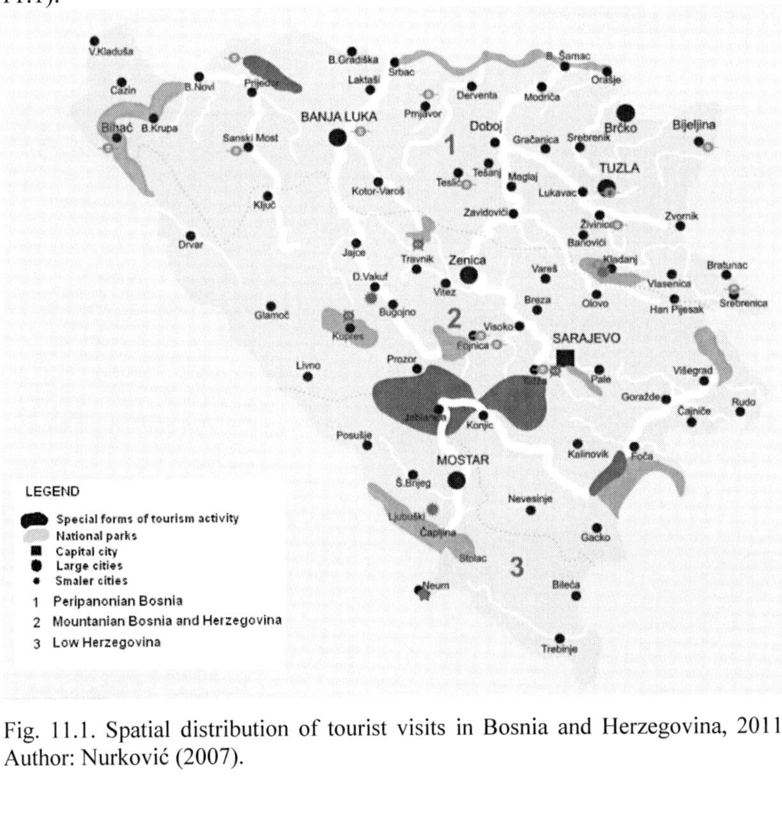

Fig. 11.1. Spatial distribution of tourist visits in Bosnia and Herzegovina, 2011 Author: Nurković (2007).

The Agency for Statistics of BiH, 2011, shows that in the period January to November 2011 tourists realised 639,746 visits, 4.5% more than the previous year, and 1,415,456 overnight stays, 5.8% more compared to the same period in 2010. In the structure of overnight stays of foreign tourists in the same period, most were from Croatia (15.0%), Serbia (13.2%), Poland (8.3%), Slovenia (8.2%), Italy (6.7%) and Turkey (6.2%), a total of 57,6%. Tourists from other countries comprised 42.4% overnight stays. However, a small number of arrivals and overnight stays in private accommodation are noticeable, respectively in the households mainly located in rural areas.

Table 11.1. Arrivals and overnight stays of tourists according to types of accommodation facilities in Bosnia and Herzegovina, 2010–2011

Types of accommodation	Arrivals		Overnight	
	I-XI/2011	Indeks I-XI 2011/ I-XI 2010	I-XI/2011	Indeks I-XI 2011/ I-XI 2010
Hoteli	522,406	105.4	1,100,152	105.8
Moteli	53,755	94.2	86,051	95.6
Pansioni	22,281	109.8	50,058	104.3
Domaćinstva (privatne sobe)	3,476	117.9	4,919	100.4
Radnička odmarališta	37,828	105.1	174,276	112.9
Domaći turisti	269,105	101.2	621,469	103.3
Hoteli	188,276	98.9	394,685	98.6
Moteli	36,896	104.4	52,501	103.8
Pansioni i prenoćišta	13,737	116.2	22,966	112.9
Domaćinstva (privatne sobe)	2,649	135.9	3,158	134.0
Radnička odmarališta i ostalo	27,547	104.6	148,159	115.6
Strani turisti	370,641	107.0	793,987	107.9
Hoteli	334,130	109.4	705,467	110.2
Moteli	16,859	77.5	33,550	85.1
Pansioni i prenoćišta	8,544	100.8	27,092	98.0
Domaćinstva (privatne sobe)	827	82.7	1,761	69.2
Radnička odmarališta i ostalo	10,281	106.5	26,117	99.7
TOTAL	639,746	104.5	1,415,456	105.8

Source: Agency for Statistics of Bosnia and Herzegovina, 2012

Cultural tourism

One of the most important trends in world tourism during the past decade is an increase of short trips generally towards easily accessible towns. This is reflected in the reduction of the average duration of tourist trips in a majority of destinations worldwide. Different correlated factors have contributed to this behaviour, such as changes in people's lifestyles, irregular distribution of work, uncertainty of jobs and a lack of time on the one hand, and the interest of consumers in cultural-historical attractions and an opportunity for consumption on the other. Therefore, people usually choose those destinations that may be reached easily and quickly. Culture, history, education, business and a visit on the way (in transit) are the most common motives for visitors to cities. Business tourism is an important and a separate segment of the tourist market which offers a huge benefit to a country's urban centres (Pepeonik 1995).

Urban or city tourism is becoming more and more important for overall tourist prosperity in Bosnia and Herzegovina. Urban centres are rich in history, culture, tradition and offer over 16,000 beds in commercial accommodation capacities (hotels, motels and luxurious apartments), recording about 320,000 overnight stays in 2009, 25% of the total in Bosnia and Herzegovina. Sarajevo, Mostar, Banja Luka, Tuzla and, to a certain degree, Brčko, contributed considerably to the urban tourism of the country as centres of socio-political and cultural life in Bosnia and Herzegovina. In addition to the major urban centres, there are other urban places rich in historical and cultural heritage. Among these are Bihać (the administrative centre of Una-Sana Canton, on the way to attractive nature, including the Una River near the border with Croatia, 20 km from Plitvice National Park); Jajce, a historical place with a World War II museum (ZAVNOBiH), old town (proposed for UNESCO inclusion), and the Pliva River and Plivsko lake (on which international sports events are held). Travnik, a historical place, and Trebinje in Herzegovina, 20 km from Dubrovnik and 30 km from Herceg Novi, are certainly among the most attractive places in Bosnia and Herzegovina (Nurković 2004).

Religious tourism

Religious tourism is an integral part of the tourist industry in Bosnia and Herzegovina. It is known that religion, either through taboos and obligations or through culture and society, affects our behaviour as consumers. Due to the country's geographic position and its location at the intersection of roads as a meeting of different civilizations and cultures over the centuries, most of the main religions can be found in Bosnia and Herzegovina. Islam, Orthodoxy and Catholicism are three dominant religions, each represented by a certain number of attractive locations which attract more and more domestic and foreign visitors. Besides the above mentioned sacred places, there are also a large number of mosques and Jewish synagogues and churches (Catholic and Orthodox) around the country which attract a number of visitors, although most of these places have not been promoted seriously for the needs of tourist industry. So, in order to be "commercialized" individual religious places in Bosnia and Herzegovina should be better integrated into an overall tourist product. In order to progress religious tourism in Bosnia and Herzegovina it should be taken more seriously, and its development must be supported by a constructive and integrated marketing strategy.

Sarajevo is the only city in which the four large religions (Islam, Catholicism, Orthodoxy and Judaism) meet in a small area, and where some of the oldest religious buildings in Europe are located. It is an area in which east, west, north and south meet, and the stratification of history and diversity of cultures are refracted. The space is full of treasures of cultural-historical and natural values; it is rich in architectural heritage, historical urban areas, nature parks, forests and highlands with flora and fauna and waters. There are also winter sport recreational centres and spas here, along with embassies, numerous diplomatic-consular offices, seats of international organizations, state institutions, branch offices of economic organizations and cultural and sports events with an Olympic heritage. The basis of natural and cultural historical heritage offers an opportunity for the development of cultural tourism, within which 891 objects of cultural heritage were registered, among which the highest concentration is in Municipality of Stari Grad (289 objects) and the Municipality of the Centre (209 objects). Organizing numerous cultural and art manifestations at the world and the European levels contributes to cultural tourism development (Lorber 2006)

Seaside tourism

Despite the change in tourist demand toward more alternative forms, holidays at the seaside still represent more than one third of the tourist demand in Europe (IPK International 2005). The increased number of cheap and low cost airline companies forming strong associations with companies for accommodation and land transport have additionally supported a permanent growth of this type of tourism. Bosnia and Herzegovina, with 21.2km of coastline, can receive about 6,000 tourists in medium and large hotels and private houses/apartments. About 46,000 visitors (mostly from Bosnia and Herzegovina) had about 280,000 overnight stays in 2009. As the main tourist product is "sun, sea and sand," and there are no other tourist activities, the revenues are limited.

Considering the above, the seaside resort of Neum and the Bosnia and Herzegovina coastline have a real possibility of transforming into a modern destination, which has benefits from their vicinity in the domestic and regional markets. By developing access from the sea (excellent conditions throughout the year) the area would attract more lucrative markets.

Ivanjica, a small place on the border with Croatia near Dubrovnik, offers real opportunities for tourism development. In the vicinity of the Adriatic coast and Dubrovnik (only 6 km and 25 km from the airport of Ćilipi), on the one hand, and close to Trebinje (15 km on a good road) on the other, ensures a good location for this "off-coast" tourist place. However, except for a couple of modern houses for holidays and touristic luxurious apartments, this village has not capitalised on its advantages. A number of other places are good prospects for larger scale tourism, such as Trebinje in Herzegovina, 20 km from Dubrovnik and 30 km from Herceg Novi, which is certainly one of the most attractive places in Bosnia and Herzegovina due to its rich cultural-historical heritage, Mediterranean climate and economic importance for east Herzegovina (www.neum.ba).

Rural tourism

Tourism based on nature and rural tourism cover a wide range of activities, including mountaineering, bird watching, camping, kayaking, fishing, hunting, nature photographing, diving, and visiting parks, farms and waterways. Bosnia and Herzegovina includes a diverse and contrasting landscape, with spectacular mountains, forests, rivers, lakes, unique national parks and beaches. Due to a lack of adequate buildings, infrastructure and creativity of the local population, and despite the favourable natural resources, this type of tourism is still insufficiently developed in Bosnia and Herzegovina. The most interesting natural localities for the development of eco-tourism are shown next.

The Sutjeska National Park-NPS (in the territory of Foča) is the oldest of its kind in Bosnia and Herzegovina, and the only one in Europe, with a thousand year old untouched ancient forest Perucica, covering 1,291 hectares, encompassing the Maglic mountain which divides Bosnia and Herzegovina and Montenegro, and is the highest in the country (2,963 m). The Kozara National Park (in the territory of Banja Luka) is "hospitable," comprising an easily accessible mountain (the highest peak is Mrakovica at 806m), and is 20–40 km from the highway Zagreb–Beograd. In addition are the Una National Park (in the territory of Bihac), the Blidinje Nature

Park, and the Hutovo Blato nature park between Mostar and the shrine of Međugorje in Herzegovina, which is a protected reserve of migratory bird colonies (protected by the Ramsar International Convention on Wetlands) (www.neum.ba).

There are also Šipovo and Plivsko lakes (near Jajce), Konjic-Jablaničko lake (on the main traffic route connecting Sarajevo and Mostar), and Nevesinje and Blagaj in Herzegovina. Bardača (in the north, in the vicinity of Gradiška) is a protected reserve of migratory bird colonies. Considering the limited available statistical data on types of tourist services, the number and structure of nature visitors/national parks are unknown. With regard to a poor tourist infrastructure (accommodation and tourist information centres) it is assumed that the local population accounts for the biggest number of visitors, comprising individuals acquainted with the locality. In the context of tourism based on natural resources, adventure tourism in Bosnia and Herzegovina distinguishes itself with destinations that can compete with many other popular places. Many natural attractions in Bosnia and Herzegovina offer exceptional environments for different adventurous activities.

Spa tourism

Generally, spa tourism has shown a strong increase in the past two decades. In many European countries spas have been transformed and developed with an increase of fitness and recreation centres and a growing demand for more luxurious hotels. Bosnia and Herzegovina is still relatively unused in a tourism sense, and still has much to offer. Spa resorts in Bosnia and Herzegovina offer a series of modern and high quality treatments at still very moderate prices. There are 16 thermal centres in Bosnia and Herzegovina at the moment, accommodated in different parts of country from Višegrad in the southeast to Mlječanica in the northwest. In 2009, spa tourism generated about 420,000 overnight stays (189,080 in RS), or 33% of tourist visits to Bosnia and Herzegovina. The thermal centres at Ilidža (Sarajevo), at Fojnica, Kiseljak, Tuzla, Olovo, Srebrenica, Višegrad and Teslic are famous for the high quality of their thermal/mineral waters.

Adventure tourism

The Una River is currently attracting more and more visitors for rafting and other activities related to water. The Tara River, with its canyon of 1,300 meters, the second largest in the world after Colorado, offers the most interesting rafting, canoeing and kayaking in Europe. The Neretva River, 100 km long, connects Zelengora Mountain with the Adriatic coast of Bosnia and Herzegovina, and offers a unique experience for rafting, swimming, kayaking and hunting (Drešković & Đug 2010). In the city of Mostar, divided by the Neretva River, the most popular adventure activity is jumping from the Mostar Bridge. The Drina River is the longest river in Bosnia and Herzegovina offering many advantages for those interested in fishing, swimming, kayaking, water skiing (two artificial lakes) and rafting. The natural environment around the river offers additional possibilities for mountaineering and climbing, walking trips, hunting and watching wild animals (Bijele Vode in the vicinity of Srebrenica) (Mikačić 1998).

Table 11.2. Spa resorts in Bosnia and Herzegovina, 2011

Name spas	Location	Number of beds
Aquterm	Olovo	90
Dvorovi	Bijeljina	117
Gata	Bihać	90
Guber	Srebrenica	96
Ilidža	Gradačac	160
Kulaši	Prnjavor	108
Laktaši	Laktaši	154
Mlječanica	Kozarska Dubica	140
Reumal	Fojnica	560
Slana banja	Tuzla	Hotel “Bristol” u izgradnji
Banja vrućica	Teslić	975
Vilina vlas	Višegrad	135
Sanska ilidža	Sanski most	Ljetni period
Ilidža	Ilidža-Sarajevo	Hotel “Terme”
Kiseljak	Kiseljak	Van funkcije

Source: Ministry for Foreign Affairs of Bosnia and Herzegovina, 2011

Mountain/winter tourism

While winter tourism includes a series of activities on snow (e.g. skiing, snowboarding, tour skiing, heli-skiing and sledging), the most common forms of mountain tourist activities are walking, hill climbing cycling, and mountaineering. In some countries, ski resorts have a small importance in summer mountain tourism. Visitors have a big choice in a very competitive market in Europe and it is not easy to attract them. Such unstable demand is not able to maintain the tourist business in many isolated areas. Skiing and winter sports have a long tradition due to favourable conditions, and first of all this refers to Jahorina and Bjelašnica (including Igman), two major ski resorts. These two destinations, with low quality infrastructure (particularly of vertical transport and ski tracks), mostly attract low budget tourists from the former Yugoslav republics (Nurkovic 2009). Tourism development in both ski resorts is unplanned and accidental in the post-war period. There are several smaller ski resorts in different parts of Bosnia and Herzegovina, including Blidinje (near Bugojno), Vlašić (Travnik) and Kozara (in the vicinity of Banja Luka), mainly aimed at the local market, including the increased number of weekend visitors (Table 11.2).

Despite the country's enormous potential (80% of Bosnia and Herzegovina is covered by big massifs), except for three major ski resorts (Jahorina, Bjelašnica/Igman and Vlašić) for winter sports-recreational activities, other types of mountain tourism are rare. Due to the inadequate or non-existent infrastructure and tourist facilities, many potential mountain areas are currently unfavourable for any regular movement of visitors. Tourism in the Bosnia and Herzegovina mountains (except skiing) is still in an initial phase, with only several agencies covering the whole country, and as a result there is a lack of information on summer mountain activities. Green Vision, the Sarajevo tourist agency, is one of the few agencies that organises tours with skilled guides in mountain areas, with an emphasis on less accessible and undeveloped localities of Sutjeska National Park, and the so-called "Hercegovački Himalaji" (Prenj-Čvrsnica-Čabulja), the heart of the Dinaric Alps in central Bosnia and Herzegovina. In the context of adventure tourism of Bosnia and Herzegovina, many natural attractions offer an exceptional environment for different adventurous activities (Đug & Drešković 2012).

Sports tourism

Thanks to natural values, Bosnia and Herzegovina has opportunities for the development of this form of tourism, and sports tourism has already acquired a reputation in this area. The existing sports infrastructure of open terrains, swimming pools, sports halls and sports activities within the hotel complexes represents a good basis for the development of this form of selective tourism, which certainly needs to be extended and modernized. While speaking of sports infrastructure in the area of Bosnia and Herzegovina, it should be said that it is spatially well distributed, and Sarajevo, Mostar, Banja Luka, Bihać, Foča, Tuzla, Zenica, Brčko Bijeljina and Goražde have conditions for such development. It would also be desirable to make recognisable centres of sports tourism, which would lead to further specialisation in golf, tennis, riding, athletics, cycling, fitness, nautical sports, extreme sports on water, mountaineering, white water sports.

Conclusion

Bosnia and Herzegovina can improve its position in the regional market by constructing tourism of a higher quality with innovative concepts. These should avoid standard models of development (such as ski or seaside resorts and other tourist products). For example, development of tourist routes at the state level would help to compensate the limited number of places with natural or historical heritage. For this purpose, the marketing strategy for tourism should primarily be oriented toward the European market. Bosnia and Herzegovina is not far from many of such destinations and its inhabitants are the targets for their marketing efforts. In order to create a competitive advantage in relation to those destinations with similar tourist resources and areas of natural beauty, Bosnia and Herzegovina should, first of all, be clearly positioned.

The review of tourist resources shows that most tourist areas and attractions have been little valorized so far. The attractiveness of the high mountains of Bjelašnica, Igman, Vlašić and others is reflected in the exisiting tourist facilities. The percentage of tourists visiting these localities is over 30%, while the percentage of tourists visiting the historical heritage in Sarajevo, Banjaluka, Mostar and Tuzla is around 60%. On the grounds of the research and geographical elements of tourism development in Bosnia and Herzegovina, it can be concluded that there is a successful connection with other economic activities in a new regional development. This, first of all, refers to the natural-geographic features and anthropogenic contents of selective tourism development in Bosnia and Herzegovina.

References

Cigale, D. (2004). Posledična navzkrižja in obremenitve slovenskega alpskega sveta zaradi turistične in rekreativne dejavnosti. PhD diss., University of Ljubljana.

Kurtović, H. (2005). Banjska mjesta u Bosni i Hercegovini – centar zdrastvenog Turizma. Paper presented at the international seminar Tourism as a Factor of Regional Development, Tuzla.

Drešković, N.; Đug, S. (2010). Potamological charactheristics of Neretva river. Proceedings. 1sr international symposium of fisheries and fishing tourism. Konjic, Bosnia and Herzegovina.

Đug, S.; Drešković, N. (2012). Nature protection in Bosnia and Herzegovina: state snd perspectives. Revij Za Geografijo,, Številka 7 – 1 / 2012. Univerza u Mariboru, Filozofska fakulteta. Maribor, Slovenia.

Lorber, L. (2005). The role of tourism in Slovenia from its admission to the Europe Union. Paper presented at the international seminar Tourism as a Factor of Regional Development, Tuzla.

—. (2006). A role of Tourism in Slovenia from its admission in the Europe Union, Tourism as a factor of regional development, Proceedings, 61-73

Pepeonik, Z. (1995). Tourism as a bearer of Croatia's development. Paper presented at the First Croatian Geographical Congress, Zagreb.

Nurković, R. (2009). Influence of tourism on the Regional development of bosnia And herzegovina. International Journal of Euro-Mediterranean Studies

—. (2004). Some new forms of regional development and basic relations of industry and tourism: The example of the Tuzla valley, (South- Eastern European Countries On Their way To Europe-Geographical Aspects, Dubrovnik, Croatia.

Mikačić, V. (1998): Tourism of Croatia-Position and Perspectives, Croatian Geographical Bulletin 60, Zagreb, Croatia.

Statistical yearbook of Bosnia and Herzegovina (2008). Sarajevo: State Institute for Statistics.

Strategija razvoja turizma Bosne i Hercegovine (2010). Sarajevo: Ekonomski institut Sarajevo.

Statistical rewiev of Federation of Bosnia and Herzegovina, 1997-2002, Sarajevo.

http\\: www.neum.ba

CHAPTER TWELVE

THE CARPATHIAN ECOLOGICAL NETWORK: GIS APPROACH TO DETECTING THE LANDSCAPE PERMEABILITY FOR PARTICULAR UMBRELLA SPECIES

FILIPPO FAVILLI, ELISA RAVAZZOLI, CHRISTIAN HOFFMANN AND THOMAS STREIFENEDER

Introduction

Road infrastructures endanger many wildlife populations by causing fragmentation and thus reducing connectivity among habitats (Forman et al. 2003; Jaeger & Madriñán 2011, Efe, 2010). Ecological corridors are "landscape elements which serve as a linkage between historically connected habitat areas" (McEuen 1993). A regional ecological network can provide connectivity between spatially separated populations, countering biological processes that lead to species extinction (Beier 1995; Bennett 1998, 2003; Taylor et al. 2006). The project BioREGIO Carpathians applies a combination of CorridorDesign (Majka et al. 2007) and Linkage Mapper (code.google.com/p/linkage-mapper/), two ESRI ArcGIS 10 spatial analysis tools to build a Habitat Suitability Model (HSM) and Network Linkages for selected umbrella species in the Carpathians. The present study aims to detect the most probable core areas and least-cost paths for the Eurasian Lynx and to highlight the role of protected areas at the Romanian/Serbian border as linkage nodes. The objectives are to answer the following research questions: Which are the most suitable landscape patches for the lynx? Are there chances that a lynx will reach another patch? If yes, using which path? Are there barriers in the identified routes? Are they surmountable? To reply to these questions efficiently it is fundamental to: (i) set the parameters for the HSM, (ii) assess the connectivity via the visualization of core areas and least-cost paths, and (iii) identify possible barriers.

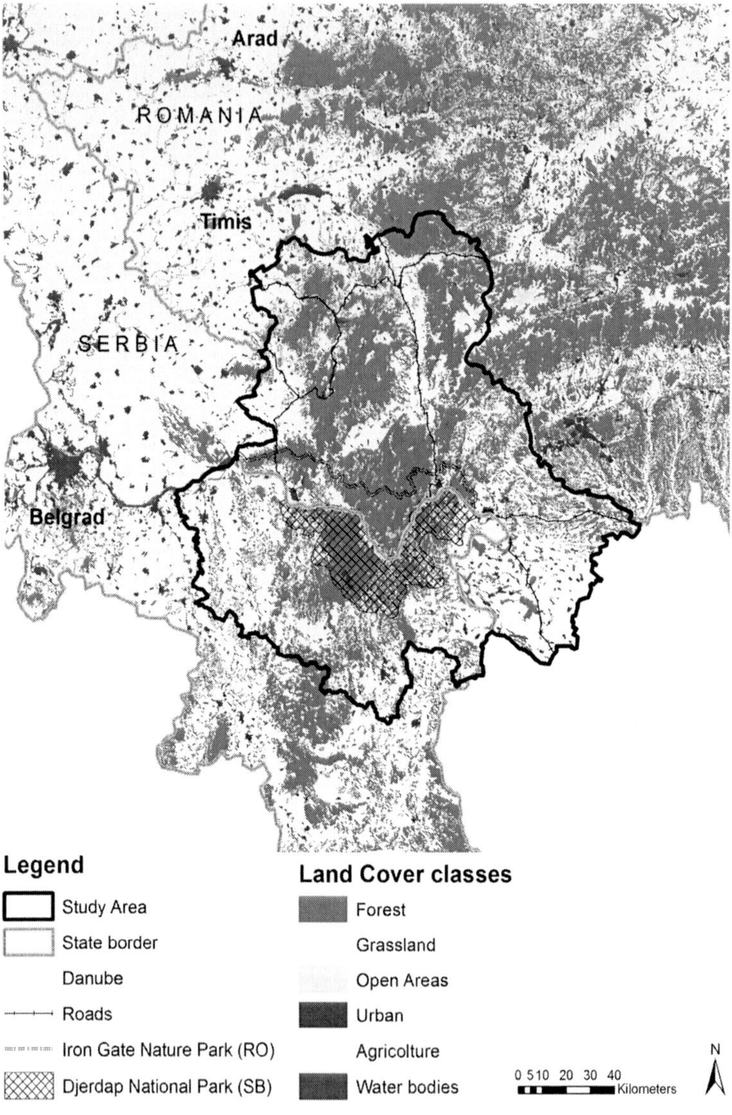

Fig. 12.1. Study Area

Material and Method

The study areas are the Iron Gate Nature Park in Romania and the Djerdap National Park in Serbia plus the surrounding landscapes (see Fig. 12.1 above).

The Iron Gates Nature Park, at the south Romanian borders with Serbia, is a site of community importance ROSCI0206 and a Ramsar site no. 1946. The National Park Djerdap presents a high level of wilderness and is situated in the north-east of Serbia, along the international border with Romania. In both areas lynxes are reported to be resident (Sommerwerk et al. 2009).

Umbrella species

The Eurasian Lynx (*Lynx lynx* L.) is a highly selective mammal which prefers forest sites with at least 60% forest cover (Fawcett 1997). It avoids human activities and disturbed habitats. The lynx has large spatial requirements which vary according to prey density and composition. Lynxes have a home range estimated between 70 (females) and 200 (males) km^2. In non-wooded habitats, they can use alternative cover such as scrub and rocks. Lynxes do not stay permanently in forest areas < 30 km^2 and the minimum core area sizes (100 km^2 breeding patch; 200 km^2 population patch) can be interrupted by open areas, but not by human infrastructures (Schadt et al. 2002; Kramer-Schadt et al. 2004; 2011).

Habitat Suitability Model (HSM)

The creation of a HSM is a two-steps process perused by applying the *CorridorDesign* GIS tool. In the first step, the habitat suitability is identified on the bases of the species' ecological preferences. According to the preferred land-cover types their suitability is valued from 0 to 100. For a complete explanation of the model refer to Majka et al. (2007). Table 12.1 below shows the selected values for the lynx used in our study.

Factor classes and weights are combined through a geometric mean. The pixel remains 0 if only one category is 0. Each pixel can then be assigned to a certain suitability class:

(1) Suitability: 50–100%=Optimal habitat
(2) Suitability: 25–50%=Sub-optimal habitat
(3) Suitability: 0–25%=Occasional habitat
(4) Suitability: 0=Avoided, barrier

The pixels with suitability values above 50% were selected, in the second step, to identify breeding (>10.000 - < 20.000 ha) and population patches (> 20.000ha) (core areas) using a moving window of 200-m radius.

Table 12.1. Factors, classes and weights used for the calculation of the first step Habitat Suitability Model

LC*	*V %*	Topo	*V %*	Elev	*V %*	Human	*V %*	Roads	*V %*
1	*100*	<30°	*50*	<500	*50*	<100	*0*	<50	*0*
2	*50*	<60°	*100*	<1000	*100*	<500	*25*	<200	*50*
3	*25*	<90°	*100*	<1500	*100*	<1000	*50*	>200	*100*
4	*0*			<2000	*100*	>1000	*100*		
5	*25*			<2500	*50*				
6	*25*			>2500	*0*				

LC *Land Cover classes (40%): 1=Forest; 2=Grassland; 3=Open Areas; 4=Urban Areas; 5=Agriculture; 6=Water Bodies; Topo: Topographic Position classes (20%); Elev: Elevation classes (10%); Human: Euclidean distance from the closest human settlement classes (15%); Roads: Euclidean distance from the closest main road classes (15%); V: Suitability Value. The weight of each factor is shown in brackets.

Lynx connectivity/Least-Cost Paths (LCPs)

The identification of probable connectivity is perused by applying the GIS tool Linkage Mapper. Resistance values for the landscape patches were necessary in order to calculate the LCPs between the identified core areas. Resistance values are defined as the opposite of the suitability and reflect the energetic cost for a certain species to move across those cells. The basic assumption of the model supposes that species move where the landscape provides less resistance (Beier et al. 2009; Conrad et al. 2012).

Dataset

Land cover and additional geographical data were acquired for the countries considered. CORINE Land Cover 2006 at a resolution of 100x100m was used as a database for the habitat suitability model and the identification of the cost-path analyses. Data were standardized to the Projected Coordinate System ETRS 1989 LCC with the projection Lambert Conformal Conic. The topographic position was generated from the command Spatial analyst/surface/slope tool in ArcGIS 10. Distance to human settlements and to roads were created as a raster image by using the Euclidean distance tool in the ArcGIS toolbox. The cell size was set to that of the CORINE land cover grid.

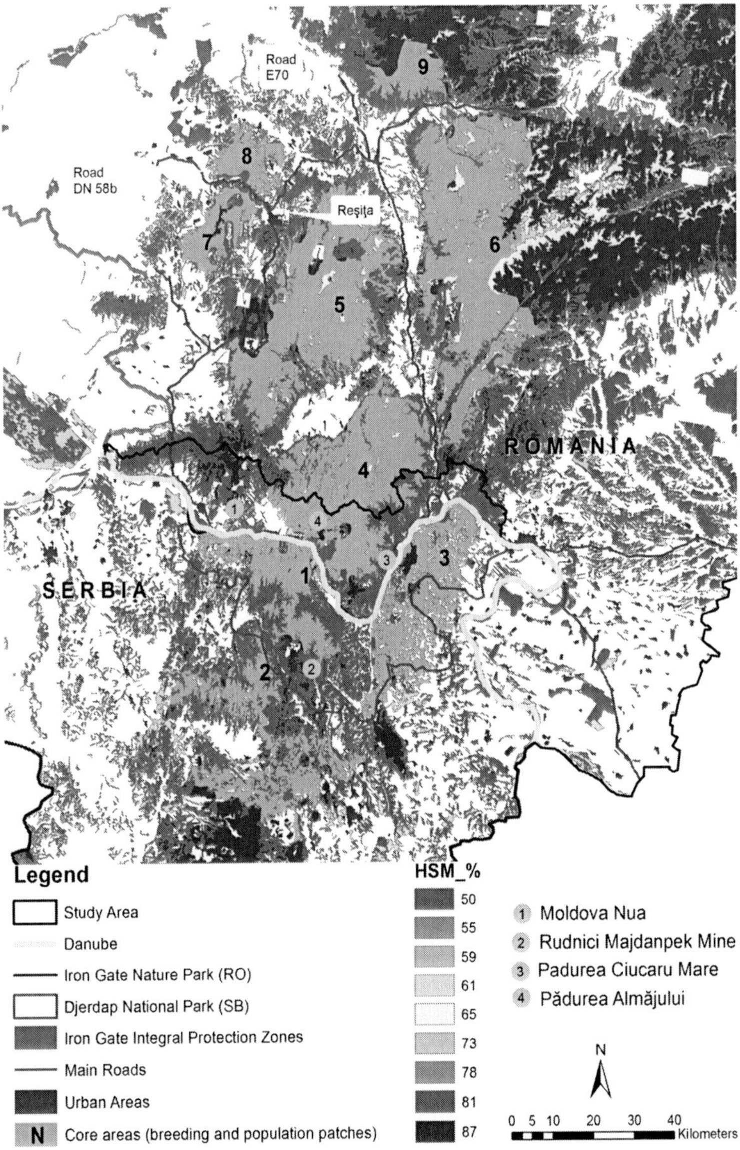

Fig. 12.2. Habitat Suitability Model and core areas

Results

The HSM steps highlighted the complex mosaics of forest landscapes which are highly preferred by the lynx. The second step of the model identified the 9 core areas displayed in Fig. 12.2, indicating a high probability that the lynx could be distributed in the study area.

In the Serbian side, the potential distribution of the lynx is spread all over the Djerdap National Park, in direct connection with the Balkan populations. As concerning the Romanian side, the lynx populations are separated in 6 main population and breeding patches (see Fig. 12.2 above).

The "integral protection zones" in the core area 4 (1, 3 and 4 in Fig. 12.2) in the Iron Gate Nature Park could represent the passage sites of lynxes emigrating from Romania to Serbia. In core area 5 the lynx is reported as residential (Stanciu 2008) but is separated from the rest of the Carpathian population by the highly frequented international road E70. Core area 6 is the biggest in the study area and is connected with the Carpathians Mountains. Breeding patches 7 and 8 are separated by the national road DN58b close to the city of Reşiţa (Fig. 12.2). Breeding patch 9 is represented only partially in the study area.

Through the use of Linkage Mapper, we selected 16 LCPs in the territories with the lower resistance as those highlighting a general view of the ecological network (Fig. 12.3). LCPs were categorized according to: the CWD (cost-weighted distance), their length, the presence of barriers that increase their mortality risk (LCP risk), and the presence of a protected area, which increases their safety (LCP safe). LCPs were then divided into 5 cut-off categories (1=Best; 5=Worst) to identify the most likely sites to which the lynx may disperse.

From the 16 selected LCPs, only 2 can be assigned to Category 1 because they are inside a forest in a protected area and do not meet any barrier. All the LCPs through the Danube have been marked with a (?) because of uncertainties as to whether the lynx has ever crossed it by swimming. It is more likely to assume the presence of the Danube as an insurmountable barrier, although sporadic lynx observations in this region were made and Serbian lynx populations have acquired some of the characteristics of the Balkan ones (Atanasov 1968; Paunovic et al. 2001). Nevertheless, the Danube has a seasonal changeability and for the lynx it could be passable during some winter months (Simeonovski & Zlatanova, 2001; Spassov 2001). The LCPs belonging to Categories 1 to 3 seem to be usable by the lynx, while 4 and 5 ones are unlikely to be used due to the less suitable landcover classes the lynx would have to pass through and the kind of barriers encountered (i.e. highways and intensive agriculture).

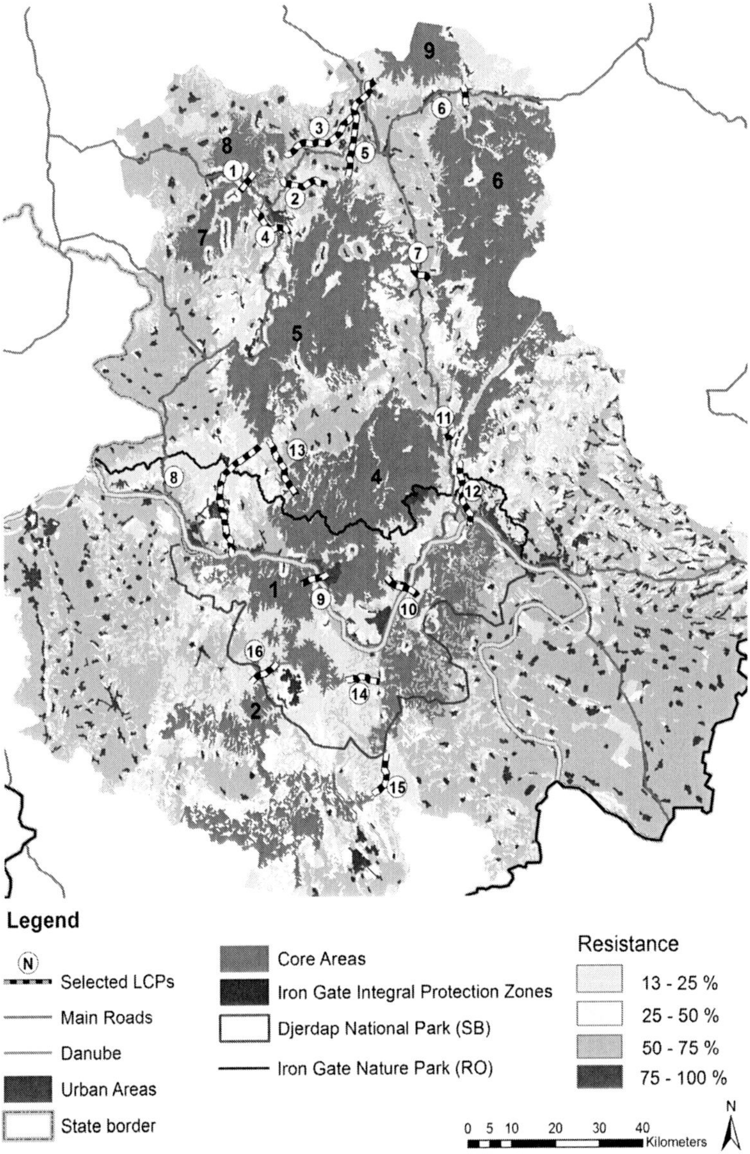

Fig. 12.3. Selected LCPs

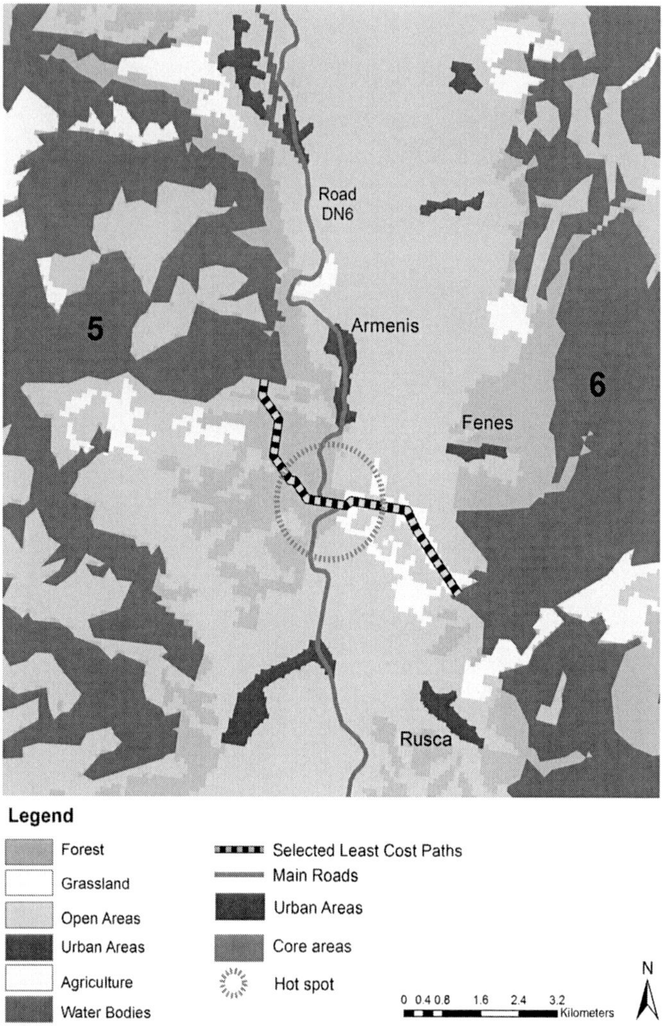

Fig. 12.4. Closer view of the LCP 7 and barrier

Discussion

The suitability and connectivity results indicate that the majority of the study area is highly suitable for the lynx. *CorridorDesign* has proven to be efficient for the detection of the lynx's most suitable habitat, only referring to ecological habits. *Linkage Mapper* has enabled the identification of the main possible connections between probable core areas to enhance the ecological connectivity of the areas. The lynx's spatial requirements are huge and the large forest habitats in the Carpathians could soon become a fragmented landscape, thus large-scale cross-boundary approaches are needed (Mladenoff et al. 1995). The south-western Carpathian lynx populations are known to inhabit the entire area north from the Danube (i.e. Rozylowicz et al. 2011). The Serbian part of the study area confirmed its being a highly suitable habitat for the lynx, due to the National Park Djerdap and the lower presence of roads and big urban areas. The Djerdap National Park includes the three main distribution areas of the lynx in this country, representing the most northern lynx distribution from the Balkans with which it is directly connected.

The Romanian study area plays a major role in protecting this species at the southern borders of its Carpathian distribution. An enhanced connectivity between core areas 4, 5 and 6 (total size 260.548 ha) could sustain more than 20 residential lynx, which is the prerequisite for the survival of a viable population (Schadt et al. 2002). The *Linkage Mapper* analysis also enabled the identification of the gaps in the potential ecological connections. Fig. 12.4 shows the LCP number 7 connecting core areas 5 and 6 in the vicinity of the city of Fenes, identifying the site where possible engineering interventions for the maintenance of the ecological network could be placed. In this case, the site selected is where the National Road DN6 runs closer to fragmented forested and open areas which could be used by the lynx as stepping stones. DN6 is a highly frequented road, but the presence of stepping stones in the vicinity, joined by a green infrastructure facility, could render the establishment of an ecological connection between the two core areas possible (Fig. 12.4).

Conclusions

The questions addressed at the beginning of the study have now been answered, showing that some of the identified routes are of highly improbable lynx use due to the kind of encountered barriers and landcover changes. The "integral protection zones" inside the Iron Gate National Park have proved their importance for wildlife protection and connectivity. The GIS approach gave us the opportunity to highlight the role of protected areas as "safe passage zones," requiring only few data and offering a good picture about landscape fragmentation and ecological connectivity. Some of the LCPs have the characteristics to be effectively used for connecting the lynx's main patches. Generally speaking, following only a visual interpretation of the results, the ecological connectivity inside the studied area seems rather good. Main roads and urban areas act as barriers for the lynx's connectivity. Nevertheless, the sites which could be important for enhancing ecological connectivity and for deriving specific recommendation can be selected. In this next phase, the role of local experts will be decisive for concrete actions.

References

Atanasov, N. (1968). Der Luchs (Lynx lynx L.) in Bulgarien. 25.32 p. In: Kratochvil, J. 1968. (ed.). History of the distribution of the lynx in Europe. Acta Sc. Nat. Acad. Pracha, Brno, 2 (4), 50 p.

Beier, P. (1995). Dispersal of juvenile cougars in fragmented habitat. Journal of Wildlife Management 59: 228–237.

Beier. P.; Majka, D.R.; Newell, S.L. (2009). Uncertainty analysis of least-cost modeling for designing wildlife linkages. Ecological Applications 19: 2067-2077.

Bennett, A.F. (1998; 2003). Linkages in the Landscape: The Role of Corridors and Connectivity in Wildlife Conservation. IUCN, Gland, Switzerland and Cambridge, UK. xiv + 254 pp.

Conrad, J.M.; Gomes, C.P.; van Hoeve, W.J; Sabharwal, A.; Suet, J.F. (2012). Wildlife corridors as a connected subgraph problem. Journal of Environmental Economics and Management 63: 1-18.

Efe, R. (2010) Biogeography, Marmara Publ. Bursa,Turkey.

Forman, R.T.T.; Sperling, D.; Bissonette, J.A.; Clevenger, A.P.; Cutshall, C.D.; Dale, V.H.; Fahrig, L.; France, R.; Goldman, C.R.; Heanue, K.; Jones, J.A.; Swanson, F.J.; Turrentine, T.; Winter, T.C. (2003). Road ecology - Science and solutions, Island Press, Washington DC, USA, 481 pp.

Jaeger, A.G.J.; Madriñán, L.F. (2011). Landscape Fragmentation in Europe. EEA Report 2/2011: 1-92.

Kramer-Schadt, S.; Kaiser, T.S.; Frank, K.; Wiegand, T. (2011). Analysing the effect of stepping stones on target patch colonisation in structured landscapes for Eurasian Lynx. Landscape Ecology 26: 501-513.

Kramer-Schadt, S.; Revilla, E.; Wiegand, T.; Breitenmoser, U. (2004). Fragmented landscape, road mortality and patch connectivity: modelling influences on te dispersal of Eurasian Lynx. Journal of Applied Ecology 41: 711-723.

Majka, D.; Jenness, J.; Beier, P. (2007). CorridorDesigner: ArcGIS tools for designing and evaluating corridors. Available at http://corridordesign.org

McEuen, A (1993). The wildlife corridor controversy: A review. Endangered species update 10. (11 & 12).

Mirić, D.; Paunovic, M. (1992). A new record of Lynx lynx (Linnaeus, 1758) (Felidae, Carnivora) in East Serbia. Bull. Nat. Hist. Mus. Belgrade, B 47: 171-174.

Mladenoff, D.J.; Sickley, T.A.; Haight, R.G.; Wydeven, A.P. (1995). A regional landscape analysis and prediction of favorable gray wolf habitat in the northern Great Lakes region. Conservation Biology 9:279–294.

Paunovic, M.; Milenkovic, M.; Ivanovic-Vlahovic, C. (2001). The lynx populations in the Fed. Republic of Yugoslavia. Kora Bericht 7: 12-17.

Rozylowicz, L.; Popescu, V.D.; Pătroescu, M.; Chişamera, G. (2011). The potential of large carnivores as conservation surrogates in the Romanian Carpathians. Biodiversity Conservation 20: 561-579.

Schadt, S.; Knauer, F.; Kaczenscky, P.; Revilla, E.; Wiegand, T.; Trepl, L. (2002). Rule-based assessment of suitable habitat and patch connectivity for the Eurasian Lynx. Ecological application 12: 1469-1483.

Simeonovski, V.; Zlatanova, D. (2001). Some notes on the systematics of the Balkan Lynx. Kora Bericht 7: 24-25.

Sommerwerk, N.; Baumgartner, C.; Bloesch, J.; Hein, T.; Ostojić, A.; Paunović, M.; Schneider-Jacoby, M.; Siber, R.; Tockner, K. (2009). The Danube River Basin. In: Tockner K, Uehlinger U, Robinson CT (Eds) The Rivers of Europe. Academic Press 2009. 700 pp.

Spassov, N. (2001). Brief notes on the status and problems of the lynx in Bulgaria. Kora Bericht 7: 26-27.

Stanciu, E. (2008). Wilderness and Wildlands in Romanian Carpathians. International Journal of Wilderness 14: 37-42

Taylor, P.D.; Fahrig, L.; With, K.A. (2006). Landscape connectivity: a return to the basics. In: Crooks KR,Sanjayan M (Eds) Connectivity conservation. Cambridge University Press, Cambridge, UK, pp 29-43.

CHAPTER THIRTEEN

PLACE NAMES IN APPREHENDING LANDSCAPES AND HUMAN NATURE INTERACTIONS IN SERIK, ANTALYA

MERYEM ATIK,
ABDURRAHMAN KANABAKAN
AND VELI ORTAÇEŞME

Introduction

Landscape is defined as an area as perceived by people, whose character is the result of the action and interaction of natural and/or human factors (Council of Europe 2000). Jessel (2006) specified that the character as an overall spatial impression is the result of a descriptive process that classifies, illustrates and characterizes areas of similar appearance.

Landscape has many symbolic meanings (Yeoh 1996) and consists of man-made products such as buildings, walls and roads, and natural features of mountains, plains and lakes which are named after and found their meanings from these names. The way humans perceive the landscape reflects on the names given to the landscape and place names, revealing aspects of landscape characteristics.

Toponymy is defined as the scientific study of place names and their origins. The term "name" refers to describing a person or an object. The "names" are known as the connotations for the acknowledgement by all (TDK 2005). Place names include valuable information about rich natural materials, local people, traditions, geography, soil, forests, animals, plants, mountains, rivers, seas, lakes, life styles and how they were perceived by people. Here, the perception of landscape becomes a major component in giving names to the landscape as a reflection of landscape character.

Landscape character is defined as the distinct and recognisable pattern of elements that occur consistently in a particular type of landscape (Swanwick 2002; Heritage Council 2006; Turner 2005). Jellema et al.

(2009) explain that landscape character can be defined as the presence, variety and arrangement of landscape features, which give a landscape a specific identity and single it out from the surrounding landscapes.

Location, people and professions as well as lakes, mountains and rivers have been taken into account in naming places in the landscape (Öz 2007). However, names in the landscape also locate human actions and their meanings in space and time (Carter 2011). Features in geography and nature as well as nodes, networks, contemporary practices, past practices and historical connections render different characters of the landscape.

Rippon (2012) remarked that understanding the origins and development of local and regional variation in landscape character, such as meanings of the place names, are important for our appreciation of human society's relationship with the natural environment.

The aim of this study is to evaluate place names in the Serik Province, Antalya and their associations with landscape characters with regard to visual and perceived aspects, biophysical features, natural processes and human activities. A number of place names in Serik have filtered through time and present examples are Çamlık, Belpınar, Deniztepesi and Akçapınar, representing different landscape characters. However, the interpretation of place names can provide a useful tool in understanding the landscape and geographic features with regard to cultural and natural associations.

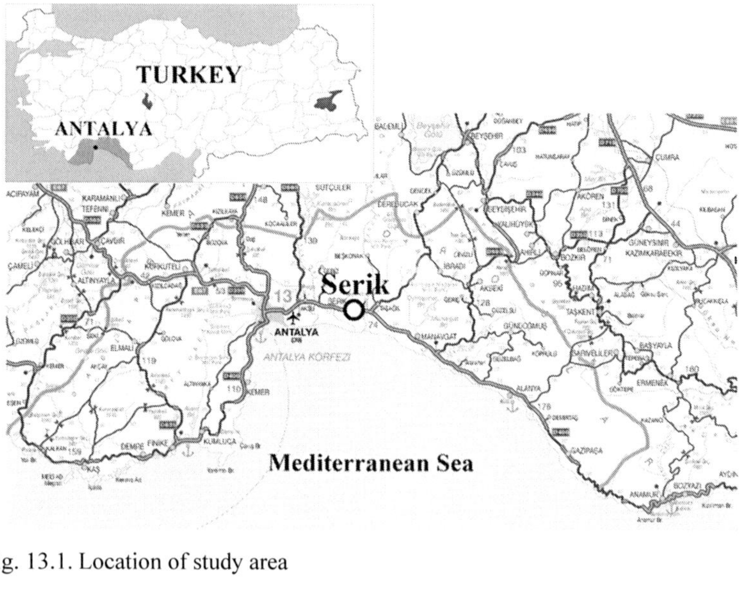

Fig. 13.1. Location of study area

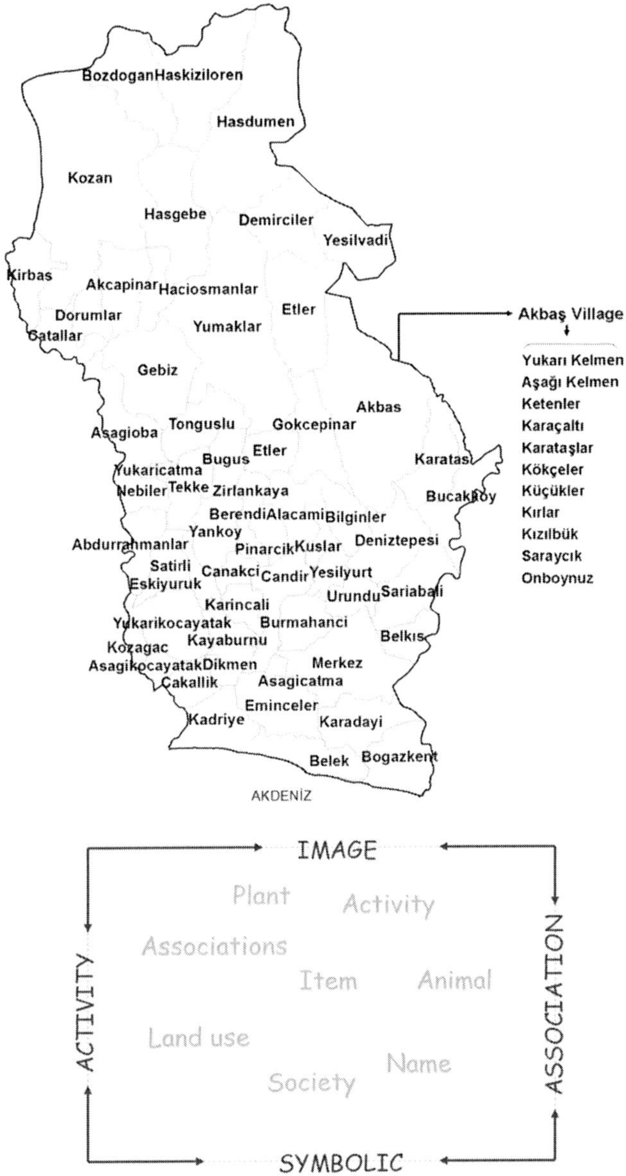

Fig. 13.2. Scale and flow of the study method

Material and Method

The Serik Province in Antalya Region was chosen for this study due to its diversity of coastal, agricultural and mountainous features and deep cultural history (Fig. 13.1). It is located between the Mediterranean Sea and Taurus Mountains on heights up to 2,500 metres. The general characteristics of the Serik Province are based on rural settlements, agricultural lands and forests as reflected in the place names since ancient times, Seljuk, Ottoman and Nomadic Yörük culture and emigrants from the Balkans, Crete and Caucasia.

The methodology of the study is based on Başkan (1971), Stephenson et al. (2004), Levinson (2008) and Murray (1995), in categorising place names on reflective landscape characters. The meanings of the place names were interpreted using Pisküllüoğlu (1995) and TDK (2005). In Serik there are 47 villages, 10 towns and a number of neighbourhoods. The study method was carried out in three scales based on the names of villages, towns and neighbourhoods. Place names were evaluated along with their connotations in visual characters, symbolic associations or human activity (Fig. 13.2).

Results

Place names are a kind of ontology of the landscape which reflect the natural, cultural, social or perceptual and visual aspect of the landscape. Van Eetvelde & Antrop (2007) emphasise that cultural themes, such as settlement types and patterns of place names, are significant but not included in the landscape typology and characterisation.

Serik was found in two locations, namely Silyon and Aspendos, in the era of Pergamons, second century AD. Following the Seljuk's arrival in 1207, they settled in Silyon. The name "Serik" was given by Ottomans in 1890 (Güçlü 2000), derived from the geography Serika from central Asia, the people living in the land extending from Kaşkar to Tibet (Selekler 2011).

There are distinct examples of place names revealing the character of landscape with regard to position, form, colour or land use. Characteristics of place names in Serik Province are given in Table 13.1 below. Here, the visual characteristics of the landscape such as colour, form and quality are the major components in place names. The cultural dimensions of the landscape are reflected in the names associated with people, society, profession or land use.

Table 13.1. Characteristics of place names in the Serik Province

Place Names	Landscape Characteristics	Place Names	Landscape Characteristics
Akbaş	White, colour	Kayaburnu	Rock, geology
Akçapınar	Whitish, colour	Kırbaş	Grey, colour
Alacami	Colour	Kozağaç	Walnut, plant
Aşağıçatma	Lower, position	Kozan	Land use
Aşağıkocayatak	Lower, position	Kuşlar	Birds, wildlife
Aşağıoba	Lower, position	Nebiler	Name, person
Berendi	Eagle, animal	Pınarcık	Spring, hydrology
Bilginler	Person, society	Sarıabalı	Yellow, colour
Bozdoğan	Grey, colour	Şatırlı	Society
Bucakköy	Location	Tekke	Space
Burmahancı	Form	Töngüşlü	Nomadic
Büğüş	Bend, form	Üründü	Food, product
Çakallık	Jackal, animal	Yanköy	Position
Çanakçı	Bowl, form	Yeşilvadi	Green, colour
Çatallar	Form	Yeşilyurt	Green, colour
Demirciler	Profession	Yukarıçatma	Upper, location
Deniztepesi	Form	Yumaklar	Round, form
Dikmen	Form	Zırlankaya	Position
Dorumlar	Wildlife	Abdurrahmanlar	Name, person
Eminceler	Society	Belek	Position
Eskiyürük	Old, quality	Belkıs	Name, person
Etler	Food	Boğazkent	Form
Gökçepınar	Blue, colour	Çandır	Quality
Hacıosmanlar	Society	Gebiz	Soil
Hasdümen	Fine, quality	Kadriye	Name, person
Haskızılören	Crimson, colour	Karadayı	Black, colour
Haspınar	Fine, quality	Serik	Food, product
Karataş	Black, colour	Yukarıkocayatak	Position
Karıncalı	Ant, wildlife		

The natural features embedded in place names are characteristic, related to plants, animals, soil and geology (Table 13.1 and Fig. 13.3). It has been argued that landscape is a visual construction and visual features often reveal both natural and cultural characteristics of the landscape. According to this study, more than 20% of place names in the Serik Province are found to be associated with colour, of fields, buildings and other elements in the landscape. It is diversified by the land use and crop pattern or season. Similarly, form literally deals with the shape of the elements in the landscape, and more than 10% of place names take this place in Serik (Fig. 13.3).

Different information is included in place names, such as where the place is located or the shape of the landscape. For example, the name of Deniztepesi indicates the position of a hill top overlooking a sea view, while Böğüş is related to the bending streams or ridges in the province. Most position statements were made regarding lower or upper, while forms were indicated with curves, bowls, and steep or round (see Table 13.1 above).

Hasgayriyası; Haskızılören, Haspınar, Hasdümen and Bozdoğan

Hasiçigayriyası is a large territory on the outskirts of the Bozburun mountains. The word refers to the quality to be "fine" and "good," while "Hasgayriyasi" pertains to the meaning of fine territory. Four villages located in and around Hasgayriyasi are connoted with the quality of "fine" and accordingly named Haskızılören, Haspınar, Hasdümen and Bozdoğan. Haskızılören has a dual meaning and multi-character: "Oren" refers to a wall often built with historic value, and "Kızıl" refers the colour crimson, associated with a native tree known as the dye tree *Alnus orientalis* and its abundance in the village.

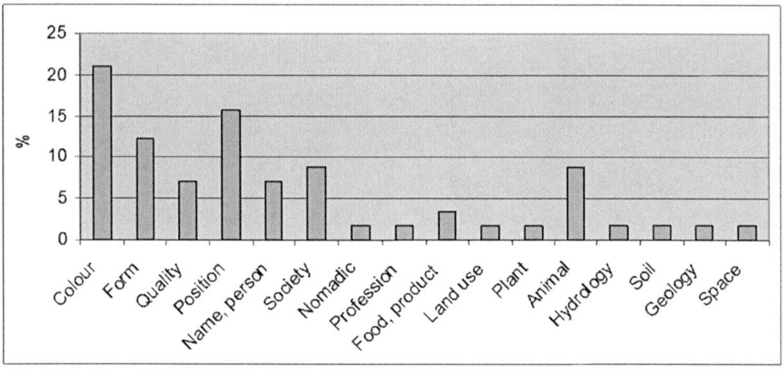

Fig. 13.3. Connotations of place names in Serik

Haspınar directly indicates the quality of water resources in the vicinity with three main springs Kocapınar, Ortapınar and Özlüpınar, related to the form and quality koca, or "big," the position orta, "middle," and the quality özlü, meaning "lush." Hasdümen, on the other hand, is related with the occupation of the land, when a military division stayed there during the war of independence.

Fig. 13.4. Villages in Hasgayriyası

Conclusion

Landscape is regarded as a combination of the interaction between man and nature. This interaction appears through our observations, perceptions and associations built with the landscape and often expressed by place names. Rippon (2012) states that the landscape around us provides a remarkable record of human achievement in exploiting, modifying, and transforming the natural environment. Place names can be appreciated as the multidimensional face of the landscape. The relationship with our environment is created through cultural filtering and perception of the natural world as well as human products in the landscape.

In this study, place names in the Serik Province, Antalya, were evaluated with their names as a reflection of landscape characters with regard to visual and perceived aspects, biophysical features, natural processes or human activities. There are a number of place names in Serik which have filtered down through time, such as Çamlık, Belpınar, Deniztepesi, Akçapınar, representing different landscape characteristics.

The study results show that visual characteristics such as colour and form were major elements in place names in the case of Serik. However, position in the landscape and names related with the society also play an important role. Names related with animals and society were expressions of the occupation and co-existence of people and their activities in the landscape.

Geographical features such as rocks, hills, streams, river beds, valleys or ridges have often been characterised with colours such as red, white, green, black and grey, forms such as long or steep, positions such as upper or lower or qualities such as fine, good or lush to create an overall definition of the landscape. Although place names are linguistic and cultural products, the interpretation of place names can be a useful tool in understanding the landscape and geographic features with regard to cultural and natural associations.

Acknowledgments

This study was supported by the Scientific and Technological Research Council of Turkey, TÜBİTAK with the Project number 112Y019.

References

Başkan, Ö. (1971). Türkiye Köy Adları Üzerine Yeni Bir Deneme, Türk Dili Araştırmaları Yıllığı Belleten: 237-251.

Carter, L. (2011). The Big 'H': Naming and Claiming Landscapes. In J. Ruru, J. Stephenson, M.Abbott (eds.) Making our Place Exploring Land-use Tensions in Aotearoa New Zealand, Published by Otago University Press, Dunedin, New Zealand, 243 pages.

Council of Europe (2000). The European Landscape Convention (STE n°176). http://www.coe.int (accessed in 20 July 2010)

Güçlü, M. (2000). 1864-1950 Yılları Arasında "Serik" İdari, Ekonomik, Sosyal, Serik Belediyesi Kültür Yayınları, Antalya, sayfa 95, (2000).

Heritage Council (2006). Landscape Character Assessment (LCA) in Ireland: Baseline Audit and Evaluation. Prepared by Julie Martin Associates in Association with Alison Farmer Associate, ISBN 978-1-906304-01-0, pp 78.

Jellema, A.; Stobbelaar, D.J.; Groot, J.C.J.; Rossing, W.A.H. (2009). Landscape character assessment using region growing techniques in geographical information systems. Journal of Environmental Management 90 (**2**): 161-174.

Jessel, B. (2006). Elements, characteristics and character – Information functions of landscapes in terms of indicators. Ecological Indicators 6 (**1**): 153-167.

Levinson, S.C. (2008). Landscape, seascape and the ontology of places on Rossel Island, Papua New Ginea, Language Sciences 30 (**2/3**): 256–290.

Murray, J.S. (1995). Unnameable Landscapes, Landscape Review 2: 30-41.

Öz, S. (2007). Evliya Çelebi Seyahatnamesi'ne Göre Anadolu'da Yer Adları. Kırıkkale Üniversitesi, Sosyal Bilimler Enstitüsü Türk Dili ve Edebiyatı Anabilim Dalı Yüksek Lisans Tezi, 207 sayfa.

Püsküllüoğlu, A. (1995). Türkçe Sözlük YKY Yayınları, İstanbul.

Rippon, S. (2012). Making Sense of Historic Landscape, Oxford University Press, ISBN:978-0-19-953378-7, Oxford, U.K. 380 pp.

Selekler, M. (2011). Yarım Asrın Arkasından Antalya'da Kemer, Melli, İbradı ve Serik. Serik Belediyesi Kültür Yayını, Kutlu Avcı Ofset Ltd., Antalya.

Stephenson, J.; Bauchop, H.; Petchey, P. (2004). Bannockburn Heritage Landscape Study. New Zealand: Department of Conservation. Science for Conservation, http://www.doc.govt.nz/upload/documents/science-and-technical/SfC244b.pdf (accessed in 02 November 2011)

Swanwick, C. (2002). Landscape Character Assessment. Guidance for England and Scotland, Cheltenham (UK);Edinburg: The Countryside Agency; Scottish National Heritage.

TDK (2005). Türk Dil Kurumu Türkçe Sözlük. Türk Dil Kurumu 4. Akşam Sanat Okulu Matbaası, Ankara, 2243 sayfa.

Turner, S.C. (2005). Devon Historic Landscape Characterisation. Devon County Council Historic Environment Service & English Heritage, County Hall, Exeter, EX2 4QW.

Van Eetvelde, V.; Antrop, M. (2007). Integrating cultural themes in landscape typologies. In: Roca, Z., Spek, T., Terkenli, T., Höchtl, F. (Eds.), European Landscapes and Lifestyles: The Mediterranean and Beyond. Proceedings of the 21st PECSRL Conference "One Region, Many Stories: Mediterranean Landscapes in a Changing Europe", Limnos and Lesvos, 2004. Lisboa: Ediçőes Universitárias Lusófonas. Ediçőes Universitárias Lusófonas, Lisboa, pp. 399-411.

Yeoh, B.S.A. (1996). Street-Naming and Nation-Building: Toponymic Inscriptions of Nationhood in Singapore. AREA 28 (3): 298-307.

CHAPTER FOURTEEN

EMENDERE THERMAL SPRINGS (SINDIRGI, BALIKESIR, TURKEY)

ALAATTIN KIZILÇAOĞLU AND ISA CÜREBAL

Introduction

Turkey is very rich in thermal resources due to the geological characteristics of the country (Bulut 1997, 69). Thermal waters in Turkey are of better quality and provide a wide range of treatment opportunities compared to those in Europe in terms of flow and temperature, quantity and various physical and chemical characteristics (Özbek 1991, 18). There are about 1,300 healing water resources in different regions of Turkey suitable for thermalism (balneology) in terms of temperature and rich mineral content (Güner-Sever 2000, 105). Visitors prefer the thermal springs for different reasons, such as treatment or recreational entertainment. Various service facilities are being established in order to meet the demands of the visitors. In order to provide a tourist attraction in thermal areas, treatment units indoors and outdoors (Turkish baths, private baths and pools) and recreational areas (walking tracks, game rooms and sports areas) should be established, and transportation and locations should be favourable. Since the demand for thermal springs has recently increased in Turkey, thermal facilities have been built around many of the thermal resources (Çekirge 1991; Kahraman 1991; Özbek 1991; Tuncel & Doğaner 1992; Taşlıgil 1995; Bulut 1997; Güner & Sever 2000; Zaman et al. 2000; Bulut & Girgin 2001; Ünal 2003; Sandıkçı 2007; İbret 2007; İlban et al. 2008; Türksoy & Türksoy 2010; Tunçsiper & Bezirgan 2010; Çetin & Özşahin 2011; Zaman & Birinci 2011; Öcal 2011; Çomak & Güncegörü 2012). The locations where these establishments are built are treatment and entertainment centres (Çetin & Özşahin 2011, 318), such as Emendere Thermal Springs. The current study aims to present Thermal Springs geographically and offer suggestions to provide more efficient and qualified services. Documents were reviewed followed by field trips and interviews with the authorities.

Geographical Location

Emendere (Ilıcalı) Thermal Springs is situated within the borders of Ilıcalı (Emendere) village in the Sındırgı District of Balıkesir Province in Southern Marmara part of the Marmara Region (Fig. 14.1) in Turkey. The thermal springs can be approached via several routes. Vehicles using the Istanbul and Bursa route take the Bursa-İzmir highway, while vehicles driving towards southeast to Bigadiç through Balıkesir using the Bursa-İzmir highway arrive at Sındırgı district after a 58 km drive. Emendere Thermal Springs is located 7 km southeast of Sındırgı district. The Balıkesir-Emendere road is asphalt with double lanes in some parts and counter-flow lanes in others.

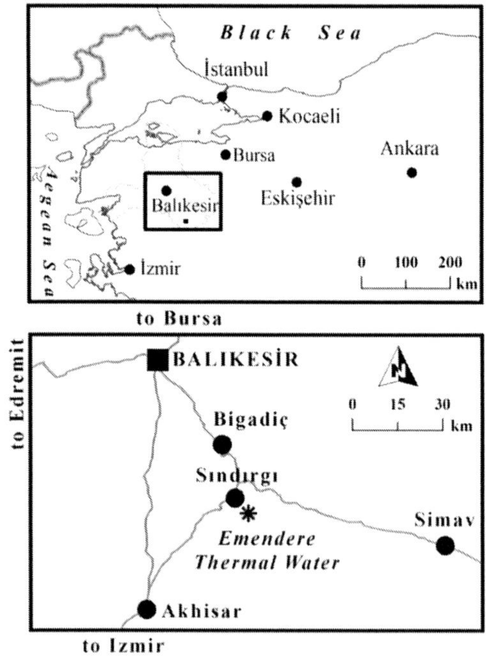

Fig. 14.1. Location map of the thermal springs area

The visitors from the Izmir direction take the Akhisar-Sındırgı road, which is asphalt and with curves at some locations. There is another road to Sındırgı from Gölcük village located on the İzmir-Balıkesir highway; however, this road is scarcely taken. There is also a route to the thermal springs through Simav. According to the statements of the authorities in the Emendere Facilities, 70% of visitors generally use the Balıkesir-Bigadiç route. The thermal springs lie at a distance of 455 km from Istanbul, 218 km from Bursa, 144 km from Izmir, 64 km from Akhisar and 94 km from Simav.

Natural and Human Environment Characteristics

The thermal springs are situated approximately 30 km from Bigadiç with a 200 m difference of elevation between these two locations. An interpolation of 37 years (1971–2007) of observation data from the Bigadiç Meteorology Station revealed that the annual average temperature is around 13.3^{0}C (Web-1) and average annual precipitation is about 560 mm. The area experiences a Marmara Transitional climate type, with hot and dry summers and cold and rainy winters (Cürebal et al. 2012; Efe et al., 2008) . The climatic conditions of the area support thermal tourism. According to the information collected from the local people, in some winters, transportation to the facilities is delayed for 1–2 days due to heavy snow.

The area is located in the south of Ilıca village with an elevation of about 400 m. Volcanic rocks such as andesite, dacite and trass/tuff, formed during the Lower Miocene period, are common in the vicinities of thermal springs (Soykan 1993). Manganese formations that have contact with the dacite on valley slopes are also observed.

The Akpınar Creek flowing through the area originates from hilly areas of up to 1,000 m and joins (from the south) the Demyan stream that flows into the Çaygören Dam built on the Simav Stream. A wavy appearance is dominant in the Plateau, fragmented by the branches of the Demyan Stream due to lithological differences (Soykan 1993). The thermal springs located on the western slopes of the Akpınar Creek valley are situated in the southern part of the two regular faults with a southeast-northwest direction which have developed parallel to each other. There are smaller faults that cross these main fault lines in the reverse direction. The thermal springs lie on the intersection of these faults. The length of the main fault is more than 10 km. The smaller fault lines which cross the main lines in Southeast-Northwest direction are 3–4 km.

Natural flora in the research area and its immediate vicinity has been destroyed due to agricultural activities and pasturing as meadow areas. Natural flora is conserved only in relatively elevated parts. Black pines are dispersed in the forest areas of the field mixed with oak species (such as *Quercus cerris*). The maquis elements of the Mediterranean vegetation like Sandalwood (*Arbutus andrachne*) and mock privet (*Phillyrea latifolia*), and hazelnut (*Coryllus* sp.) and cornelian cherry (*Cornus mas*) from the Black Sea vegetation are observed side by side in the Akpınar valley. Hackberry (*Celtis sp.*) is another interesting plant, alongside which we can also find plants like terebinth (*Pistacia terebinthus*), ivy (*Hedera helix*) and sycamore (*Platanus orientalis*).

The thermal springs, located within the borders of Emendere (Ilıcalı) village, are at a distance of 1.5 m to the village with a population of 581 (Web-2) as per data from 2011. The main sources of income for the villagers are agriculture and animal breeding.

Characteristics of the Thermal Spring Water and Facilities

The temperate of the thermal water is 32°C. The water has a positive effect on the P 53 gene identified as controlling the aging and creasing of human skin. It has radioactive qualities and belongs to acrotothermal water group (Web-3). Thermal source is a mineral water rich in radon (8284 Pci/lt) (Web-4). There is a multitude of minerals in the healing water of the Emendere thermal springs which is good for skin diseases such as wounds, burns, psoriasis, mange, mycosis and eczema (Web-3). Thermal water has healing effects on kidney stones, gout and stomach diseases (Web-5).

The springs facility was operated by Emendere villagers in the past and was known as "Mange Baths." The visitors from the vicinity and the neighbouring provinces used to take baths for treatment in derelict cabins that included bathtubs. Later facilities were built by the Sındırgı Municipality with the financial and architectural support from the Istanbul Metropolitan Municipality. The facilities were opened for service in June 2008 and leased to the village legal entity to operate for 49 years by Sındırgı Municipality.

The Sındırgı Emendere Thermal Springs facilities were established on 9 acres of land, 2 of which were the indoor area. Accommodation facilities are located on the west of the valley in the southwest-northeast direction of the thermal area. Excavation was undertaken in this section to flatten the land in the form of steps. Accommodation facilities were built on three steps built during the excavation (Fig. 14.2). Walking tracks were also built in the area which slopes down to Akpınar Creek Valley (Fig.14.3). Along the walking tracks terraces with benches have been set up for rest and observing the scenic view (Fig. 14.6).

There are six separate one storey buildings spread over the area which include adjacent apart rooms (Fig. 14.7). The number of rooms for accommodation is 28 (Fig. 14.8), with a total bed capacity of 70. One-room apartments can accommodate 2 persons and 2 room apartments can accommodate 3 or 4. Rooms have satellite TV, air conditioning, mini bars, bathtubs and telephones. Room service is available 24 hours a day. Underfloor heating is used in all rooms and thermal water is reheated in the boiler in winter to be routed to the rooms. A web site is available for the facilities and online reservation is possible. Wireless internet is also available.

Fig. 14.2. Apart rooms in the front of building with reception, restaurant and indoor pools at the back

Fig. 14.3. A view of the reading room with an oriental decoration where local clothes are displayed

There are two restaurants (Fig. 14.9) (indoor and outdoor), and cafeteria and market facilities. Open buffet and à la carte services are available in the restaurant that can seat 250 visitors. It is also used as a conference hall with minor arrangements. Car parking is available for visitors. There is a two storey cafeteria by the car park and the second floor is arranged as an authentic scenic viewing area and resting location, normally used in the summer. Visitors can entertain themselves with several games in the cafeteria (table tennis, rummikub, chess, backgammon, card games, checkers etc). The lower section of the cafeteria is used as the market which sells souvenirs and personal hygiene materials used in the baths. Local products are not sold in the market. There is a reading room with an oriental corner decorated with local Yağcıbedir carpets and a display of local clothing (Fig. 14.3).

The facilities also include four baths for daily use, Turkish baths, a sauna, a massage room, a fitness centre and a playground, with two of the daily baths (Fig. 14.4) used for treatment purposes. The 24-hour open sauna has a capacity of seven persons. One personnel is available for massage. Mud baths are offered in the facilities and the mud is generated by sifting the clay material obtained from the source of the spring. The mud obtained in this manner is transferred to the mud bath pool and mixed with thermal water to be used in treatments.

In total there are three pools, two of which are located indoors. The pools are not used for treatment although they are filled with the healing water of Emendere. Since the temperature of the thermal water is low, water is heated in winter for pool use. The size of the unisex outdoor pool is 12x16 m with a depth of 140 cm (Fig. 14.9). There are two indoor pools for males and females in the main service building which houses the restaurant and reception. The sizes of the indoor pools are 8.5x6 m with a depth of 140 cm, and the pools also include 2x2 m children's pools.

Necessary precautions are taken to ease the climb from the facilities to the Akpınar Creek valley. Scenic view terraces, walking tracks and rest areas have been well planned. Adjustments are made in the facilities for the disabled, and entrances are built to allow independent movements with a wheel chair. A total of 17 personnel served in the facilities during the summer of 2012.

Fig. 14.4. A view of the day baths by Akpınar Creek Valley.

The number of personnel decreases to 10 during the winter due to the less-busy season. The employees are generally from the nearby Ilıca village. College students from Sındırgı do their internships here in the facilities. The main expenses of the facilities comprise personnel salaries, purchase of provisions, electric bills and heating expenses.

Emendere Thermal Springs in Terms of Tourist Activities

The Emendere Thermal Springs are a point of attraction due to the healing effects of the water and the relaxing characteristics of the environment. Suitable climate conditions, fresh air and the relaxing atmosphere of the area make the thermal springs appealing to healthy individuals as well as patients seeking treatment.

As per the rates given in November 2012, two people can stay in one bedroom apartments for 100 TL (half board), and one can stay for 75 TL daily. The price for two bedroom apartments for half board accommodation for one person is 95 TL and for two persons 120 TL. The price of daily use of the baths is 15 TL, with free use for 0–6 years and 50% reduction for the 7–12 age group. A discount is provided for groups or tour-based organizations.

According to the managers of the facility, the number of average daily accommodations in the facilities is 35. About 12,500 individuals stay at the facilities annually, while 50% of the visitors are day customers who come from Sındırgı, Balıkesir and Izmir, benefitting from the restaurant, pool, sauna and the market; 40% of the customers prefer the facilities for treatment whereas 60% prefer it for purposes other than health, such as rest. The facilities are mostly preferred by seniors during the winter and the youth during summer. The share of foreign tourists is 1%. Visitors from abroad are generally EU citizens, and 99% of the customers are domestic tourists. Approximately 40% of domestic tourists come from Istanbul, 30% from Izmir, 10% from Balıkesir, 10% from Bursa and the rest from other provinces in Turkey. Although it is very close, the number of visitors from Balıkesir is relatively limited, probably due to the abundance of thermal springs around that area.

Fig.14.5. A view of walking tracks

Fig.14.6. Rest and scenic view terraces that include benches and the steps leading to Akpınar Creek Valley

Fig.14.7. View of the apart rooms

Fig.14.8. Interior of the apart room

Fig.14.9. View of indoor and outdoor restaaurant and outdoor pool

Fig.14.10. View of the 2-storey cafeteria

Nearly 50–60% of tourists come to the facilities with the help of tour organizations and the majority of the remaining tourists prefer to drive with their own vehicles. A limited number of customers arrive at Sındırgı by taking buses in Balıkesir and switching to taxis in the winter and the service buses in the summer to reach the facilities. Occupancy in the thermal springs increases in summer season, and during school breaks and holidays such as New Year's and religions holidays.

Weekends are generally busier throughout the year. The lowest occupancy is observed during the winter at about 60–70%. However, this rate can reach 80% during weekends in the winter. Apartment rooms in the facilities are completely booked during the summer. Agreements with tour operators ensure the increase in occupancy rates day by day. According to the managers, 70% of customers arrive from the Balıksir-Bigadiç road, 15% use the Akhisar route and 5% come through Simav. The number of visitors who arrive by using the Gölcük route is rather limited. The facilities contribute to the Sındırgı Municipality and the village budget. Tourists purchase items in Sındırgı and contribute to the local economy. The most popular items are local rugs, carpets and colognes.

Conlusion and Suggestions

The Emendere Thermal Springs are located in a rich environment in terms of natural attractions. The visitors to the facilities can be alone with nature in addition to the treatments they receive. Service units have been established to meet the demands of customers in accommodation, catering and social areas. The number of customers has increased in time and occupancy rates have gone up.

There is a need to establish new service units and organize effective presentation activities in the Emendere Thermal Springs. Some suggestions are given below.

Since occupancy rates expand during the summer, bed capacity should be increased. There are areas available in the field to build new facilities. The lower section of the current facilities should be excavated to create flat land where a ten-room accommodation facility can be built. Therefore, the bed capacity can be roughly increased by thirty.

The current facilities have undertaken almost all measures to create comfort for the disabled. However, a pool for the use of the disabled can be built as well. A children's pool can be opened by the outdoor pool. A seam bath can also be established.

A play area can be built for winter use and indoor entertainment areas for children can be established.

Visitors can trek to Ilıca village, about 15 km from the facility. They can observe nature and communicate with the local people. They can also explore the nearby tourist areas (Hisaralan Thermal Springs, Çaygören Dam, Cüneytçayı Bridge) with their private vehicles or with tour buses.

Introductory videos that present the natural environment features of the facilities should be uploaded to the website of the facility and the number of photos should be increased.

Presentation of the facilities should be given priority. Leaflets can be published to present the facilities. Stands can be erected in fairs for presentation. Connections can be made with more tour operators to increase occupancy rates. Large signboards can be placed at a few locations along the Bursa-İzmir highway to present the thermal spring facility to capture the attention of travellers, and advertisement billboards in Istanbul, Izmir, Bursa and Ankara can be used to present the facilities.

References

Bulut, İ. (1997), Yozgat Province Thermal Springs in Terms of Their Touristic Potential. Eastern Geographical Review. Volume:3, Issue:2, p.69-114.

Bulut, İ.; Girgin, M. (2001), Geographical Survey of Bingöl Kös Thermal Springs. Eastern Geographical Review. Volume:7, Issue:5.

Cürebal, İ.; Kızılçaoğlu, A.; Soykan, A.; Efe, R.; Sönmez, S. (2012). Precipitation Characteristics in Balıkesir and its Vicinity. 8.Türkiye Romanya Akademik Seminar. Proceedings.

Çekirge, N. (1991), Thermal Spring Architecture and Our Thermal Springs. Anatolia: An Int J. of Tourism and Hospitality Research. Volume:2, Issue:3.

Çetin, B.; Özşahin, E. (2011), Gönen (Balıkesir) Thermal Resources in terms of their Effects on Tourism and Spatial Transformation. Turkish Studies-International Periodical for the Languages, Literature and History of Turkish or Turkic Volume 6/2 Spring 2011, p. 317-340, Turkey.

Çomak, N.; Güncegörü, B. (2012). Integrating Thermal Springs to Tourism. Marmara Geographical Review. Issue:26, p. 55-70.

Efe, R., Sönmez, S., Cürebal, İ., Soykan A. (2008). Balıkesir'in Geoturizm Yöreleri: Marmara Geoturizm Yöresi Örneği. Üçüncü Ulusal Balıkesir Turizm Kongresi. 17-19 Nisan 2008. Bildiriler Kitabı, s. 323-330, Balıkesir.

Güner, İ; Sever, R. (2000). Less Known Mineral Springs: Malatya-Aşağı ispendere Mineral Springs. Turkish Geographical Review. Issue:35, p.105-122, Istanbul.

İbret, Ü. (2007). A Recently Developing Thermal Tourism Center in Turkey:: Çavundur Thermal Springs. Eastern Geographical Review. Vol:12, Issue:18.

İlban, M.O.; Köroğlu, A.; Bozok, D. (2008). Destination Image in Tourists Travelling for Thermal Tourism Purposes: Gönen Sample. Istanbul Commerce University, Social Sciences Journal. Year:7, Issue:13, p. 105-129.

Kahraman, N. (1991). Thermal Tourism Phenomenon and Yalova Thermal Springs. Anatolia: An International Journal of Tourism and Hospitality Research. Volume:2, Issue:3.

Öcal, T. (2011). A Sample of Thermal Tourism Resources Planned in Niğde: Nargöl. Marmara Geographical Review. Issue:34, p.287-311.

Özbek, T. (1991). Significance of Thermal Tourism in the World and in Turkey. Anatolia: An Int. Journal of Tourism and Hospitality Res. Volume:2, Issue:3.

Sandıkçı, M. (2007). Assessment of Customer Satisfaction and a Field Study in Sandıklı Hüdai Thermal Springs. Afyon Kocatepe University İ.İ.B.F. Journal, Volume:9, Issue:2.

Soykan, A. (1993). Geomorphology of Sındırgı and Bigadiç Depressions and Their Immediate Vicinity, Unpublished Doctorate Thesis, Ist. University, Istanbul

Taşlıgil, N. (1995). Thermal Tourism Potential in Manisa. Turkish Geographical Review, Issue:30, p.299-317. Istanbul.

Tuncel, M.; Doğaner, S. (1992). Thermal Tourism in Kütahya. Aegean Geographical Journal, Issue:6, p.47-60.

Tunçsiper, B.; Bezirgan, M. (2010). Perception of Economic Effects of Thermal Tourism, A Survey in Emet District. Gümüşhane University. Electronic Journal of Social Sciences. Issue:2.

Türksoy, A.; Türksoy, S.S. (2010). Assessment of Çeşme District Thermal Resources in the Framework of Development of Thermal Tourism. Ege Academic Review. 10 (1), p. 699-725.

Ünal, Ç. (2003). Geographical Survey of Pasinler Thermal Springs. Eastern Geographical Review. Volume: 8, Issue:10.

Web-1: http://www.mgm.gov.tr/index.aspx#sfU

Web-2: http://www.tuik.gov.tr/Start.do

Web-3: http://www.balikesirkulturturizm.gov.tr/dosya/1-220179/h/balikesirtermalbrosurleri.pdf

Web-4: http://www.sindirgi.bel.tr/index.php?option=com-_cantent&task=view&id=26& Itemid=39

Web-5: http://www.balikesirturizm.gov.tr/belge/1-66435/saglik-ve-termal-turizm.html

Zaman, M.; Polat, S.; Özdemir, M. (2000), Diyadin Thermal Springs. Eastern Geographical Review. Volume:6, Issue:4.

Zaman, M.; Birinci, S. (2011), A new sample to centers where thermal tourism can be developed in Eastern Black Sea: İkizdere Thermal Springs. Atatürk University Journal of Graduate School of Social Sciences, 2011 (1):405-429.

CHAPTER FIFTEEN

GEOTOURISM POTENTIAL OF TRAVERTINE CONES IN THE SURROUNDINGS OF BOLLUK LAKE (CIHANBEYLI-KONYA), TURKEY

AYLA BOZDAĞ AND GÜLER GÖÇMEZ

Introduction

Geotourism is sustainable tourism with a primary focus on experiencing the earth's geological features in a way that fosters environmental and cultural understanding, appreciation and conservation, and is locally beneficial (Dowling & Newsome 2006, Atalay et al., 2008; Cürebal et al. 2013). It is about creating a geotourism product that protects geoheritage sites, helps build communities, communicates and promotes geological heritage, and works with a wide range of different people. According to James & Hose (2008), geotourism is a form of special interest tourism and focuses on geology and the formation of landscapes. Tongkul (2006) defines it as the utilization of geological heritage resources for education-based tourism.

In the last decade, geotourism has been introduced as a new niche market that shows travellers how to increase their awareness about natural resources and ways for preserving them through outdoor and recreational activities. In geoparks, tourists can see something different from other tourism activities. Today, geoparks are ideal destinations for tourists because, as a living outdoor museum and messenger of geotourism, they apply innovative strategies which not only improve the local economy but also develop tourists' knowledge. Additionally, involving local people in local geotourism marketing helps to increase cultural communication between geotourists and local communities (Farsani et al. 2012).

Travertine cones with different sizes and the Bolluk Lake are great geotourism locations, and natural structures with well-qualified features. The purpose of this study is to introduce and protect this naturally marvellous and geologically rare area for tourism purposes.

Material and Method

The study area is located approximately 10 km southeast of Cihanbeyli town and 90 km north of Konya in Central Anatolia (Fig. 15.1). The region is very important because of its proximity to the Salt Lake and its hosting of so many travertine cones, which are not common in the world.

A geological map of the area was prepared at 1:25000 scale (Fig. 15.2). Locations and positions of travertine cones were established using a global positioning system (GPS). In order to determine the physical characteristics of the travertine cones, measurements (height, outer diameter, inner diameter, the inner/water the depth) were made using a steel tape measure. In addition, the water content of the travertine cones was determined.

In order to determine the chemical characteristics of the water within the travertine cones and the Bolluk Lake, water samples were collected. Major anions were determined in the hydrochemistry laboratory at Selcuk University. The major cations and trace element concentrations of the water samples were analyzed in the ACME Laboratory, Canada.

Fig.15.1. Location of the study area

Geological Setting

Upper Cretaceous Ophiolitic melange (Koçyaka formation) forms the basement of the study area, which is composed of serpentinite, gabbro, quartzite, chert, metasandstone and Middle Triassic-Upper Jurassic limestone, dolomitic limestone and crystalline limestone blocks (Ulu et al.

1994). Upon these units and without unconformity comes the Middle Miocene-Lower Pliocene İnsuyu formation, which is composed of alluvial deposits, conglomerate, sandstone, tuff, claystone, marl and limestone. Upper Miocene-Pliocene Çataltepe andesite, which is located in Karadağ in the southeast of this area, cut these units. The İnsuyu formation is covered by a Plio-Quaternary Cihanbeyli formation in a non-uniform manner. This unit is composed of conglomerate, limestone with gravel, sand, clay and gypsiferous clay, which is the lateral vertical transition with the Tuzgölü formation and is composed of gravel, sand, clay, silt and gypsum levels. Gypsum layers extend kilometres in the lateral directions (Ulu et al. 1994). Quaternary-Recent travertines overlie the older rocks (Fig. 15.2).

The crystalline limestone blocks belonging to the Koçyaka formation, sandstone, conglomerate and limestone of the İnsuyu formation form the reservoir rocks of hot and mineral water. The altered serpentinite and metasandstone of the Koçyaka formation are impervious base properties of the latter units. Also, clay and marls of the İnsuyu, Cihanbeyli and Tuzgölü formations form the cover rocks. Upper Miocene-Pliocene Çataltepe andesite is considered to be the heat source.

Traverten cones and Bolluk Lake, which is full of salt and sodium sulphate, developed in the control of the NE-SW trending Altinekin Fault Zone (Dirik & Erol 2000). Another important fault in the study area, the NW-SE trending dip-slip normal fault, extends along the Bozdağ, Bolluk Lake, Küçüktütün Hill and Karadağ.

Formation of Travertine Cones

The formation of travertine begins when water interacts with the soil zone, carbonate aquifers, organic material, or regional geothermal activity to produce H_2CO_3 (Deines 1992; Pentecost 1995). The H_2CO_3 increases the dissolution of carbonate rocks, resulting in elevated dissolved CO_2 concentrations in the water. The concentration gradient between atmospheric CO_2 and dissolved CO_2 in the water emerging from the aquifer initiates outgassing (Jacobson & Langmuir 1970; Langmuir 1971, Atalay et al., 2008). As CO_2 concentrations move toward equilibrium through outgassing, the water becomes supersaturated with $CaCO_3$, eventually reaching a critical level. When this critical level is exceeded, a kinetic barrier is surpassed and $CaCO_3$ precipitates to form travertine deposits (Stumm & Morgan 1981; Dandurand et al. 1982).

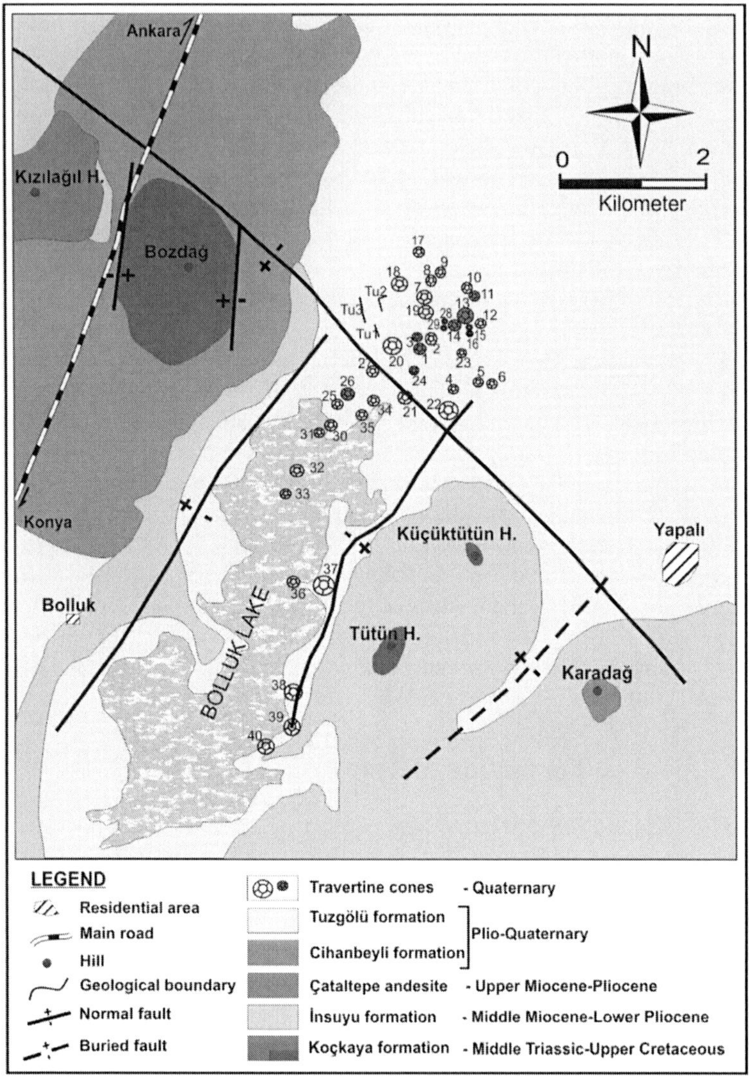

Fig. 15.2. Geological map of the study area

There are about forty travertine cones of different sizes in an area of 12 km^2 around Bolluk Lake. These travertine cones were formed by the precipitation of $CaCO_3$ from hot and mineral waters rising from slightly confined limestone aquifers at a shallow depth (Canik 1987). Hot and mineral waters in the area emerge from the low-pressure aquifer through fractures and faults. Dissolved CO_2 in the water in conjunction with a decrease in pressure disappeared over time and the water became supersaturated with $CaCO_3$, and as such the $CaCO_3$ precipitation from these waters began. Due to low pressure and discharge conditions, $CaCO_3$ in the water started to precipitate around the source. Precipitation from these waters first appeared as a thin layer and later formed a wet surface covered with plants (Canik 1987). Cones have grown with the rising of their sides by the deposition of the travertine around the spring over the course of thousands of years. The height of the cone is closely related to the surface pressure of water (Pasvanoğlu & Canik 2007).

According to the mineralogical analyses of samples taken from the travertine cones, travertines contain 10–88 % micrite, 3–50 % sparite, 0–65 % intraclast, 0–5 % ooid, 5–30 % porosity, 0–10 % fossil and 0–8% pellet. Although micrite in some of the samples is found at very high levels, in some of the samples sparite, micrite, calcite, intraclasts, fossils, pellets and porosity are all observed.

Travertine Cones and Bolluk Lake

There are about forty travertine cones of different sizes in an area of 12 km^2, and three of the travertine cones are within Bolluk Lake. Distributions and sequences of these travertine cones are parallel to the general direction of northeast-southwest trending Altinekin Fault Zone.

Cones with different shapes, such as crater, elongated and ring structured, offer a spectacular view (Pasvanoğlu & Canik 2007). The outer basement diameters of the travertine cones are much larger than the mouth diameters and look like an overturned plate. The size of the overturned plate-shaped cones differ from one to another. Most are semi-circular, while a few are ellipsoidal.

The heights of beige, grayish-white coloured travertine cones range from 0 to 23 m, the long and short inner axes range from 5.4 to 47.1 m, and 3.8 to 46.45m., respectively, the outer edge lengths range from 0 to 170.5 m., and the water depths or inner depths range from 0 to 9.9 m. Only T1, T3, T11, T13, T14, T24, T26 and T27 travertine cones include water. On the basis of AIH (1979) classification, travertine waters are Ca, Mg, Na, SO_4, HCO_3 and Cl type. The physical properties of the travertine cones are given in Table 15.1 above. According to the literature, there is no other place hosting so many travertine cones in the world. Also, three vein type travertines formed along the fault zone (NW-SE direction).

Cone number 13: It has a height of 8.5 m, the depth of the water is about 9.90 m. The long and short inner axes of the cone are 12.5 and 18.9 m, respectively. The outer edge lengths range from 13 to 18.2 m (see Figs. 15.2 and 15.3a, and Table 15.1).

Table 15.1. Physical properties of the travertine cones in the study area

Travertine ID	Inner Mouth Lenghts (m)		Height	Water / Inner Depth	The Outer Edge Lengths (m)			
	Long axis	Short axis	(m)	(m)	N	S	E	W
T1	13,10	6,65	2	2,5	14	11	11,3	12,4
T2	11,55	7,30	1,5	-	4,18	6,30	2,80	2,40
T3	9,80	7,10	-	1,65	31,8	31	28	33
T4	16,40	13,30	2,5	0,58	42,4	37,2	49,3	43,5
T5	13,90	12	0,5		10	10,3	9,8	6
T6	19,63	12,70	1,5	-	22,8	22	22	22,5
T7	24,10	22	5	-	35	35	33	33
T8	6,90	5,30	1	-	8,2	6,85	7,8	6,5
T9	5,40	4,70	3	-	5,90	7,30	6	6,2
T10	10,2	8,7	1,5	3,3	10	8,2	8,5	8,4
T11	7,80	5,40	2	3	3,40	3,30	3,40	2,60
T12	10	6,8	2,4	3,2	9,8	6,7	7,65	14,7
T13	18,90	12,50	8,5	9,90	13	15,5	13,5	18,2
T14	9	9	4,5	4,56	7,20	6,30	7,4	4,8
T15	5,70	4,10	3,5	5,10	-	-	-	-
T16	6,10	3,80	3,5	5,10	-	-	-	-
T17	24,45	21,9	5	0,11	136,8	88,2	150	75,5
T18	18,6	16,9	6,5	0,28	147,5	165	170,5	150
T19	9,8	8,2	3	-	27,10	18,4	21,1	32,15
T20	47,10	46,45	6	-	136	152	148	120,5
T21	15,40	9,8	5,1	5,40	34,5	30,1	25,3	30,3
T22	22,95	20	3	-	30,3	33,8	33,7	32,5
T23	12,8	9,65	0,3	-	8,9	6,7	6,65	10,7
T24	7,6	6,4	-	2,5	-	-	-	-
T25	13,4	11,4	2,6	2,6	9,5	11	15,9	10
T26	13,5	11,1	1,5	2,5	1,8	4,20	3,9	1,5
T27	15,4	15,3	0,4	0,20	8,1	13,9	8,6	15,9
T28	8,9	8,9	-	2,7	-	-	-	-
T29	9,44	8,3	-	2,25	-	-	-	-
T30	7,5	7,2	-	-	12,4	5,6	6,4	10,4
T31	15,2	14,4	3,5	0,53	10,9	11,1	12,75	10,3
T32	29,4	24,42	1	-	5,05	5	4,9	4,8
T33	8,9	8,1	0,5	-	24,8	18,4	26,2	32,2
T34	20,9	19,2	1	0,65	13,8	15,1	11,1	19,2
T35	12,8	10,9	1	2,1	2,1	1,8	1,42	1,3
T36	30,4	31,3	9	1,25	34	35,7	35,1	34,8
T37	46,65	38,2	23	1,15	92,3	88,6	98,9	91,8
T38	31,10	29,6	3	0,24	55,7	58	39,9	51
T39	11,6	9,8	5	-	25,5	17,5	17,6	23,5
T40	40,6	25,95	14	1,30	55	21,8	47,6	11,3

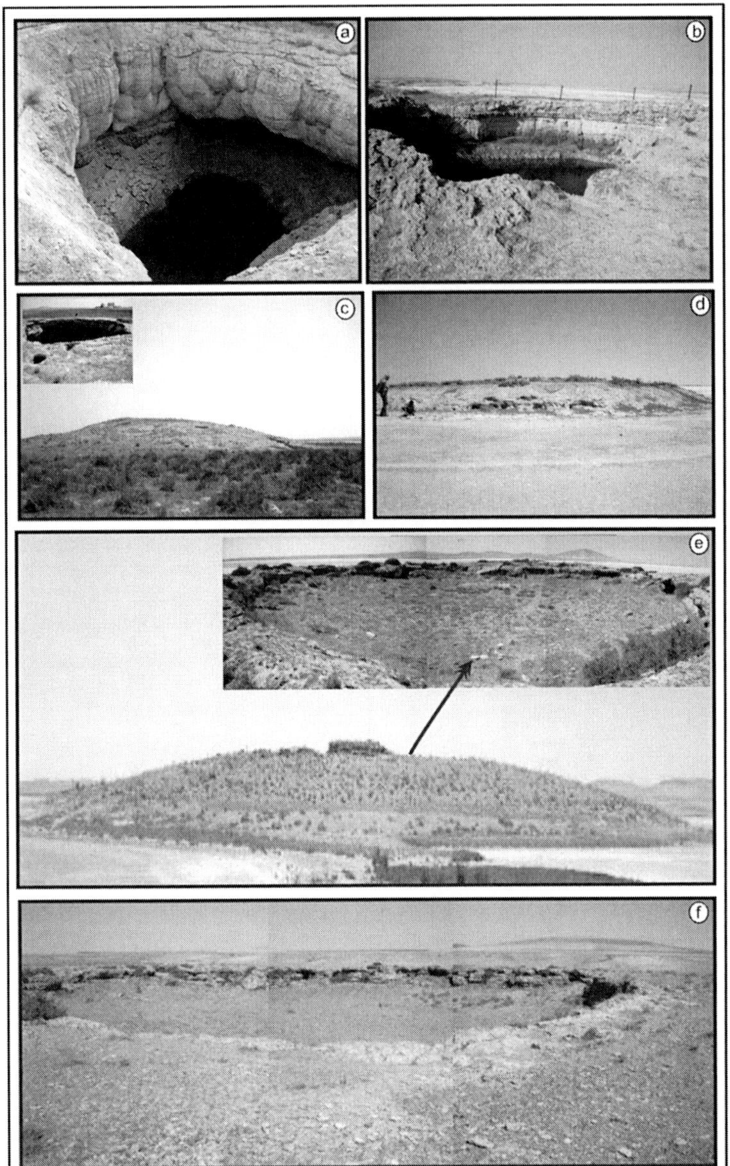

Fig. 15.3. View of some cones in the area: (a) the cone number 13, (b) the cone number 26, (c) the cone number 21, (d) the cone number 32, (e) the cone number 36, (f) cone number 37

Fig. 15.4. View of some cones in the area: (a) cone number 8, (b) cone number 28, (c) cone number 7, (d) cone number 10, (e) cone number 31, (f) cone number 14, (g-h) vein type travertines

Cone number 26 (Telli cone): It has a height of 1.5 m and the depth of the water is about 2.5 m. The long and short inner axes of the cone are 13.5 and 11.1 m, respectively. The outer edge lengths range from 1.5 to 4.2 m (see Figs. 15.2 and 15.3b, and Table 15.1).

Cone number 21 (Güvercin ini cone): It has a height of 5.1 m, and the inner depth is about 5.4 m. It is an ellipsoidal cone and is completely dry. The long and short inner axes of the cone are 15.4 and 9.8 m, respectively. The outer edge lengths range from 25.3 to 34.5 m (see Figs. 15.2 and 15.3c, and Table 15.1).

Cone number 37 (Tütün cone): It has a height of 23 m and the inner depth is about 1.15 m. The long and short inner axes of the cone are 38.2 and 46.65 m, respectively. The outer edge lengths range from 88.6 to 98.9 m. It has the appearance of a hill (see Figs 15.3 and 15.3f, and Table 15.1).

Cone number 14 (Girik cone): It has a height of 4.5 m, and the depth of the water is about 4.56 m. It is an ellipsoidal cone and the inner axis is 9 m. The outer edge lengths range from 4.8 to 7.4 m (see Figs. 15.2 and 15.3f, and Table 15.1).

Bolluk Lake is one of 76 wetlands in Turkey according to the inventory of the Ministry of Forest and Environment of Turkey (Erdem 1995), and was accepted as National/Historic Site in 1992 according to Ramsar Convention Criterias (2004). The lake developed within a graben formed by dip-slip normal faults (approximately in the NNE–SSW direction) at the eastern and western edges of the lake (Dirik & Erol 2000). The lake's surface area is $11km.^2$ The lake has a Na-Cl water type and contains large amounts of sodium sulphate. Since 1963, the Alkim Company has produced sodium sulphate and sodium chloride from two saltpans. The pH value of the lake is mostly alkaline in nature, 9.06. T31, T32, and T33 travertine cones are within the lake and these cones appear as islands.

The lake is also a living area for many bird species, such as Mediterranean gulls, terns, spoonbills, and avocets, as well as endangered flamingos in Europe. These bird species brood on the travertine cones within the lake. The area also contains endangered and endemic plant species.

Geotourism Potential and Protection of the Area

Today, people come to the area in search of the beautiful natural environments, recreation and tourism activities. The geological and geomorphological structures of the unique elements of the area are both scientific, educational and interesting examples and the focus of scientists and nature lovers. The natural and cultural resource evaluation of geoparks and geotourism is a current and valid approach (Koçan 2012).

The protection of natural heritage is the process of discovering the importance of evolution and natural history. Geological heritage values many documents containing the history of the Earth's crust. Their conservation and protection has the potential to transfer all that holds geological value to future generations (Koçan 2011).

Erol (1967–1968), in his study conducted in this area, stated that there are 63 travertine cones and Canik (1987), and Pasvanoğlu & Canik (2007), in their study of the area, mentioned that some cones have a low water discharge, and other cones are continuing to form. In this study, about forty travertine cones have been found in the area. Most of the cones are dry, and have completed their formation, and only eight include water.

The climate of the study area is hot and semi-arid. The rate of precipitation decreased gradually during the last 20 years (Fig. 15.5). The twenty-year (1989–2009) average precipitation in the Cihanbeyli basin is 312.8 mm and the average temperature is 11.69 °C. According to the Thorntwaite (1948) method, 93.67 % of the precipitation is subjected to evapo-transpiration (Bozdag 2010). Water levels in the cones, which are of the same aquifer as the hot and mineral waters in the area, have lowered over time due to low precipitation, excessive evaporation and the overexploitation of groundwater for irrigation in recent years. In addition, some cones have been destroyed for use as building blocks. Travertine cones, which are not common in the world, are currently facing a period of corruption and extinction due to human exploitation.

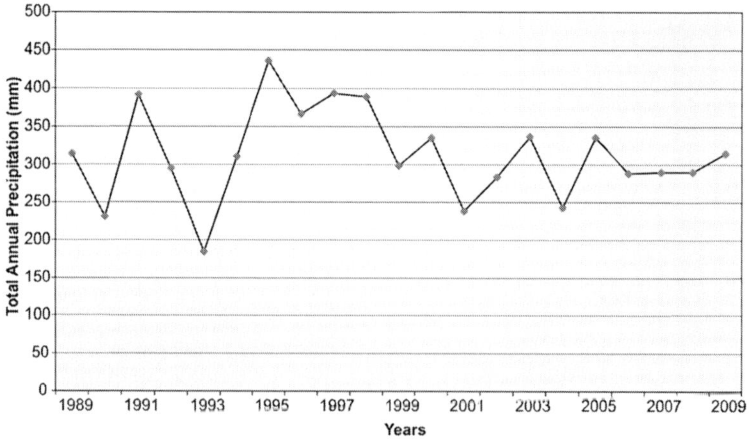

Fig. 15.5. The graph showing annual precipitation of Cihanbeyli (Konya) (over twenty years).

In recent years, the Bolluk Lake has also faced completely drying out due to groundwater depletion, lack of precipitation, shallowness of the lakes, excessive evaporation, and inadequate recharge sources. According to chemical analysis results of the water in Bolluk Lake, As, B, Se, Pb and P values exceed tolerance limits for CWQG (Canadian Water Quality Guidelines for The Protection of Aquatic Life) 2005 and ÇOB (Republic of Turkey Ministry of Environment and Forestry) 2004 (Bozdağ 2010).

The water mixing into the lake from the main discharge channel, which carries sewage from the Konya province, and the waste water exiting the salt production plant are major threats to the area and are leading to negative consequences in terms of tourism.

Conclusion

Travertine cones and the Bolluk Lake, which is full of salt and sodium sulphate, developed in the control of the NE-SW trending Altinekin Fault Zone. In a flat area of 12 km^2 around the Bolluk Lake, Cihanbeyli (Konya), about forty travertine cones with different shapes, such as crater, elongated and ring structured, offer a spectacular view. Also, there are three vein type travertines forming along a fault zone (NW-SE direction).

The area is home to many bird species, including endangered flamingos in Europe, and also contains endangered and endemic plant species. Bolluk Lake and the travertine cones, which are great geotourism attractions, are natural structures with unquestionable appeal, and their spectacular morphologic features have to be protected. These cones, which are not common in the world, are currently facing a period of corruption and extinction caused by human beings. In recent years, Bolluk Lake has also faced the danger of drying out and pollution. It is very important to protect and utilize this naturally marvellous and geologically rare area for tourism purposes.

Because of its determined properties, the study area is recommended as a Geopark. The area should be dedicated to geotourism via a conservation status for the sustainable development of local communities and the region.

References

Atalay, İ.; Efe, R.; Soykan, A., Cürebal, İ., Sarı, C. (2008). Formation of Antalya Travertine Deposits and Karstic Ground Water Systems. In: Efe, Cravins, Ozturk, Atalay (Eds). Environment and Culture in the Mediterranean Region. Part I, Chapter Six, pp. 93-108. Cambridge Scholars Publishing. ISBN (13): 9781847186584, 2008, Newcastle, UK.

Bozdağ, A. (2010). Investigation of Hydrogeological, Hydrochemical and Isotopic Characteristic of the Cihanbeyli (Konya) and Envıronment. PhD Thesis, Institute of Science, Selçuk University, Konya, 299p.

Canik, B. (1987). Bozdağ-Yapalı Toprakkale dolayındaki (Cihanbeyli) sıcak ve mineralli sular ve oluşukları. Ulusal I. Hidrojeoloji Sempozyumu, A.Ü. Fen Fak. Jeo.Müh. Böl. S. 111-123, Ankara.

Curebal, I.; Efe, R.; Soykan, A.; Sönmez, S. (2013). Factors that Affect the Formation and Development of Gölcük Uvala in Western Turkey. 8th IAG International Conference on Geomorphology, Paris, August 27-31, 2013, Abstract Volume, p.299.

CWQG (2005), Canadian Water Quality Guidelines for The Protection of Aquatic Life. http://www.ccme.ca/assets/pdf/wqg_aql_summary_table.pdf

ÇOB (Republic of Turkey Ministry of Environment and Forestry), (2004). Water pollution control regulations. Official Newspaper, 31 December 2004, Issue 25687, Ankara (in Turkish).

Dandurand, J.L.; Gout, R.; Hoefs, J.; Menschel, G.; Schott, J.; Usdowski, E. (1982). Kinetically controlled variation of major components and carbon and oxygen isotopes in a calcite-precipitating spring. Chem. Geol. 36, 299–315.

Deines, P. (1992). Mantle carbon: concentration, mode of occurrence and isotopic composition. In: Schidlowski, M., Golubic, S., Kimberley, M.M., McKirdy, D.M., Trudinger, P.A. (Eds.), Early Organic Evolution. Springer, Berlin, pp. 133–146.

Dowling, R.K.; Newsome, D. (eds.) (2006). Geotourism. Elsevier.Oxford.

Dirik, K.; Erol, O. (2000). Tuzgölü ve civarının tektonomorfolojik evrimi Orta Anadolu, Türkiye, Haymana-Tuzgölü-Ulukışla Basenleri Uygulamalı Çalışma. T.P.J.D. Bülteni, Özel sayı 5. s:27-46.

Erdem, O. (2004). Sulak Alanlar: Önemi, Temel Sorunları, Türkiye'nin Uluslararası Öneme sahip sulak alanları. Haber Ekspress: İzmir gediz Deltası ve Kuşları (28 February 2004).

Erol, O. (1967-1968). Cihanbeyli güneyinde, Bolluk Gölü Çevresindeki Traverten Konileri. Türk Coğrafya Dergisi, 24/25, 64-87.

Farsani, N.T.; Alves Coelho, C.O.; Martins da Costa, C.M.; Neto de Carvalho, C. (2012). Geoparks and Geotourism: New Approaches to Sustainability for 21th Century.Brown Walker Press. Boca Raton, Florida. USA.

Jacobson, R.L.; Langmuir, D. (1970). The chemical history of some spring waters in carbonate rocks. Ground Water 8 (3),5–9.

James, H.C.L.; Hose, T.A. (2008).. Are We in Danger of Losing the"Geo" in Geotourism? An assessment of the geological potential of selected sites in southern-Central Britain. In: Dowling R.K. and Newsome, D. (eds.) Geotourism. Proceeding of the Inaugural Global Geotourism Conference, "Discover the Earth Beneath our Feet", Fremantle, Western Australia, 17-20 August, 199-208.

IAH (International Association of Hydrogelogists) (1979). Map of Mineral and thermal water of Euarope. Scale 1:500,000. International Association of Hydrogelogists. London, UK.

Koçan, N. (2011). Kızılcahamam-Çamlıdere Bölgesi'nde (Ankara) Jeolojik Mirasın Korunması. Iğdır Üni. Fen Bilimleri Enstitüsü Dergisi. 1(4): 63-68

—. (2012). Ekoturizm ve Sürdürülebilir Kalkınma: Kızılcahamam-Çamlıdere (Ankara) Jeopark ve Jeoturizm Projesi. Karadeniz Fen Bilimleri Dergisi.3/2(6) :69-82

Langmuir, D. (1971). The geochemistry of some carbonate groundwater in central Pennsylvania. Geochim. Cosmochim. Acta 35, 1023–1045.

Pasvanoglu, S.; Canik, B. (2007). "Formation of the Spring Cones in the Vicinity of Cihanbeyli Bolluk Lake, Turkey",Cairo 10th International Conference on Energy and Environment, March 11-15, Luxor, Egypt, 1-13

Pentecost, A. (1995). Geochemistry of carbon dioxide in six travertine-depositing water of Italy. J. Hydrol. 167, 263–278.

Ramsar Convention (2004). The Ramsar Convention Manuel: A Guide to the Convention on Wetlands (Ramsar, Iran, 1971), 3rd edition, Switzerland.

Stumm, W.; Morgan, J.J. (1981). Aquaticchemistry. JohnWiley & Sons, New York.

Thorntwaite, W.C. (1948). An approach toward a rational classification of climate. Geogr Rev. 38: 55-94.

Tongkul, F. (2006). Geotourism in Malaysian Borneo'. In: Dowling, R.K. & Newsome, D. (eds.), Geotourism, Elsevier-Butterworth Heineman, Oxford, UK, 26-41.

Ulu, Ü.; Bulduk, A.K.; Ekmekçi, E.; Karakaş, M.; Öcal, H.; Abbas, A.; Saçlı, L.; Taşkıran, M.A.; Adır, M.; Sözeri, Ş.; Karabıyıkoğlu, M. (1994). İnlice-Akkise ve Cihanbeyli-Karapınar alanının jeolojisi: M.T.A. Der. Rap. No: 9720, Ankara.

CHAPTER SIXTEEN

AN ASSESSMENT OF SEASCAPE VISIBILITY FOR COASTAL LAND USE PLANNING IN THE MEDITERRANEAN REGION OF TURKEY

HAKAN ALPHAN AND FIZYON SÖNMEZ

Introduction

Valuing natural land uses, especially in an urban context, can be useful for policy planning and decisions. There is growing pressure on natural land uses to satisfy the need for additional housing and commercial spaces in developing cities. As urban sprawl is increasingly challenged, emphasis is now placed on city densification. Several movements have stimulated the debate over reorienting development practices to support neighbourhood diversity, accessibility of public spaces, and urban design accounting for pedestrians. The competing demands need to be balanced by city planners and effective land use regulation policies should integrate the benefits of preserved open spaces (Baranzini & Schaerer 2011).

Landscape as a visual experience holds considerable value in society. It is often the main attraction for tourism and can add significantly to the quality of life of residents. The most important point is that the values associated with landscape are represented and robust systems are established for conceptualizing landscape. Landscape scenery is inherently difficult to conceptualize because of its perceptual nature, yet landscapes are an extremely important resource for tourism and quality of life so there is a need to classify and manage them (Brabyn & Mark 2011).

Certain land use types are more desirable than others and have corresponding impacts on property values. For instance, views of urban green spaces (Jim & Chen 2006), water (Benson et al. 1998; Luttik 2000), and forests (Tyrväinen & Miettinen 2000) all significantly increase property values, while views of buildings and industrial areas decrease values (Bishop et al. 2004). Assessing the visual impacts of changes in land cover has become increasingly important to natural resource and land use planners. Viewshed is able to provide information on visibility and view structure that is useful in studies of scenic quality (Sander & Manson 2007).

The viewshed approach, commonly used outside the urban and architectural fields, calculates inter-visibility. A viewshed is a two-dimensional spatial layer that records the inter-visibility between points in a tessellated landscape and accounts for the location, height and angle of the view of an observer located in three dimensions. The viewshed procedure computes locations in an elevation model connected by a line-of-sight to a viewpoint location within a specified distance of an observation point. A digital elevation model (DEM) is used most often in this procedure, although triangular-irregular-networks (TINs) can also be used (Sander & Manson 2007).

The automated generation of landscape viewsheds and visible areas was one of the earliest applications of digital elevation models and Geographic Information Systems (GIS) (Amidon & Elsner 1968). Subsequently, the ability to compute viewsheds from elevation data has been incorporated into GIS packages. This function is particularly relevant to landscape assessment, given that landscape is first and foremost the appearance of the land. Viewsheds can be calculated from the point locations using line of sight calculations and digital elevation models. The analysis is dependent on having accurate surface elevation information, as poor quality will result in viewshed uncertainty (Fisher 1991).

Viewshed calculation can play an important role in urban and natural resource planning. However, difficulties exist in incorporating built structures into viewshed computation. These features are rarely reflected in available elevation maps and adding them can be complicated as their locations, shapes and heights are often uncertain (Baranzini & Schaerer 2011).

The viewshed from any one location is measured by the number of cells that can be seen from that location. Archaeologists often examine viewsheds because they are concerned whether one specific cell, one hill fort, long barrow or site of rock engraving (located in one cell) can be observed from another (Lake et al. 1998).

Predicting whether one point is visible from another (inter-visibility analysis) and predicting the total area visible from a single point (viewshed analysis) are standard tools in GIS software (Kim et al. 2004). Determining the viewshed of a chosen point can be very computer intensive, and there are several algorithms available. Kim et al. (2004)

describe some commonly used algorithms (Cooper 2005). The quantitative analysis of viewsheds requires a raster digital elevation model of the landscape. In other words, the landscape is divided into a series of squares, or cells, each of which is characterized by its height above sea level (Lake et al. 1998). The main objective of this study is to demonstrate the utility of landscape visibility information for landscape planning and management.

Materials and Methods

The study area is located within the boundaries of the Erdemli district of the Mersin Province in the Mediterranean region of Turkey. The Erdemli district lies between 36° 30' 13" and 37° 08' 18" Northern latitudes, and 32° 54' 57"- 34° 23' 43" Eastern longitudes, covering an area of 2078 km2, 62% of which is occupied by forests. The study area lies on the west coast of Erdemli town (see Figs. 16.1 and 16.2). Limonlu (Lamas) is one of the major elements of a complex network of streams that flows into the Mediterranean Sea through deep valleys carved from limestone. The climate is typical of the Mediterranean—winters are mild and rainy, whereas summers are hot and generally humid on the coastal zone (Alphan & Derse 2013).

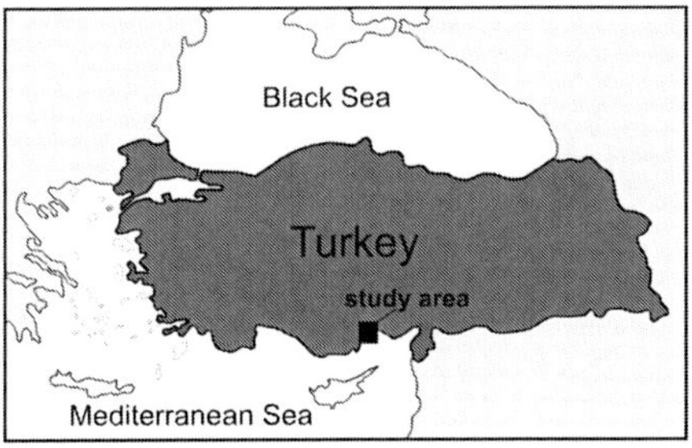

Fig. 16.1. Location of the study area

Built-up development mostly occurs along the coastline and around the town centres. Areas near the coastline mainly occupied by multi-storey buildings are designed to serve as summer apartments. Tall buildings nearest to the coastline take advantage of sea scenery. However, the scenery is partially or completely obscured for most of the buildings in the built-up zone due to the fact that the coastal margin of this zone is occupied by these tall buildings.

Agricultural crop production in the study area can be considered under four categories:

(1) open-field agriculture
(2) citrus plantations
(3) greenhouses
(4) vineyards and orchards.

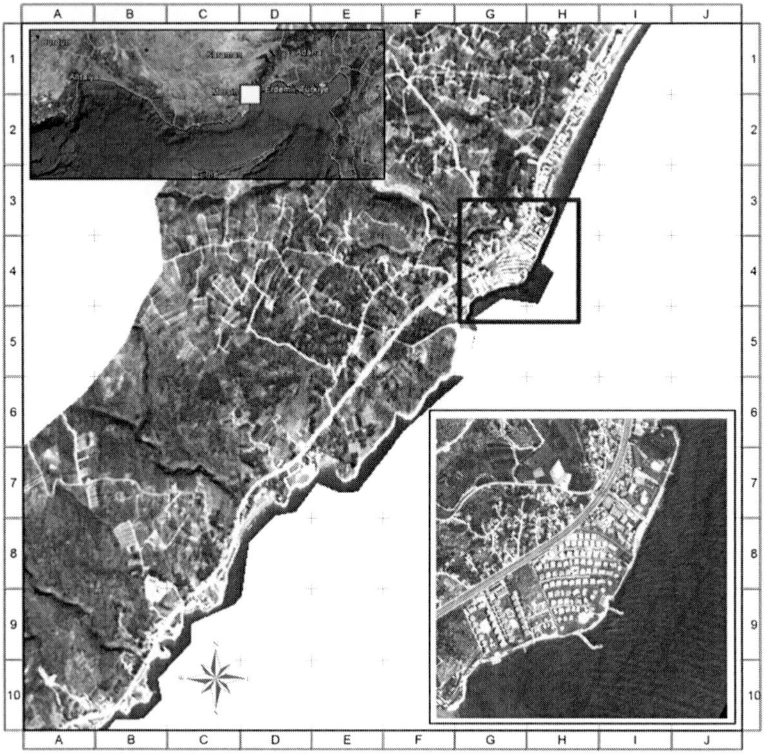

Fig. 16.2. Satellite images of the study area

Open field agriculture serves for growing a rich variety of fruits and vegetables year-round. It occupies relatively flat areas on the coastal plain, where citrus plantations are also extensive. Plastic and glass greenhouses generally exist on mild slopes of the foothills of Taurus Mountains, while orchards and vineyards create a complex mosaic with forest patches and other semi-natural vegetation at relatively higher altitudes (Alphan & Derse 2013).

The coastal zone of the study area has narrow alluvial plains formed by the streams flowing into the Mediterranean Sea between the Mediterranean Sea and Taurus Mountains, providing highly fertile soils for crop production. However, their destruction by multi-storey buildings created a trend by which agriculture was forced to move to upper lands. The study area has undergone rapid and extensive changes during the study period due to unprecedented tourism development on prime agricultural land on the coastal zone.

Methodogy

Viewshed analysis is described as a tool for identifying and classifying the positive and negative visual characteristics of a land surface. It addresses:

(1) Which areas of a land surface can be seen by one or more observers?

(2) How many observers can see a particular point?

Viewshed is the area on the ground that is visible from a specified location or locations.

A viewshed approach allows an analyst to perform two different analyses that produce the following outcomes:

(1) a viewshed map, which is a set of binary raster maps that represent the visible areas from the observers

(2) a viewshed network, which is a spatial viewshed matrix generated using GIS (Montis & Caschili 2012).

The main procedures in the methodology are as follows:

- Building heights are recorded and transferred into GIS environment
- A digital elevation model (DEM) of the study area is created
- Sample points are selected to represent sea panorama
- Buildings are incorporated into DEM
- Visibility of sample points (i.e. sea scenery) is calculated
- Quality of sea scenery is classified for each of the candidate pixels in DEM.

Results and Discussion

Assessing Seascape Visibility: Case Study

Many studies have shown that development on coastal areas mostly takes place on the areas with a sea view. The quality of sea view (i.e. partial or panoramic views) is also an important factor that determines spatial extent and the type of development that occurs. Visibility of seascape was assessed in the case of Erdemli located 40 km west of the city of Mersin, one of the large cities located on the Mediterranean coast of Turkey.

A total of 13 observation points were selected to represent the sea panorama. More observation points can be selected to represent the sea view better; however, this would be much more computationally demanding. Buildings were recorded using their spatial attributes (i.e. UTM coordinates) using GPS. Building heights were calculated as the function of the number of storeys.

A digital elevation model was calculated using digital elevation data of 10-meter contour intervals. Seascape visibility was assessed using land cover (i.e. buildings) and elevation data. Buildings were digitized on a high spatial resolution Quickbird image. The flow of this procedure is given in Fig. 16.3 below, while the procedure that resulted in a digital 3D model of the study area is given in Fig. 16.4.

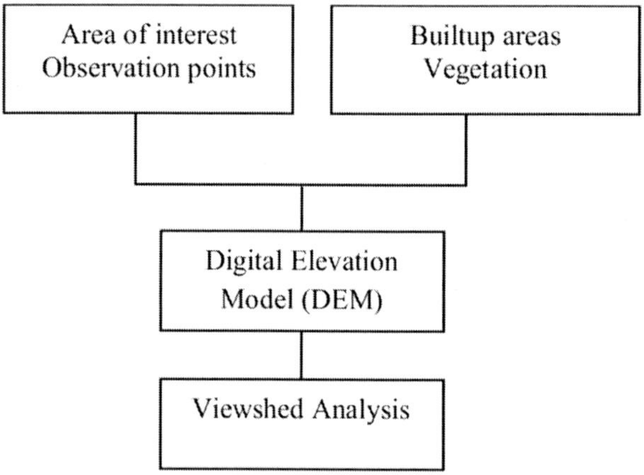

Fig. 16.3. A simple diagram depicting inputs for viewshed analysis.

Outcomes of the Analysis

Visibility analysis was done and a visibility map created for the study area (Fig. 16.5).

Analysis results included the following:

Only a narrow band of the coast has sea scenery.

Almost 100% of the areas with panoramic views are located in a narrow zone between the frontline of the buildings and the coastline, meaning potential new development areas on the coast (e.g. non-urban areas) may only have partial sea views.

Only a few narrow inland patches in close proximity to the coastline have partial sea scenery.

None of the settlements showing a scattered pattern of built-up features have sea scenery.

Sea view becomes available as elevation increases northwards. However, in this case proximity to the coastline also increases, adversely affecting suitability for development.

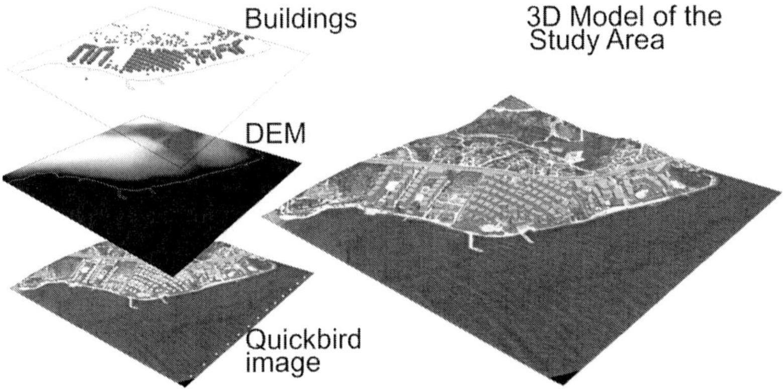

Fig. 16.4. Combination of buildings, DEM and Quickbird image for creating a digital 3D model of the study area

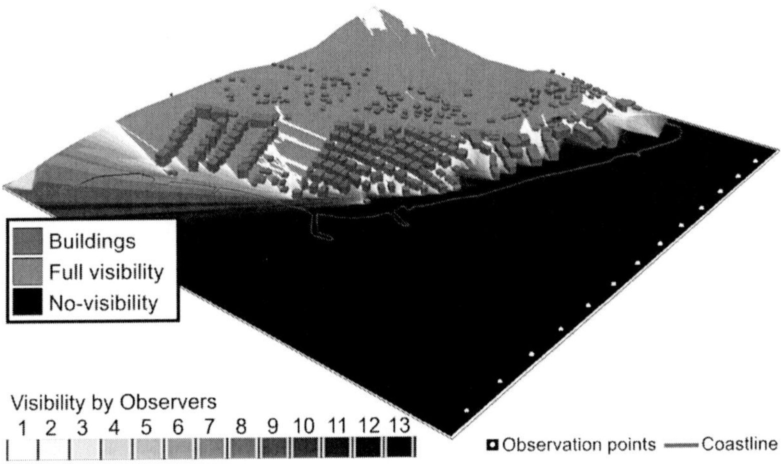

Fig. 16.5. Visibility map showing degree of visibility

Conclusions and Prospects

Digital analysis of visibility in natural, semi-natural or built environments and landscapes provides an important means for environmental planning and management. This analysis approach relies on viewshed analysis which requires 3D modelling of landscapes. The main advantage of this technique is to understand the visibility of surfaces from a particular observation or a set of multiple observation points. The visibility of seascapes in the case of Erdemli, a Mediterranean coastal town, was assessed using viewshed analysis.

Information on the availability and quality of sea scenery information provides an important means for development planning. However, suitability analysis for development also requires extra information, such as slope of the area, aspect and proximity to the coastline. This study demonstrated a methodology for analyzing sea scenery. Precision may be increased by increasing the number of observation points and analysing the impact of building heights on the areas with sea scenery that are potentially available. The increased number of observation points may lead to more complex multiple viewshed calculations, while keeping the height of target locations variable may result in repeated calculations of multiple viewsheds corresponding to different building heights that will potentially be realized through development.

Acknowledgements

This project is supported by the Turkish Scientific and Technological Research Council (TUBITAK) and the Çukurova University Department of Research Projects under the grant numbers ÇAYDAG 111Y253 and ZF2013YL05, respectively.

References

Alphan, H.; Derse, M.A. (2013). Detecting Landscape Changes on an Altitudinal Gradient in the Mediterranean: The Case of Mersin, Turkey. Natural Environment and Culture in the Mediterranean Region-II, Efe, R., Ozturk, M., Atalay, I., (Editors), p 153-163. Cambridge Scholars Publishing-UK, ISBN: 978-1-4438-2986-1, 518 pp.

Amidon, E. L.; Elsner, G. H. (1968). Delineating landscape view areas - a computer approach. Berkeley, California: USDA Forest Service.

Baranzini, A.; Caroline S. (2011). A sight for sore eyes: Assessing the value of view and land use in the housing market. Journal of Housing Economics, 20:191-199.

Benson, E.D.; Hanson, J.L.; Schwartz Jr. A.L.; Smersh, G.T. (1998). Pricing for urban environments. Computers, Environment and Urban Systems, 34:518–531.

Bishop, I.D.; Lange, E.; Mahbubul, A.M. (2004). Estimation of the influence of view components on high-rise apartment pricing using a public survey and GIS modeling. Environmental Planning, B 31, 439–452.

Cooper, G.R.J. (2005). Analysing potential field data using visibility. Computers & Geosciences, 31: 877–881.

Jim, C.Y.; Chen, W.Y. (2006). Impacts of urban environmental elements on residential housing prices in Guangzhou (China). Landscape Urban Planning, 78,422–434.

Kim, Y.H.; Rana, S.; Wise, S. (2004). Exploring multiple viewshed analysis using terrain features and optimisation techniques. Computers & Geosciences, 30:1019–1032

Lake, M.W.; Woodman, P.E. and Mithen, S.J. (1998). Tailoring GIS Software for Archaeological Applications: An Example Concerning Viewshed Analysis. Journal of Archaeological Science 25, 27–38.

Luttik, J. (2000). The value of trees, water and open space as reflected by house prices in The Netherlands. Landscape and Urban Planning. 48, 161–167.

Montis and Caschili (2012). Nuraghes and landscape planning: Coupling viewshed with complex network analysis. Landscape and Urban Planning, 105 (3): 315-324

Sander, H.A; Manson, S.M. (2007). Heights and locations of artificial structures in viewshed calculation: How close is close enough? Landscape and Urban Planning 82:257–270

CHAPTER SEVENTEEN

RELATIONSHIP BETWEEN THE CONCEPT OF SUSTAINABLE AGRICULTURE AND AGRI-TOURISM IN TURKEY

HASAN SAYILAN

Introduction

Today, humanity is faced with the problems of global warming, climate change scenarios, ozone depletion, destruction of forests and species, the degradation of natural resources such as land and water, poverty and rural development, all of which is the result of a determinant economic system. This approach reduces the welfare of humans to the level of the quantity of the goods and services that they consume and makes them believe that every problem can be solved with technological developments, ignoring relationships except for the consumption between humans and other creatures. The solutions of the problems that humans meet today require leaving this approach and appearing more humane and environmentally conscious. One of the spearheads of this approach is sustainable agriculture.

Sustainable agriculture means environmental health, economic profitability and sustaining a better social and economic equality (IFAS 2008). It also means that agriculture for most people should allow for the preservation of natural resources, protect the environment indefinitely, improve the safety and health of the community and be profitable for people working in agriculture, which will enable the production of enough food for everyone. It also has some social values such as revitalization of rural areas, providing a rich life to farmers, and has the ability to obtain healthy food for everyone (ATTRA 2005).

In this context, the concept of sustainable agriculture is providing an alternative understanding and requires the reduction or complete removal of the negative effects created by modern agriculture applications. It provides an overall structure that mentions all aspects of the agricultural sector with economic, social and environmental dimensions. The establishment of this structure is important for the agricultural sector which needs more necessities today than in the past. Today, agriculture undertakes an important function in the protection of the social structure, cultural and rural heritage, natural balance and the environment as well as the nutritional requirement of humanity. Other functions are improving the quality of life through high-quality production and protection of human health, increasing farmer income and providing a basis for rural development. Sustainable agriculture, with these functions as an income-generating element in the rural areas, also has a positive impact on the tourism sector.

The tourism sector is one of the fastest growing sectors of the global economy, but in today's world important changes are observed about the tourism consumption patterns in parallel with economic, social and technological changes. Developments in the tourism sector and its dynamic structure have affected the diversification of the tourism activities. Participation in the luxury tourism activities is decreasing gradually and the trend is moving away from the usual tourist centres.

The tourist consumers, who are more conscious today, require new approaches for the products based on nature or traditional products (Rastrollo & Alarcon 2000, 210). Agricultural tourism (agro-tourism) is also a product of the new understandings and pursuits of the visitors regarding nature.

Agricultural tourism is gaining its place among the rural tourism activities with its special values. Generally, it is a kind of tourism that has common targets with sustainable agriculture, provides sustainability in tourism, shows respect for the environment, and values the cultural heritage as well as authentic features of the region. In addition, it has a purpose of providing additional income to the rural population and decreasin the rural-urban migration and the economic, social and cultural issues accordingly.

Sustainable agriculture and the process of integration of tourism are gaining momentum dependent on agricultural tourism. The local people living on agriculture in the regions where tourism has developed are leaving the productive agricultural areas to the touristic companies. They are using rationally the productive areas for agriculture and the partly unproductive areas are used for tourism and recreation. Thus, a planned and rational integration of tourism and agriculture is realized. This tourism is also very important in terms of realizing the maximum efficiency of agricultural areas.

In this study agro-tourism, a special area of rural tourism, is discussed. The concept of agro-tourism is mentioned based on the relationship related to sustainable agriculture, and suggestions are made about developing this kind of tourism in Turkey.

Materials and Methods

Turkey is in the northern hemisphere in the middle section, between 36° to 42° north latitudes and 26° to 45° east longitudes. There are meridians between the east and west parts of Turkey and this results in a time difference of about 76 minutes. Turkey's total land area is 814,578 km², 97% of which is in Asia with the rest in Europe. Turkey is on a point connecting Asia, Europe and Africa. The Canakkale and Istanbul Straits separate Asia from Europe and make a natural connection between the Mediterranean and Black Sea basins. This gives a very important and strategic position for the country in the world in terms of political, economi and military aspects.

Turkey's geographical location leads to a variety of geological features and notable topography, climate, wildlife and vegetation. Its average height in Europe is 330 m, in Asia 1,050 m, and on the whole is 1,132 m. The country is a peninsula surrounded on three sides by the seas lying on mild temperate zone. The climate, type and amount of precipitation and vegetation are affected by the proximity and distance to the sea and elevation. Generally, the north and south mountains extend parallel to the sea. The sectors most affected by this situation are agriculture and tourism.

Table 17.1. Average number of parcels and parcel sizes according to Farmer Registry System (CKS) (The Ministry of Food, Agriculture and Livestock-GTHB)

Years	Number of Parcels (P)	Number of farmers (C)	Area (decares) (A)	P/C	A/C
2002	15.332.976	2.588.666	164.960.378	5,9	63,7
2006	16.457.203	2.609.723	164.930.261	6,3	63,1
2011	15.856.663	2.292.380	156.287.667	6,9	68,1

Source : Prepared by using GTHB CKS

While we look at the existing land-use situation, about three million hectares of land in Turkey is covered by fruit orchards and olive groves. The area of other agricultural lands is about 21.4 million hectares, and grassland covers 14.6 million hectares. The total land used for agricultural purposes in 2011 was 38.2 million hectares (Gunlu 2012, 10).

If we examine the data given in Table 17.1, after the year 2002 it is seen that the parcels per farm are 5.9, but became 6.9 parcels per farm in 2011. In 2011 land size was 68.1 acres. Generally in Turkey small-scale agricultural enterprises are extensively involved in small size companies, which is one of the Initial factors restricting the efficiency of production and reduction in profitability, and it also restricts the ratio of the lands where agricultural tourism can prove successful.

In this study, agricultural tourism (agro-tourism), which is a special area of the rural tourism applications, has been evaluated theoretically in order to enlighten this potential. There is a high level of agro-tourism potential, and attempt has been made to discuss the relationships between sustainable farming applications and agricultural tourism.

For data collection, comprehensive literature reviews were studied to determine the applications of the agricultural tourism (agro-tourism), the main target of this study. Other materials were collected from all available published records from the world and Turkey.

General Agricultural Census Results and published statistical data of The Ministry of Food, Agriculture and Livestock were used. The averages, percentages, accounts and statements were recorded and suggestions made with the aim of developing agricultural tourism in Turkey.

Findings and Discussion

The Concepts of Sustainable Agriculture and Agricultural Tourism (Agro-Tourism)

Sustainability is derived from the Latin word sustainer, which means protecting the existence with continuity or support in the long term (Rigby & Cáceres 2001, 22). In the simplest terms, sustaining means the presence of something by itself (Meadowcroft 1997, 168). The concept of sustainability has qualitative features and characterizes the concept used in conjunction with itself. One of these concepts is agriculture.

Agriculture could be described as producing plant and animal products, raising their quality and yield, arranging appropriate preservation of these products, handling them and evaluating and their marketing. Today, the agriculture is a holistic and multi-dimensional structure with such aspects as biological, economic, environmental, social and political. Therefore, to make evaluations on the agricultural sector there is a need to consider its different dimensions, which could be classified as bio-physical environment, socio-political environment and economic and technological environment (Yunlong & Smit 1994, 300).

The dimensionality of the agricultural sector and some problems with the modern agricultural applications require new studies with approaches based on evaluating these dimensions all together, and those which are minimizing or eliminating the negative effects of the modern agriculture applications. One of these new approaches is sustainable agricultural practices.

Sustainable agriculture aims at long-term efficiency and can be described as efficient food production activities without exhausting the world resources and without polluting the environment. Another description defines it as an agriculture which follows the main principles of nature in growing agricultural crops and livestock breeding systems in a way that allows nature to renew itself. For most people, sustainable agriculture means the agriculture that will provide protection of the natural resources, protect the environment indefinitely, increase the safety and health of the public and provide enough food for everyone and be profitable for the farmers. This concept also includes some features like social justice and the protection of animal welfare. The sustainable agriculture described above is existing as an alternative to the traditional and modern agricultural practices, which means the agricultural applications reliant on other non-farm inputs and are highly specialized, using heavy synthetic materials. The rule which needs to be accepted here is that the sources for the agricultural production in the world are not limitless, and continuous improvement cannot be provided by disrupting the natural balance (Schaller 1993, 89–90; Turhan 2005, 13–14).

An important purpose of sustainable agricultural applications is also to minimize the gap in the development between the towns and cities by improving the living conditions of the locals. In addition to the above, it also has the effect of social transformation, creating employment opportunities for the locals and thus decreasing migration by integration of the rural development programme with agricultural as well as other activities required in rural areas. It is related to the aspect of rural tourism and agricultural tourism.

The concept of rural tourism has many common targets within the sustainable agriculture concept. It is known by different names by professionals, such as farm tourism, village tourism, highland tourism, agro-tourism, and ecotourism. The reason for these different names is that there is no common idea about the scope of the rural tourism (Esengun et al. 2002, 32). Agro tourism is gaining a place inside the rural tourism activities with the special activities it includes (Fig. 17.1). Rural tourism is a kind of tourism that integrates with culture, the natural environment and agriculture and also adapts to other types of tourism. The main purpose of rural tourism is staying in a village, farm or a chalet, for example, and having a holiday by meeting and socializing with the rural culture (Soykan 2003, 3). Rural tourism is not seasonal, and rural settlements can be visited in every season. Due to this feature, rural tourism includes several recreational activities. Traditional agriculture creates a unique cultural landscape and lifestyle, and a tight relationship exists between agriculture and rural tourism (Ozguc 1998, 275).

Some descriptions of agricultural tourism have been put forward, such as the "incorporation of a farm house by combining it with the components of tourism" (Weaver & Fennell 1997, 467). It is a phenomenon where people observe and live within daily farm conditions (Mitchell 2006), combining tourism and recreation activities with farm activities (Prezezborska 2003) in places where agricultural production is rich. This feature of agro tourism creates additional sources of income for the tourism producers. Thus, we can talk about combining agricultural production with tourism instead of creating displacement and shattering the agricultural areas doing tourism business (Kucukaltan 2002, 151). Agro-tourism is important for the aspects of diversification of products and provision of additional income for the farmers. The tourists, satisfied with nature by participating in the agricultural activities and gaining sensitivity and awareness to our living environment (Fig. 17.1), are real agro-tourists.

Generally agricultural tourism enables the development of sustainable tourism, respecting the environment and giving value to the cultural heritages and authentic features of the regions. The authentic features are multi-functional, which provide an opportunity for participation and information about agricultural areas, agricultural occupations, local products, traditional cuisine and the daily lives of the people. This provides tourists with rural activities in which they can have fun in travelling, learning and discovery.

According to Woo & Yeon (2006, 68), agricultural tourism is the forming and mixing of agricultural, ecological and cultural products, which include relaxing and therapeutics in social areas and education in the environmental field. The intention of agricultural tourism is not displacement of agricultural production and tourism, but combining agricultural production with tourism (Kucukaslan & Ak 2008). Agricultural tourism can be described as participating in the activities of a farm, farm plant or similar operations with the purpose of training, active production or leisure.

One of the targets of the Rio-Agenda 21 is the creation of non-agricultural employment opportunities, such as rural industry, textile industry and rural tourism in rural areas (Kucukaltan 2002, 151). This way, the agricultural areas are protected and a different economic income is created together with the concept of tourism.

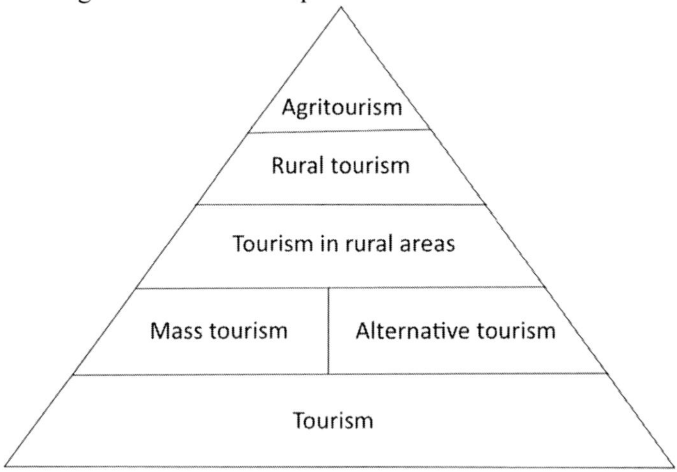

Fig. 17.1. The Pyramid of Tourism Related Concepts
Source : Sznajder, Przezborska & Scrimgeour (2013, 3)

The potential of agro-tourism in the world and in Turkey

Agro-tourism started at the end of the 1800s. The first activities in this connection included short visits for urban dwellers to help them escape the city and visit their relatives. After the 1920s the rural area visits became easier for the people. Following the great depression and World War II during the 1930s and 1940s, interest in rural recreation increased. Horse riding and feeding became farm animals became popular in the 1960s and 1970s. Farm vacations, overnight stays with breakfast and commercial farm tours became popular during the 1980s and 1990s (Hatch 2008, 2).

This shows that rural tourism, together with agricultural tourism, became widespread during in the last twenty years. Lots of factors have contributed towards its increase. For example, the shorter holiday times of workers, increased car travelling, combining business with holidays, diversifying experiences, and desires to search for lineages by travelling as families (Maetzold 2004, 84).

In "Tourism Vision of 2020," it is mentioned that the products presented to the visitors in rural tourism are still relatively limited, but will increase in the next 5–10 years to an important level. Although an influx to rural tourism is not expected during this period, it is clear that there will be a high increase in the levels of rural tourism (WTO 2004).

Although in the national tourism management of most countries, rural tourism is developing as an important sector, it is really difficult to get information on the dimensions figures in agricultural tourism due to the scarcity of countries which collect statistical information on rural tourism. Therefore, the results of only some studies on rural areas can be mentioned.

A study carried out by the European Commission in the mid-1980s on European tourism habits attempted to find tendencies in rural tourism. According to this study, at least one third of people go to rural places for their holidays in the countries where rural tourism has developed. In the Netherlands, 39% of people going on holidays chose the rural places as their first choice, and in England three quarters of the population went to rural areas at least once a year. In the rural tourism concept, when we look at the statistics on accommodations in Europe during 1987, we can see that tourists mostly look for accommodation in rural boarding houses and hotels (30%), friend's houses (21%) and in rural camping areas (16%) (Soykan 2000, 21).

Rural tourism has been considered a strategy in regional socio-economic development since the 1950s in most countries of the central and north Europe and also in southern countries since the 1970s. The local authorities have seen that rural tourism serves for the common interest of the visitors and local public and they have created rural tourism by considering the multi-faceted effects of agro tourism.

The EU has also accepted the importance and addition of tourism in reducing regional disparities and promoting employment. The structural funds and especially the European Regional Development Fund (ERDF) provide the most important EU financial assistance for tourism. These are the protection of the rural environment, sustaining the local cultural traditions and skills, renewing the villages, and support for the farmers entering the tourism sector and diversifying their activities etc. (Commission Européenne 2000). In Europe today, the effective countries in agricultural tourism are the UK, France, Germany and Austria (Busby & Rendle 2000, 635).

The agro-tourism operators in the world came together and organized an international arena. In this way, agricultural tourism became a sector serving more consciously, regularly and in high-quality, which also includes nature and eco-tourism. The International Agricultural Tour Operators Association (ATOI) serves the interested people in this area.

In the Tourism Strategy of Turkey 2023, reference is made to the agricultural tourism with an article on "making awareness of the public, private and non-governmental organizations about eco-tourism, rural tourism and agro-tourism by introducing sustainable tourism" under the title of the envisioned principles. The "action plans will be prepared for agricultural tourism (agro-tourism), adventure tourism, cave tourism, sports tourism with the cooperation of public, private sector and civil society organizations" under the title of eco-tourism.

The first investment made for agricultural tourism was for the Adatepe Olive Oil Museum, which is on the Canakkale-Izmir road in the district of Kucukkuyu. The high-potential places for agro-tourism in Turkey are Bozcaada in Canakkale, Sirince Izmir, Cesme (Alacatı) and Foca, Cappadocia in Nevsehir, Sarkoy and Murefte in Tekirdag, Pamukkale in Denizli, Avsa in Balıkesir, and Iznik in Bursa. We can see that there are many more places which have good potential due to their geographical diversity.

The agricultural tourism in Turkey was started in 2004 by the Wheat Association for Supporting Ecological Living with support from The United Nations Development Program. The project of Ta-Tu-Ta (Agro-Tourism-Trade) drew the attention of domestic and foreign tourists who want something different from the classic sea-sand-sun holiday, and who have consciousness of ecological agriculture. This project is a movement which connects volunteers and organic farms to build a sustainable global society by promoting the cultural and educational experiences based on confidence and exchange. The movement is connected to the organizational network of WWOOF (World Wide Opportunities on Organic Farms), within the frame of the Ta-Tu-Ta Project. Detailed information about 88 farms in different parts of Turkey is given on the web site of the association (bugday.org) .

Marmara, the Black Sea and the Aegean and Mediterranean regions of Turkey are important not only with their coastal tourism, but are also very important because of their high potential in rural tourism due to their geographical locations. The Central and Eastern Anatolian regions are also important for rural tourism with their specific agricultural structures. In addition, the lakeside villages in Bursa are very convenient for agricultural tourism (Ak 2006). The Toros Yoruk villages on Taurus Mountains are very convenient for the integration of agricultural and cultural tourism (Kaya & Arıkan 2008, 67).

The relation of sustainable agriculture, rural areas and agricultural tourism (agro-tourism) in Turkey

In recent years all over the world, efforts to increase agricultural land and water resources are encountering difficulties. Improvements in agricultural areas have decreased due to some unfavourable environmental impacts and soil erosion, as well as air pollution. Every year, millions of hectares of agricultural land are lost because unplanted lands are severely eroding and other areas are used for non-agricultural purposes, such as building houses and factories.

The important topics in evaluating the concept of sustainable agriculture for our country are: erosion, the use of more inputs, pollution, irrigation, industrialization, the effects created by urbanization and the impact of tourism, and the degradation of pastures and meadows and their gradual decrease in usage. Moderate and splitting erosion is seen in approximately 80% of the land of our country and the impact of erosion can affect the sustainability of agriculture (Gunaydın 2005). The agricultural applications, and especially the form of processing and techniques, destroy vegetation, leading to erosion. The primary reasons for such development are insufficient investments for soil conservation and reclamation and avoiding the application of soil protection measures. The prevalence of small-scale agricultural enterprises and low agricultural income can be considered as the reasons for this. Therefore, it is important to follow sustainable agriculture in rural areas where it is intensive in Turkey.

When we look at the equivalent meanings of kırsal (rural) in foreign languages, the word is identified with the case of villages, but in reality it is used as the the opposite meaning of word of kent (city) (Kucukaltan 2002, 152). Gaining new meanings of socio-cultural, demographic, economic, environmental and spatial diversity, the rural areas include the changing conditions of the century, making it difficult to describe rural areas (Ceken et al. 2007, 3). According to Soykan (2006, 73), the rural areas are described as a concept based on low population density, as small residential units with a population of generally less than ten thousand. The description of rural areas developed by the OECD and used by the EU is the most used, which is also used for determining the regional typologies. According to this description, rural areas are places with administrative boundaries considered as rural and with a population density of less than 150 people per square kilometre. The description accepted by TUIK as per the statistics related to the field of rural areas mentions that the residential units with a population of more than twenty thousand are considered cities, and other residential units with less than this are considered rural areas; this is also generally used by other associations in our country (Rural Development Plan 2010, 39).

In many rural and regional development plans developed in Turkey, some activities are carried out under the title of rural tourism, eco-tourism, and agro-tourism to create new business opportunities and increase the incomes of rural people. These projects include the education intended to increase the sense of taking more care of their areas by creating awareness about environmental protection and natural resources.

In sustainable agriculture, realization comes through the tools which can provide environmental, social and economic sustainability. These concepts are included within the concept of sustainable agriculture and should include the four elements on which sustainability is based, namely: food safety and assurance, employment and income growth, protection of natural and environmental resources, and the more active participation of people in decisions and management.

The sources created by nature, culture and people are at the same time tourism sources, because they are also used by the tourism industry and tourists. These sources are a collection of environmental values. Most of the environmental values belong to society and generally cannot be re-created or re-priced. While we think about the expectations of the tourists for a successful tourism policy, a protected and "developed" environment is necessary (Kahraman 1997, 103). Within the scope of the Turkish Tourism Strategy (2023), eco-tourism, rural tourism and agro-tourism are mentioned under sustainable tourism.

Considering the climate, natural resources, vegetation, seas, lakes, terrain, soil resources and irrigation areas of Turkey, all these are providing a diversity of production in agriculture and also giving opportunities for culture and alternative tourism activities (Efe et al 2008) . There are some relations between some sectors in the economy, tourism and agriculture—mainly that they complete each other. Tourism is a service sector where basic food, drinks as well as fruits, vegetables are consumed intensely intensively. A large part of the basic inputs of the tourism sector are provided by the agricultural sector. The sector plays a major role in supplying the basic eating and drinking needs of domestic and foreign tourists. The diversity in the agricultural products of the country can also positively affect the tourism demand for the country. The climate and natural resources depend on the structure of both sectors and require building the structure in both sectors by considering their dynamics.

The recent additional requirements for increasing tourism activities in Turkey also resulted in the development of agricultural sector. The interaction between tourism and agriculture featured the agricultural tourism in relation with the geographical and climatic features of our country (Cıkın, Ceken & Ucar 2009, 4–6).

Agricultural tourism has a common purpose with sustainable agriculture (agro-tourism) and has been shaped as a new kind of tourism which does not damage the agricultural areas and targets the people who want to be away from city life and who miss rural life. In this kind of tourism, tourists join the agricultural activities directly and can satisfy their longing to commune with mother nature. This kind of tourism is also important for getting the maximum yield from agricultural areas. Agricultural tourism is also a multi-functional kind of tourism that provides tourists a way for participation in the agricultural areas, agricultural occupations, local products, traditional cuisine and daily lives of the people. It provides some sort of rural activity to the tourists so that they can enjoy travelling, learning and discovering.

The Importance of (Agri-Tourism) and its benefits

Agricultural tourism is beginning to use a marketing of agricultural agricultural products-oriented structure which is the purpose of rural development in various national developmental programs. It is possible to use the sources more efficiently in the rural areas where not enough importance is given to the development of the national economy. Agricultural tourism is similar to eco-tourism but the attractiveness of the area is not only the natural aspect, it is also cultural scene. If the attractiveness aspects are provided to the tourists the number coming to the region will increase and this will add to the income levels of the locals. Agricultural tourism will increase the regional development, and agricultural diversity is also considered important in this connection.

The agricultural environment includes the tourism products which are directly linked to the agricultural production and agro-housing, and accommodation on a farm through staying in a closed area or camping, educational excursions, meals, recreational activities, and the sale of farm products or handicrafts are included in this (Gunduz 2004, 39). The local people who provide the possibility to use the tourist services in the villages or farms, dependent on agricultural tourism, are using their empty lands as camping areas for tourists. The possibility of using such areas for touristic purposes by the local people adds more and more to their income by bearing the low costs from the agricultural production. After an increase, on the request of the people, village tourism in the places where agriculture and tourism are abundant can provide accommodation for domestic and foreign tourists. The basic life conditions of the rural areas make it possible that farm houses can respond to the tourist needs with a low capital. Today, the developments on the supply on agricultural tourism are increasing the charm and attractiveness of the tourism activities in this sector (Olalı & Timur 1988, 157).

In the international meeting in 1992 of the "Rio-Agenda 21," attention was drawn towards making profitable operations of the policies in countries which are suitable for national development plans and targets. One of the decisions was developing rural tourism "agricultural tourism." According to this, the poor people in rural areas should be taken away from the obligation of using marginal lands, and instead new non-agricultural employment opportunities should be planned and industries constructed, such as rural industry, textile industry, fisheries and rural tourism (Yıldırak & Gulcubuk 1998, 188). All these facts are highlighted in the agricultural tourism which includes not only protecting the agricultural areas and using agricultural products in tourism, but also providing an alternative type of tourism for development with an economic dimension. The point that should be highlighted here is realization of agricultural tourism in agricultural areas (Kucukaltan 2002, 151).

Table 17.2. The classification of agro-tourism activities

Special Activities and Festivals	Recreational Activities	Farm Activities
Music festivals	Bird watching	Farm visits
Agriculture festivals	Horse riding	Sale of agricultural products
Holiday celebration	Animal nutrition	Agricultural exhibitions
Seasonal activities	Wildlife	The farm accommodation
Youth camps	Photography	The activities of collect-on-your-own
School tours	Wildlife-watching	Educational activities
	Wine producing	Tree rental
	Camping	
	Ancient village trips	

Today, the different agricultural tourism alternatives are seen as depending on the international tourism demand. In the regions where agricultural activities are rich, collection of fruits by tourists or harvesting are the alternatives. The ethnographic museums, public markets and traditional festivals are other elements which can support the development of agricultural tourism. The tourists renting trees from the farmers make the payment at the beginning of the season and can optionally join the harvesting and maintenance works. The festivals carried out connected with the agricultural tourism (strawberry festival, peach festival, etc.) and village bazaars play an important role for the revitalization of the local economy and in generating more and more interest. Such kinds of events for the local people who are sponsoring and hosting are providing opportunities for the promotion of local culture and also providing the flow of the domestic and foreign tourists to the bazaar and small villages (Gunduz 2004, 42). The tourists go to villages and are socializing with the villagers in the cafes, collecting cotton in the cotton fields and milking the cows, which lead to the development of agricultural tourism. Various agricultural tourism activities depend on the different needs of the tourists, and agro-tourism is finding its place in the rural tourism activities with special activities (Soykan 2003), as shown in Table 17.2 above.

There are benefits of agricultural tourism in many ways, in that it contributes to the diversity in the agricultural sector and advancement of industry. The benefits of agricultural tourism in terms of the agricultural sector are (Michigan Advisory Commission 2007; Williams et al. 2001, 4):

- it enables the growth of agricultural enterprises
- it enables the development of different agricultural products
- it provides increased farm incomes
- it enables the creation of a new consumer markets
- it provides the promotion of local agricultural products
- it helps the agricultural enterprises to be more functional and administrative
- it ensures the sustainability of the agricultural sector.

The benefits of the agricultural tourism in terms of countryside or the community are (Agriculture and Forestry 2000, 2; Williamset et al. 2001, 4):

- opportunities for additional income rises for the local businesses
- help in the development and revival of the infrastructure and local activities in the region
- help for better protection of the environment in rural areas
- allowing an introduction of manners, precedents and handicrafts of the region
- helping people to have agricultural education and awareness in the region
- providing local people additional income and business resources
- guiding the investors to make some kinds of investments in rural areas
- employment for all family members in the farms
- additional income for the families.

The benefits of the agricultural tourism in terms of tourism are (Williams et al. 2001, 5):

- different tourism products and services for different visitors
- number of visits to rural places increases

- the promotion of many different destinations
- provision of an extension of the tourist season in rural areas
- provision of inflow of foreign currency.

Agro-tourism can be evaluated as a new income source for the farming communities which supports the protection of the environment and continuity of small-scale enterprises. On the other hand, some problems like migration, poverty, hunger and malnutrition occur due to a deterioration of structure in the natural resources, problems primarily affecting the rural areas. To overcome these problems, the resources of rural areas should be managed and planned adequately and appropriately.

Other economic contributions of the sustainable agriculture are:

- it provides protection and development of the agricultural lands
- it creates opportunities for marketing of agricultural products
- it helps to increase the level of welfare of local people
- it provides increasing importance and respectability of agricultural activities for the urban public
- it provides promotion of agricultural activity and contribution of agriculture regarding the economy and quality of transport, thus supporting the education of public,
- employment, additional income, personal income, standard of living, rural migration and public revenues are positively affected
- the load of traditional holiday stations (coastal tourism) has lost the feature of relaxing so spreading holidays into agro-regions becomes possible
- participations to cultural festivals, weddings and similar events in the region are increased
- job opportunities and ownership develops in favour of the local community
- irreversible sources can be consumed at the lowest level
- it supports the protection of the biological diversity.

General Considerations

The development of sustainable agriculture and agro-tourism in the rural areas due to respect for environmental values is creating a possibility for sustainable development in rural areas by increasing the variety of economic activities. The development of environmental protection awareness, protection of biodiversity and ecosystems and the non-deterioration of natural beauty create sources of attraction for consumers in tourism and provide an important alternative possibility for Turkey.

Turkey has appropriate geographic and climatic diversity for the growth of different agricultural products. The agricultural sector has assumed a very important role in the country's development from the past to present. Even though the share of the agriculture sector in the economy is lower when compared to the past, it still holds importance for the improvement. In the tourism sector, the agriculture and animal husbandry activities are very important as alternative sources of tourism. Especially in the rural areas where traditional agriculture and livestock activities are going on, agro-tourism can increase the level of income of the locals and help them maintain their livelihoods instead of migrating from the villages to the cities. The basic logic of agro-tourism is to carry out the activities without damaging agricultural areas and affecting the sustainability of agricultural activities. Hence, while hosting the visitors making use of agro-tourism, this will mean a harmonization of visitors to the lifestyles of local people instead of giving special services and opportunities to them. In this process, the volunteering of local people and visitors should be well organised.

When we look at various applications, Turkey does not have a large enough share from agricultural tourism although it has a rich geography, culture and nature potential in comparison with most other countries in Europe. Suggestions about integrating and applying agricultural tourism (agro-tourism) with sustainable agriculture are:

- greater sensitivity should be shown to preserving the ecological balance
- importance should be given to promotion of agricultural production and increasing the sensitivity of sustainability in agricultural production
- touristic sources of supply should be commercialized with the idea of use protection and the capacity of rural areas should be considered, otherwise there will be irreversible losses for nature and local people
- farmers dependent on agricultural tourism should have the awareness that agricultural biological diversity is valuable and worth protecting
- unused lands appropriate for agricultural activities should be used by considering both agricultural diversity and suitability for agro-tourism
- in tourist areas, tourism should be integrated with the agriculture and the production should be guided to the market conditions
- an environment should be created where locals are not negatively affected, the tourists are not conned and the environment is protected from the tourist activities
- for applying agro-tourism, people who live with agriculture should be willing to work together and cooperate with farmers

- in the locations where the supply potential of agricultural tourism is high, enough importance should be given to the areas in introducing foreign languages into the education programme for tourism and for the local people
- there should be promotion of local agricultural tourism products and services
- locals farmers should function as a bridge between nature and urban people and should share their information about nature with visitors
- in the locations where supply of agricultural tourism is sufficient, the physical infrastructure services like rural roads, drinking water, sewage and water treatment plants should be expanded
- investments should be accelerated in the places where agro-tourism activities are abundant, as the increase in investments will result in the increase in demand of agricultural products, thus both investments for special sector and public sector should be accelerated
- local people living as farmers should create better quality products or have more efficient production, and should be encouraged to continue natural agricultural production
- in the areas where agricultural tourism is rich, credit and support should be given by the government to the people who are doing business in agro-tourism and to those using their lands for accommodation or camping
- in the areas where the agricultural tourism is rich, support should be given to the housewives to develop hand skills and those women should go to courses to produce local food and drinks.

Turkey should use the potential of agricultural tourism in the best way. However, tourism in not possible in all places where agricultural activities are available. The success of application will increase and results will be good by looking for truly suitable areas for agro-tourism. Turkey has some advantages in comparison with other countries in the market regarding the aspects of natural, seasonal, and socio-cultural tourist activities. Due to these advantages, it is possible to get more market share in the developing world by developing tourism and it is also possible to encourage development in the agricultural sector where the agricultural activities are rich in rural areas.

References

Agriculture and Forestry (2000). Agricultural Business Profile on Agro-Tourism. PEI Department of Agriculture & Forestry.

Agricultural Tour Operators International "Members of ATOI, 2012", Access: 22.12.2012, http://www.atoi.org/media/atoi_profiles_12-08-2012-a4.pdf

Ak, İ. (2006). A new option in tourism, eco-tourism, agro-tourism and agricultural tourism. The 3rd Organic Farming Symposium of Turkey , 1-4 November 2006, Yalova

ATTRA, (2005). Sustainable Agriculture: An Introduction, <www.attra.ncat.org>, (Access date: 04.01. 2013).

Wheat Association, http://www.bugday.org/bugdaygil/Tatuta/?p=0&lang=tr (Access: 23.01.2013)

Busby, G. and Rendle, S. (2000). The Transition From Tourism on Farms to Farm Tourism, Tourism Management, 21(6), 635-642.

Civelek, M.; Dalgın, T.; Ceken, H.; Ekiztepe, B. (2013). Uluslararası Sosyal ve Ekonomik Bilimler Dergisi. International Journal of Social and Economic Sciences 3 (2): 01-07

Commission Européenne (2000). Pour un tourisme rural de qualité, Direction Générale des Entreprises Unité "Tourisme", Bruxelles.

Cıkın, A.; Ceken, H.; Ucar, M. (2009). The effect of tourism on Agriculture Sector , Agro-Tourism and Economic Consequences. Journal of Agricultural Economics, 15(1):1-8.

Ceken, H.; Karadag, L.; Dalgın, T. (2007). A New Approach on Rural Development,a Theoretical Study intended to Rural Tourism and Turkey, Artvin Coruh University Journal of the Faculty of Forestry, 7(1), 1-14.

Efe, R., Sönmez, S., Cürebal, İ., Soykan A. (2008). Balıkesir'in Geoturizm Yöreleri: Marmara Geoturizm Yöresi Örneği. Üçüncü Ulusal Balıkesir Turizm Kongresi. 17-19 Nisan 2008. Bildiriler Kitabı, s. 323-330, Balıkesir.

Esengun, K.; Akca, H.; Saygılı, M. (2002). The Role of Rural Tourism on the Development Administration of rural Areas. Standard Jounal, 470, 29-35.

Gössling, Stefan (2002). Global Environmental Consequences of Tourism, Global Environmental Change,(12/4),283 – 302.

Gunaydın, G. (2005). Turkey's Agriculture and tendencies to change, TMMOB Chamber of Agricultural Engineers,Ankara.

Gunlu, H. (2012). Land Consolidation Applications: Turkey, The expertise thesis of planning, unpublished, Ankara.

Gunduz, S. (2004). A research about the creation of Agricultural Tourism Model and detection of the Suitable Areas for the Agricultural tourism in the Provincecof Ankara , the District of Kalecik. PhD Thesis, Ankara University Institute of Science and Technology The depament of landscape architecture Ankara.

Hatch, A.D. (2008). Agri-Tourism: A New Agricultural Business Enterprise, LSU Ag Center; Research & Extension, USA.

IFAS (2008). Institute of Food and Agricultural Sciences, University of Florida, http://entomology.ifas.ufl.edu/mcsorley/agroecology/ (acc. date: 04.12.2012).

Kahraman, N. (1997). The Development of Sustainable Tourism. The Discussion Meeting of the Implementation of sustainable development. Ankara: Environmental Foundation of Turkey, s:103-113.

Kaya, Ece and Arıkan, İrfan (2008). A study on the acquiring of the village of Yoruk to tourism As cultural heritage, III. National Tourism Symposium, 17-19 October 2008, Egridir. The book of noticements, ss. 65-78.

Rural Development Plan 2010-2013, Ministry of Agriculture and Rural Affairs, Ankara.

Kucukaslan, N.; Ak, İ. (2008). A model about the evaluation of Misi Village in Bursa province in terms of Agriculture tourism. III. Balikesir Tourism Congress, The book of noticements , 17-19 April 2008, ss.40-44)

Kucukaltan, D. (2002). Agro Tourism and The Agricultural Tourism management in Turkey, II. Tourism Council Proceedings, The publication of Ministry of Tourism, Ankara, (1)152.

Maetzold, J.A. (2004). Nature-Based Tourism & Agritourism Trends: Unlimited Opportunities, National Alternative Enterprises and Agritourism Leader, USDA/NRCS, Washington, DC.

Meadowcroft, J. (1997). Planning, Democracy and the Challenge of Sustainable Development. International Political Review, 18(2): 167–189.

Michigan Advisory Commission (2007). Agricultural Tourism Local Zoning Guidebook and Model Zoning Ordinance Provisions. http://www.michigan.gov/documents/mda/MDA_zoneguide_185763_7.pdf

Mitchell, J.T. (2006). "Conflicting Threat Perceptions at A Rural Agricultural Fair", Tourism Management, 27(6), 1298- 1307.

Nilson, P.A. (2002). Staying on Farms- An Ideological Backround. Annals of Tourism Research, 17 (8).

Olalı, H.; Timur, A. (1988). Tourism Economics. Office trade press, Izmir.

Orhan, A. (2010). The usage of the -geographical indicators- on converting the local values to tourism products : The Pismaniye(turkish fairy floss) of Izmit The example , Anatolia: Journal of Tourism Research, Volume 21, Number 2, Guz: 243-254,

Ozguc, N. (1998). Geography of Tourism. Istanbul, Der Publications.

Przezborska, L. (2003). Relationships Between Rural Tourism and Agrarian Restructuring In a Transitional Economy: The Case of Poland, Directions in Rural Tourism, Aldershot: Ashgate. pp. 203-222.

Rastrollo, M.A. and Alarcón, P. (2000). The Competitiveness of Traditional Tourist Destinations in the Knowledge Economy. Ed. D. Fesenmaier, S. Klein and D. Buhalis, Information and Communication Technologies in Tourism 2000, (pp. 209-217), Springer Verlag, Wien.

Rigby, D. ve Cáceres, D. (2001). Organic farming and the sustainability of agricultural systems, Agricultural Systems, 68: 21–40.

Schaller, N. (1993). The concept of agricultural sustainability, Agriculture, Ecosystems and Environment, 46: 89–97.

Soykan, F. (2000). Rural Tourism and the experience gained in Europe, Anatolia: Journal of Tourism Researches, s.21, Yıl: II, Turkish-September-December.

—. (2003). The rural Tourism and the Importance of rural tourism for Turkey, Aegean Journal of Geography, 12, 1-11.

—. (2006). Perspective of rural tourism in Europe and the experience gained, II. Balikesir National Tourism Congress, 20- April 22, 2006, Balıkesir, 71-87.

Turhan, S. (2005). Sustainability in Agriculture and Organic Farming, Journal of Agricultural Economics, 11(1): 13–24.

Weaver, D.B. ve Fennell, D.A. (1997). The Vacation Farm Sector in Saskatchewan: A profile of Operations, Tourism Management, 18(6), 357–365.

Williams, P.; Paridaen, M.; Dossa, K. ve Dumais, M. (2001). Agritourism Market and Product Development Status Report,Centre for Tourism Policy and Research Simon Fraser University Burnaby, BC, Canada.

Woo, L.S. ve Yeon, N.S. (2006). Agro-Tourism As A Rural Development Strategy in Korea, Journal of Rural Development, 29(6).

WTO (2004). Trade and Environment at the WTO, World Trade Organization, April, 2004.

Yıldırak, N. ve Gulcubuk, B. (1998). The Restructuring on Rural Areas of sustainable rural development For the realization of sustainable rural development, 3rd Congress of Agricultural Economics, Ziraat Bank Publication, Ankara.

Yunlong, C. ve Smit, B. (1994). Sustainability in agriculture: a general review, Agriculture, Ecosystems & Environment, 49(3): 299–307.

(http://www.tarim2023.com/kutuphane/turkiye-topraklari.105/) (access date : 04.12.2012

CHAPTER EIGHTEEEN

ECOTOURISM MANAGEMENT PLAN FOR IŞIKLI LAKE ÇIVRIL-DENIZLI IN TURKEY AND ITS ENVIRONS

MEHMET ALI ÖZDEMIR AND
SEVGI TATAR (DÖNMEZ)

Introduction

Ecotourism is a type of tourism that can prove beneficial to both local economy and at the same time create tourism awareness. We have decided to conduct this study in and around Işıklı Lake with the aim of enabling tourism consciously and preventing unconscious consumption in a world where the natural resources are rapidly diminishing. Işıklı Lake is worth studying with its rich flora and fauna and its numerous tourism alternatives (see Fig. 18.1 below).

Materials and Methods

Işıklı Lake is located 12 km southeast of Çivril County in the province of Denizli in the West Central Anatolia part of the Aegean Region in Turkey (see Fig. 18.2 below). In this study, physical-based maps were formulated by using numeric topographical data to reveal the current situation of Işıklı Lake and its environs. The accuracy level of these maps was enhanced by fieldwork and these were reinforced by photographs and figures. By using the 9.3 and 10 version of the CBS program ArcGIS, a location map, digital elevation model, a river system map and a map for ecotourism alternatives of Işıklı Lake were prepared. The topographic maps that formed the basis of the study were digitalized with the help of ArcGIS 9.3 and 10 and edited in terms of their geometry in accordance with the UTM projection system.

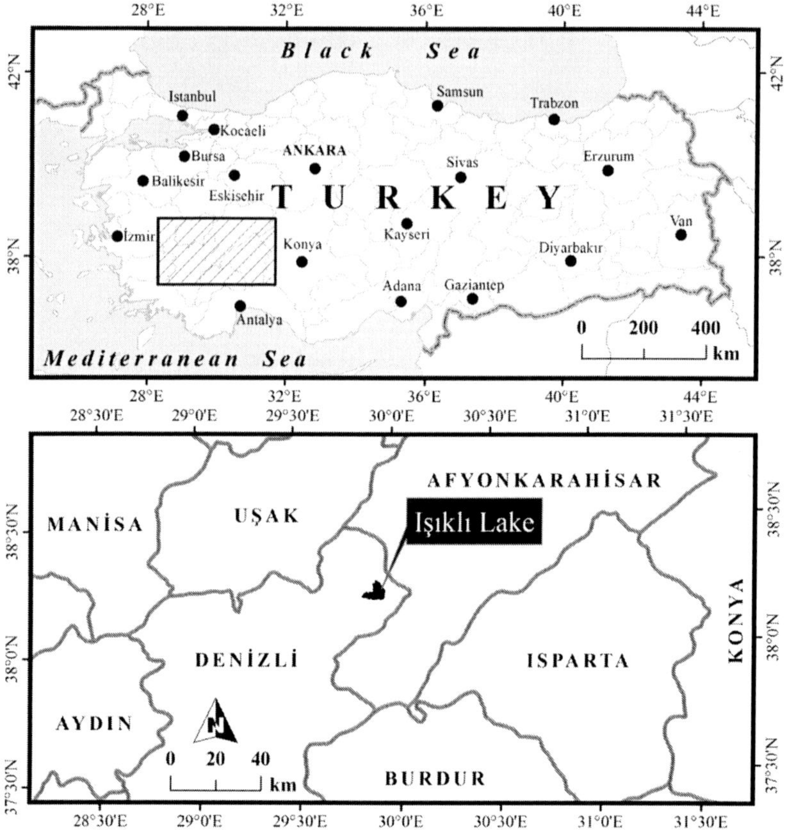

Fig. 18.1. Location of study area.

Fig.18.2. Işıklı Lake

Observations and Discussion

Study area and climate

According to the data published by the "General Directorate of State Meteorology," the average temperature in the field of study was recorded as

13.2°C in Çivril and 12.6°C in Dinar (Fig. 18.3). The annual average precipitation is 444.1 in Çivril, and the summers are distinctly dry. There is an irregular precipitation system. The area thus lies within the continental type of climate with a Mediterranean precipitation system (see Fig. 18.4).

Işıklı Lake and Gökgöl Lake are surrounded by Akdağ (2,449m) to the east, Irgıllı Plain and Bozdağ (2,421m) to the south, Gümüşsu Plain to the southeast and Çivril Plain to the west and north. The boundaries between the elevated parts and the basin floor have usually been determined by tectonic lines. These areas began to rise during the Neocene period and continued rising during Quaternary period (see Fig. 18.5 below).

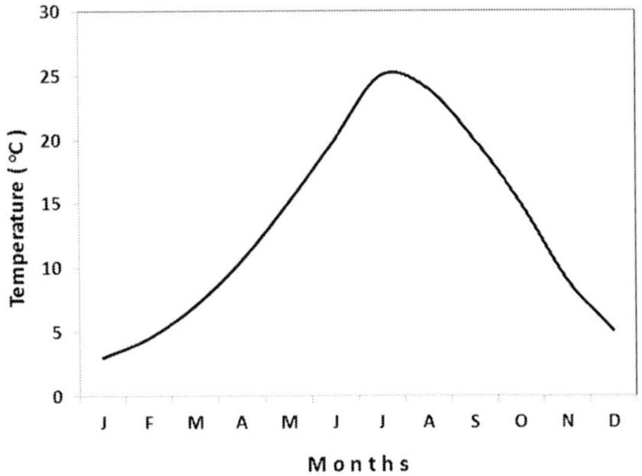

Fig. 18.3. Monthly Average Temperature Range of Işıklı Lake

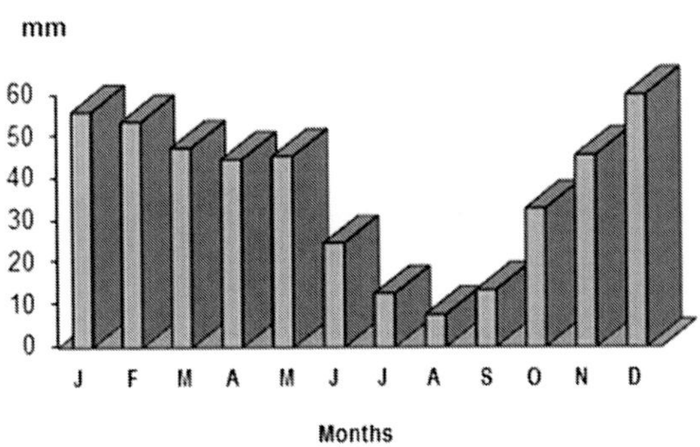

Fig. 18.4. Monthly Average Rainfall Range of Işıklı Lake

Hydrological features

The field of study is located in the feeding ground of the river basin of Büyük Menderes. The Büyük Menderes river, which is the most prominent river in the study area, has its source in Akdağ in the north of Işıklı Lake and flows into the Aegean Sea. The river runs across the Çivril Plain, which is in the northwest part of the study area, forming meandering streams (see Fig. 18.6 below). There are irrigation canals in and around Işıklı Lake built by the General Directorate of State Hydraulic Works (DSİ) to provide irrigation water. There are also drying canals in the area of the irrigation canals to remove the water that returns from irrigation.

There are several springs, such as Işıklı, Yuva and Gökgöl, feeding Işıklı Lake along the border that separates Akdağ from the basin floor. The total surface area of Işıklı Lake, which forms the basis of our study and Gökgöl Lake, with a connection to Işıklı Lake, is 62.89 km^2 and the surrounding area is 56.4 km^2. A canal and a regulator were built on the lower course of the southern edge of the Işıklı and Gökgöl Wetland by 21st District Administration of DSI. According to their measurements, the lowest water level observed in Işıklı Lake was 814 m, and the highest is 821 m, while the average is 817.8 m (G.I.G.S.A.Y.P. 2011).

Biogeographic features

Gemici (1986) determined 1,600 taxa belonging to 106 genera and 124 reported as endemics in the area.

Plants found in the wetland of Işıklı Lake are cyclamen (*Cyclamen mirabile*), tülü (*Typha angustifolia*), hyacinth (*Butomus umellatus*), white and yellow water lily (*Nymphaea alba*, *Nuphar lutea)*, water chesnut (*Trapa natans*), duckweed (*Lemna minor*), kuştutan (*Phragmites australis*), utricularia (*Utricularia australis*), iris (*Iris pseudocorus*), and bilberry (*Solanum dulcamara*) (Gemici 1986).

In addition, there are black pine forests (*Pinus nigra*), forest or sub-forest formations of pubescent oaks (*Quercus pubescens*), prickly junipers (*Juniperus oxycedrus* subsp. *oxcycedrus*), red pine (*Pinus brutia)*, trembling poplars (*Populus tremula*), acı çehri (*Rhammus rhodopeus* subsp. *rhodopeus*) on the west and southwest foots of Akdağ (on the northeast of Işıklı Lake) overlooking Işıklı Lake (Aksoy 1996).

A total of 327 taxa belonging to 71 genera have been identified within Işıklı Lake and the surrounding wetland area, and 22 are endemics (G.I.G.S.A.Y.P 2011).

Natural fauna species in Işıklı Lake are horned pikes (*Esox lucius*), carps (*Cyprinus carpio*), freshwater grey mullets (*Leuciscus cephallus*), chondrostoma nasus (*Chondrostoma meandrense*), goatfish (*Barbus pectoralis*), gobio (*Gobio gobio*), dwarf siraz (*Hemigrammocapoeta kemali*), yağ fish (*Pseudohoxinus maeandricus*), sailton pupfish (*Aphanius anatoliae*), sabanejawia aurata (*Nemacheilus angorae*) and stone-eater (*cobitis taenia*). In addition, the introduced species are: the tench (*Tinca tinca*), pond fish (*Carassius gibelio*) and gambusia affinis (*Gambusia holbrooki*).

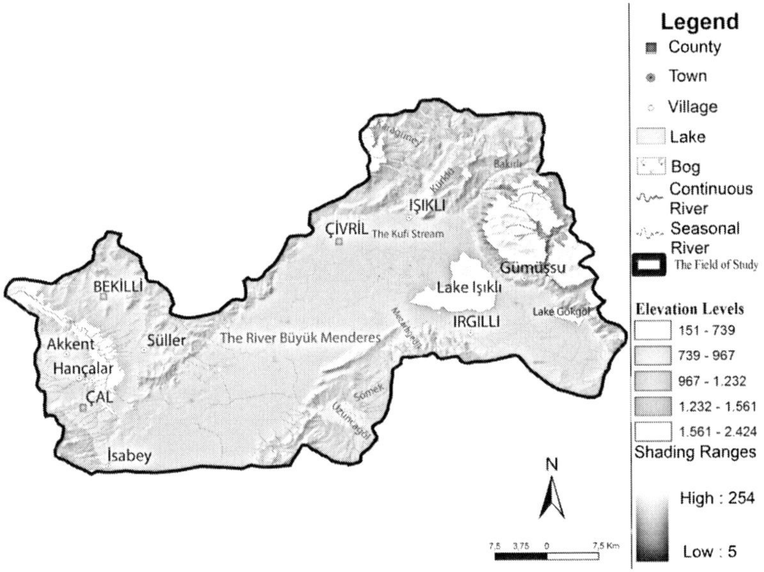

Fig. 18.5. Elevation Levels Model of Işıklı Lake Basin

The terrestrial faunal diversity around Işıklı Lake include: Red deer (*Cervus elaphus*), wolf (*Canis lupus*), fox (*Vulpes vulpes*), pine marten (*Martes martes*), badger (*Meles meles*), wild boar (*Sus scrofa*) and jade (*Equus przewalskii*). Of these, *Cervus elaphus* is under protection.

The ecosystems of Işıklı and Gökgöl provide an important wintering destination for water birds (Yarar & Magnin 1997). The data published reveals that there are 153 bird species in the area. Nearly 100 species are native to the area during the summer months. The results show that more than 700 trout herons use the area. The Dalmatian pelicans (wintering, feeding after breeding and quite likely breeding) and greater spotted eagles (feeding and wintering) reside in the area but have the status of "vulnerable," according to the IUCN red list. In addition are pygmy cormorants (wintering), ferruginous duck (breeding, wintering), black vulture (feeding), corn crake (stay), curlew (wintering), coracias (breeding), the latter with the status of threatened (NT). The species other than these are in the Low Risk status (LC) (G.I.G.S.A.Y.P. 2011).

Hunting: The wetland of Işıklı and Gökgöl is the wintering, feeding and breeding area for numerous water birds. Hunting activities are carried out in the area within the framework of the Land Hunting Code number 4915 and Article 3 of this code. For this reason, there are several hunting associations in the county of Çivril and the town of Işıklı. There are around 200 hunters registered in these associations. A great threat to the bird species in the region is caused by the approximately 800 hunters that come to Işıklı from the surrounding provinces, resulting in uncontrolled hunting. Indeed, 211,496 individual birds were counted in the water bird census carried out in the mid-winter in 1996, 190,690 of which were wild geese. On the other hand, only 28,225 individuals were counted in the mid-winter census conducted in 2010. This dramatic decrease in the number of water birds indicates that the area is under a great threat from uncontrolled and illegal hunting (G.I.G.S.A.Y.P. 2011).

Ecotourism Management Plan for Işıklı Lake and its Environs

Development of ecotourism and participation: Well thought out planning is a must for the ecotourism activities in a given area to attain success. Development and management requires planning at every stage. When we look into every kind of tourism activity carried out across the world, we see that long-term tourism plans are extremely successful and satisfying. In addition, planning an approach to ecotourism gains more importance with a view to preventing natural and cultural wealth from being damaged (WTO 1994). A good planning rests on the effective participation of all the communities of interest Efe et al. 2008; (Drumm & Moore 2002). In this study we have projected the participation of various communities of interest and their types of participation have been explained in the table (see Fig. 18.7 below).

Distinct stages in the process of planning

Analysis of current situation: When Işıklı Lake and the surrounding area are examined in terms of ecotourism, certain positive and negative results have been obtained.

Positive results: The wetland of Işıklı Lake is regarded as Class A wetland category according to the Ramsar Convention and embodies many important bird species as a wintering and breeding area. There are rich landscape features in and around the lake, and the area on the whole possesses the required conditions to become one of the most important ecotourism areas in the region and Turkey with the help of projects to be conducted. The local residents are willing to take part in the solutions to the problems.

Negative results: There are hunting and fishing activities going on in the area, but poaching in particular damages the area. There has been a decrease in the number of birds around the lake in recent years. The contamination levels in the lake have increased due to domestic and industrial wastes coming from Sandıklı and Dinar and the waste water entering from the Gümüşsu region. The occasional agricultural facilities carried out on the lake bed damage the lake.

Aim of the management plan: To evaluate the ecotourism potential that exists in the area with the help of a management plan prepared for the lake and its surrounding area, and to enable the area to attain the status it deserves among the most important ecotourism areas in Turkey with the work to be carried out.

Identifying alternatives: The study area is in the category of Class A wetlands as per the Ramsar Convention. It has the potential to offer a wide range of ecotourism activities with its hydrological features, favourable climatic and geomorphological conditions, and natural and historical sites. The ecotourism alternatives that are connected to this area need be handled within the framework of Ecotourism Management Plan Activities for Işıklı Lake and its environs.

Ecotourism Management Plan Activities for the Study Area

Establishing an ecological village: An ecological village is a place where economic, social and cultural activities are in harmony with the environment. The fundamental objective of the activities conducted in such places is to allow people to lead harmonious and reconciled lives with the environment. In 1998, the "100 Best Applications" list of the United Nations selected the eco-village model as the least damaging. The features of an ecologic village are:

- Growing organic food within the bio-area of the community as much as possible
- Building houses from local materials
- Combining and using village-based renewable energy systems

- Protecting the nature and preserving wildlife areas (webnaturel.com).

During the field studies conducted by us, it has been decided that it is suitable to establish an ecological village around the Akgöz Spring with coordinates of 38° 22′62′′north and 29º 86′02′′east, which is within the borders of the town Işıklı and is one of the most important sources feeding Işıklı Lake.

Creating botanical parks (arboretum): An arboretum is a collection of living plants arranged in a way that reflects the relationships between plant groups. The main objective of such areas is to create rich collections, stating the homelands, scientific names and colloquial names of the plants in their natural environment (botanikbahcesi.com).

We have proposed to create botanical gardens in three spots around Işıklı Lake during the field observations conducted. These areas were identified as pastures that belong to the state treasury and thus there is no hindrance in turning them into botanical gardens. The area that has been identified has the coordinates of 38º 31′33′′ north and 29º 82′35′′ east, 38º 29′20′′ north and 29º 85′50′′east, 38º 23′82′′ north and 29º 87′12′′ along course of the Kufi Stream.

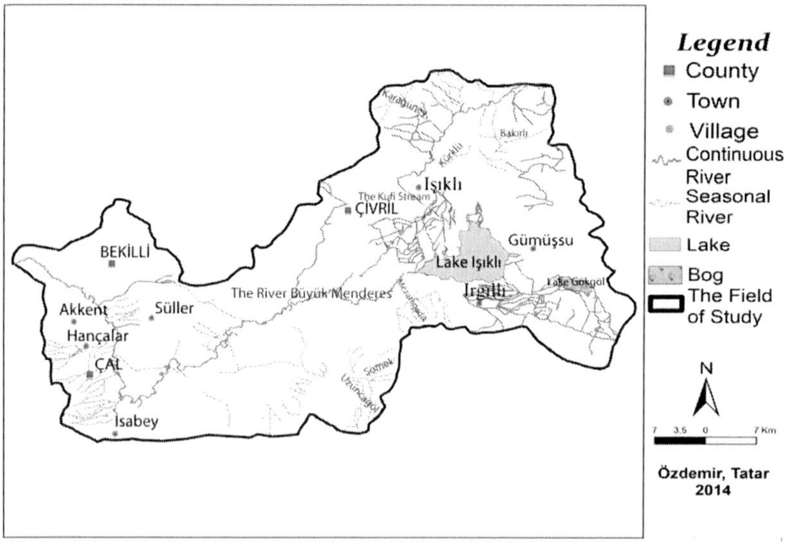

Fig. 18.6. The River System of Işıklı Lake Basin

Creating a natural aquarium: Natural aquariums are made by imitating the natural environment of a given region as much as possible. In these aquariums, water animals, water values and water movements of a given region are linked to the region itself in terms of quality and quantity. Creating a natural aquarium in the coordinates of 38º 22′62′′ north latitute and 29º 86′02′′ east longtitude in the south-eastern part of the Işıklı Lake would be very attractive in terms of ecotourism.

Building bird watching spots: Bird watching is a pursuit which aims at observing and defining wild birds in their natural habitats. This can be done as a hobby as well as with scientific ends. There are 153 species of birds in the Işıklı Lake and Gökgöl Ecosystems according to the avifauna inventory (G.I.G.S.A.Y.P. 2011). The suitable spots for building bird watching spots were identified after the study conducted around Işıklı and Gökgöl lakes.

Bird watching spots can be created in the coordinates of 38º 23′95′′ north and 29º 86′53′′ east, 38º 22′72′′ north and 29º 83′71′′ east, 38º 19′97′′ north and 29º 87′57′′ east across the islands Kocaada and Gencelli, which is a shelter for the birds and which is within Bucak village, Seraserli Village and Işıklı Lake in the quarters of the western and northern parts of Işıklı Lake.

Creating a walking track and photographic safari on bikes: Trekking is an ecotourist activity which has spread widely in recent years in Turkey. These are generally low speed walks intended to go from one place to another under completely natural conditions and in places that are away from residential areas. Trekking is a field within the scope of ecotourism that has steadily gained more economic value (akademiorganizasyon .com).

Photographic safari, on the other hand, is a trip in which one takes photographs in natural habitats, rich in flora and fauna. These trips can be done while trekking as well as in the form of a bicycle tour.

A suitable route for trekking and photographic safari trips on bikes was identified during the study in and around the Işıklı Lake. The 20 km track that starts in the area to the southeast of the town of Işıklı and continues to the village of Bucak in the northwest of Işıklı Lake has been found suitable for building a walking track and a track for photographic safari on bikes along the two banks of Kufi Stream. The route of the track emerges as a favourable location in terms of ecotourism. The coordinates of the starting point of the track have been determined as 38º 31′61′′ north and 29 º 82′13′′ east and the coordinates of the end of the track have been determined as 38º 23′81′′ north and 29º 87′03′′ east.

Organizing rowing tours on Işıklı Lake: Işıklı Lake offers great opportunities with regards to ecotourism with its natural habitat. Rowing tours on gondolas on the lake will be a great experience for nature lovers. During the field study we found that the most suitable starting point for such tours is the village of Beydilli, located in the northeast of Işıklı Lake with the coordinates 38° 26′59′′ north and 29° 87′12′′ east. It was also found that other rowing tour routes could be identified in the western and north-western parts of the lake.

Table 18.1. Ecotourism management plan for Işıklı Lake and the surrounding area and communities of interests (Özdemir, Dönmez 2012)

Settlement Place	Ecotourism Activity Places And Classes	Government Agency
Çivril County	Accommodation, Beycesultan Mound	The Municipality of Çivril, Çivril County Special Administration, County Offices of Government Ministries, Local Representatives, Tour Operators
Işıklı Town	Accommodation, Fishing, Ecologic Village, Ancient City of Eumenia	The Municipality of Işıklı Town
Gümüşsu Town	Accommodation, Building of A Rafinement Plant, Gümüşsu Falls	The Municipality of Gümüşsu Town
Irgıllı Town	Trekking	The Municipality of Irgıllı Town
Yuvaköy Village	Photosafari	Local Administrators of The Village of Yuvaköy
Seraserli Village	Building Bird Watching House	Local Administrators of The Village of Seraserli
Bucak Village	Building Bird Watching House, Botanical Garden	Local Administrators of The Village of Beydilli
Beydilli Village	Rowing Tours, Restaurant, Fishing	Local Administrators of The Village of Beydilli
Çetinler Village	Photosafari	Local Administrators of The Village of Çetinler
Süngüllü Village	Photosafari	Local Administrators of The Village of Süngüllü
Beyköy Village	Photosafari	Local Administrators of The Village of Beyköy
Sundurlu Village	Photosafari	Local Administrators of The Village of Sundurlu

Dinar County	Building of A Rafinement Plant, Ecotourism Activities	The Municipality of Dinar, Dinar County Special Administration, County Offices of Government Ministries, Local Representatives, Tour Operators
Sandıklı County	Building of A Rafinement Plant, Ecotourism Activities	The Municipality of Sandıklı, Sandıklı, County Special Administration, County Offices of Government Ministries, Local Representatives, Tour Operators
Denizli Province	Ecotourism Activities, Advertising Activities	The Governorship of Province of Denizli, Province Offices of Government Ministries, The Municipality of Denizli, Local Representatives, Tour Operators
Afyonkarahisar Province	Ecotourism Activities, Advertising Activities	The Governorship of Province of Afyonkarahisar, Province Offices of Government Ministries, The Municipality of Afyonkarahisar, Local Representatives, Tour Operators
Aegean Region	Ecotourism Activities, Advertising Activities	The Governorships of Provinces,Province Offices of Government Ministries, The Municipalities of Provinces, Local Representatives, Tour Operators

Cultural activities

Mound Beycesultan, the ancient city of Eumenia and the battleground of the Battle of Myriokephalon, which hold the remains of historically significant antique cities in the vicinity of Işıklı Lake, have important features with regards to cultural tourism.

Organizing tours to Mound Beycesultan

Mound Beycesultan is among the biggest and the most important prehistoric centres of Anatolia. It is in the northwest of Işıklı Lake and 5 km from the centre of the county of Çivril and in the south of the county. The first scientific excavation in the region was carried out between 1954 and 1956 by the excavation team of the British Archeology Institute. It was concluded that the site is one of the best examples of the original city formations in Anatolia (yesilcivril.com). The items excavated in Beyce Sultan are exhibited in the Pamukkale Museum in Denizli. The second round of excavations in Beycesultan was initiated on August 6, 2007 and are ongoing.

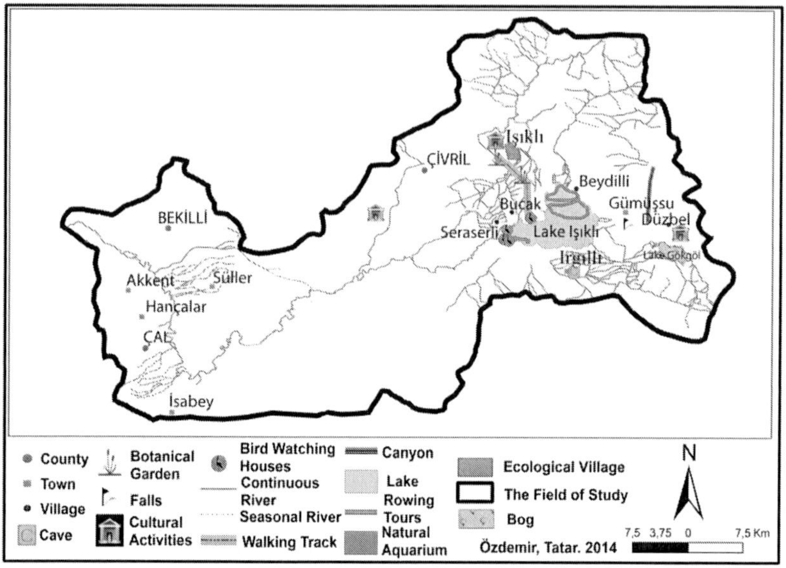

Fig. 18.7. The Ecotourism Alternatives Map of Işıklı Lake and Surrounding Area

Organizing tours to the ancient city of Eumenia: Eumenes II of Pergamon gave his name to the city he founded in the area where the modern day town of Işıklı is located, which is on the Çivril-Dinar highway within the borders of Denizli. Eumenia, which was an important Phrygian city, was also recognized as a religious centre during the Roman period. Of the remains of the city is the parliamentary building located in the nearby area of the Akgöz Spring, in the southeast of the town of Işıklı. The plain today on the mountainous area known as Sarıbaba Hill was used as a fortress in the Byzantine times, on which stands the necropolis of the Eumenia on the slopes of the hill in the area (yesilcivril.com).

Organizing tours to the battleground of the Battle of Myriokephalon: After the victory in the Battle of Manzikert that opened the doors of Anatolia to the Turks on August 26, 1071, the Battle of Myriokephalon was fought in 1176 around the valley of the Kufi Stream, which was known as "Tzibritzi" (Wikipedia_entry on_Çivril). The war fought in the vicinity of the village of Düzbel has great importance with regard to the spread of Turks in Anatolia as well as cultural tourism.

Nature Activities

Organizing tours in Teresuyu and Akkale caves: Teresuyu Cave is on an elevation of 1,600 to 1,700 m within the borders of Çivril, in Akdağ. Its length inside is approximately 150 m and it has stalactites and stalagmites. Akkale Cave is in the vicinity of Akkale in the town of Gümüşsu in the county of Çivril. It is 40 m long (pamukkale.gov.tr).

Recreational activities and organized tours in Gümüşsu Falls: Although Gümüşsu (Homa) Falls, located in the town of Gümüşsu in the northeast of Işıklı Lake, is one of the waterfalls with great tourism potential, it hasn't been utilized to its fullest. The falls came into being as a result of the process of the water following its source from the Pınarbaşı Spring, with a tectono-carstic origin flowing for 150 m, falling from two knick points of 15 and 30 m high formed from fault lines. There is a picnic area around the falls which cannot meet demand, and there is a stabilized walking track to the west of the falls. In addition, an observation terrace has been built in the place where the knick points are separated (Polat et al. 2012). However, the facilities aimed at tourism are not sufficient. Facilities in harmony with the landscape should be built in the area.

Organizing tours through the canyon of Akdağ Karanlıkdere Tokalı: The Tokalı Canyon, formed by the river of Karanlıkdere in the southeast of the mass of Akdağ in the northeast of Işıklı Lake, has an altitude of 1,600 m and stretches across an area of 20 km and ends in a settlement on an elevation of 900 m in the Gümüşsu (Homa) town in the county of Çivril (pamukkale.gov.tr). Camping tours through the canyon, which was discovered a few years ago, are organized by the nature organizations, based particularly in and around Denizli. The passage through the canyon takes about 7 to 8 hours.Evaluation of the ecotourism infrastructure of the study field transportation

Işıklı Lake is 100 km from the provincial centre of Denizli, 68 km from the provincial centre of Uşak and 155 km from the provincial centre of Afyonkarahisar. It is located within the borders of the town of Işıklı, which is 10 km from the county of Çivril. It is 30 km away from the county of Dinar within the city borders of Afyonkarahisar. The transportation around the lake is not problematic as it is on a well-built road between the counties of Dinar and Çivril.

Water and energy resources: There is electricity in the nearby villages of Işıklı Lake and the surrounding area. The potable water and the utility water in the county centre of Çivril is provided by means of the three caisson wells drilled by the municipality. The potable water in other towns and villages is provided by the municipalities or drinking water associations.

Waste: Işıklı and Gökgöl lakes, which feed Işıklı Lake with their surplus waters, are facing severe pollution. When the waters of Işıklı and Gökgöl lakes are compared, it can be said that the waters of Gökgöl are a little colder, the alkali and dissolved oxygen levels are higher in Işıklı Lake, and the levels of salinity and electric conductivity are different in both lakes. The primary reason for this is the polluting potential of the Stream of Büyük Menderes, which runs from Dinar to the area, a notable fresh water source that feeds Gökgöl Lake. The Büyük Menderes Stream collects the waste water in the county of Dinar as well as the drainage water from the plantations nearby and dumps them into Gökgöl (G.I.G.S.A.Y.P. 2011). This contamination causes adverse effects on the water life, principally fish. This also affects Işıklı Lake because surplus waters from Gökgöl feed Işıklı Lake, especially during the overflows.

Services: The closest banks to Işıklı Lake are located in the county of Çivril, 10 km from the lake. There is a fully equipped public hospital, a family health centre and several hotels in the county. However, there are no hotels or lodgings around the lake that can provide accommodation for the tourists.

Potential local workforce within the ecotourism area: We believe that when the targets with regard to ecotourism are attained in the area, the local residents will give their support in terms of workforce because it has been found that local residents have the required awareness to utilize the lake more efficiently without damaging the ecosystem, according to the studies carried out.

Vehicles and facilities to be diverted to ecotourism: It has been observed that there are not enough vehicles that can be diverted to ecotourism in the area, yet it is possible to make use of the tractors and trailers that local residents use in agricultural activities, especially in summer. The number of rowboats used in fishing in the region is not registered and there is a loophole in this regard. There are more than 500 rowboats used on the lake.

Identifying the communities of interest: An effective planning rests upon the participation of all the communities of interest (Drumm & Moore 2002). It is possible to take the communities of interest into account in the various stages of the planning process and incorporate them in to the decision-making process within the planning. It is also advisable to organize planning teams with the participation of groups with different interests and specialties. At this point we should answer the questions of Who are the communities of interest? At what stages should they participate? In what way should they participate? According to Rahemtulla & Wellstead (2001), the parties involved in ecotourism are: (a) participants (visitors), (b) tour operators (persons or organizations that are the suppliers and (c) resource managers (those responsible for planning, preservation and management). According to Drumm & Moore (2002), the administrative staff of the area where the ecotourism will be carried out, representatives from public organizations, experts and scientists, local representatives, tour operators and representatives from non-profit NGOs should take part in the planning process. The units that should take part in the administrative staff constituted within the framework of the Ecotourism Management Plan for Işıklı Lake are as follows in order of priority.

Fig. 18.8. Three-Dimensional Model of Işıklı Lake Basin

Number one priority: The Governorship of the province of Denizli, Denizli Provincial Directorship of Forestry and Water Affairs, Denizli Provincial Directorship of Environment and Urban Planning, Denizli Provincial Directorate of Culture and Tourism, Denizli Provincial Directorate of Nature Conservation and National Parks, the Municipality of Denizli, Denizli Provincial Special Administration, DSİ 21st District Administration, Scientists from Dokuz Eylül University, Ege University, Pamukkale University, Afyon Kocatepe University.

Number two priority: District governorship of Çivril County, the Municipality of Çivril, Çivril County Special Administration, Forestry Operation Directorate of Çivril.

Number three priority: The municipalities of the towns Işıklı, Gümüşsu and Irgıllı, local administrators of the villages of Yuvaköy, Seraserli, Bucak, Beydili, Çetinler, Sundurlu, Beyköy and Süngüllü.

Number four priority: Non-profit NGOs, local representatives, tour operators.

In our study, the communities of interest and the fields of interest have been formulated in a table (Fig. 18.8). If the proposed board of directors is organized in accordance with their respective priorities, it will be possible to realize the ecotourism management plan for Işıklı Lake.

SWOT Analysis

Strengths (S)

- Işıklı Lake is situated within the county borders of Çivril
- Beycesultan, the ancient city of Eumenia and the Gümüşsu Falls are located in the vicinity of Işıklı Lake
- Rich natural landscape in and around Işıklı Lake
- Presence of wildlife and wildlife habitats
- Presence of invaluable areas with regard to ornithology
- Location of the lake on one of the important bird migration routes in Turkey
- It has the potential for water based and various other recreational activities (see Fig. 18.9)
- Natural and cultural wealth in the region which could be utilized in ecotourism activities
- Willingness of the local residents to take part in the ecotourism activities
- Presence of the areas that are suitable for photographic and jeep safaris.

Weaknesses (W)

- Inadequacy of the promotion and marketing aimed at ecotourism and their failure to meet the demand
- Underdeveloped tourism service sector
- Inadequacy of the promotion of handicrafts and traditional products in the region
- Lack of qualified personnel trained in ecotourism activities
- Lack of tour organizations
- Lack of knowledge about and awareness of ecotourism.

Opportunities (O)

- High potential for alternative tourism types, and its suitability for various ecotourism activities
- Attractiveness of Işıklı Lake and the surrounding area with regard to ecotourism
- Cultural and natural texture of the area which could be made convenient with the help of conservation and restoration works
- Çivril Apple, Culture and Agricultural Festival held annually to promote local foods, dishes and handicrafts.

Threats (T)

- Damages done to the natural environment, floods and landslides caused by erosion
- Sediment accumulation and the increase in vegetation in Işıklı Lake
- Environmental pollution and contamination in the lake (disposal of domestic and industrial waste from the counties of Dinar and Sandıklı and the town of Gümüşsu into the lake through the Menderes River)
- Reckless hunting damaging the wildlife
- The decrease in the number and types of birds
- Reckless agricultural and stockbreeding activities
- Reduction in the wintering and breeding areas of the birds due to the damage done to the vegetation on the floating islands in the lake
- Occasional over-draw downs
- Disposal of the solid wastes from the settlements in the vicinity of the foothills of Akdağ Nature Reserve which is near the lake around the county of Dinar
- Risk of unplanned ecotourism activity upsetting the natural structure of the area.

Conclusions

Ecotourism is a fairly novel kind of tourism, and could be a means for the conservation and the management of Işıklı Lake. The lake and its environs have a great ecotourism potential due to its natural, cultural and recreational wealth. While carrying out ecotourism activities, the priority should be not to damage the ecosystem in and around the lake, and good planning is required for this.

Certain measures should be taken in order for ecotourism activities to be conducted in and around the lake to avoid upsetting the ecological balance.

The carrying capacity of Işıklı Lake and the surrounding areas in ecological and social terms should be determined together with other ecotourism activities in accordance with this capacity (ecological villages, trekking facilities, etc.)

An Ecotourism Management Plan for Işıklı Lake and the surrounding area has been carried out in our study, and the compatibility of the management plan to the area supported by a SWOT analysis. When the Ecotourism Management Plan for Işıklı Lake and the surrounding area is realized in accordance with its contents, it will enable Işıklı Lake and the surrounding area to become one of the most notable ecotourism areas in our country.

References

Akdağ Tabiat Parkı Geliştirme Projesi (2006). Denizli Orman Bölge İl Müdürlüğü, Denizli

Aksoy, N. (1996). Denizli – Çivril Bölgesi'nin Flora ve Vejetasyonu'na Genel Bir Bakış,

Karadeniz Teknik Üniv. Orman Fakültesi Orman Mühendisliği Bölümü Orman Botaniği Anabilim Dalı, Yüksek Lisans Tezi, Trabzon.

Drumm, A.; Moore, A. (2002). Ecotourism Development, An Introduction Ecotourism Planning. Vol. I. The Nature Conservancy, USA.

Efe, R., Sönmez, S., Cürebal, İ., Soykan A. (2008). Balıkesir'in Geoturizm Yöreleri: Marmara Geoturizm Yöresi Örneği. Üçüncü Ulusal Balıkesir Turizm Kongresi. 17-19 Nisan 2008. Bildiriler Kitabı, s. 323-330, Balıkesir.

Gemici, Y.(1986). Çivril (Denizli) , Sandıklı ve Dinar (Afyon) İlçeleri Arasındaki Akdağ ve Çevresinin Flora ve Vejetasyonu, Ege Üniv.Fen Fak.Biyoloji Bölümü Botanik Anabilim Dalı, İzmir.

Gökgöl ve Işıklı Gölleri Sulak Alan Yönetim Planı (2011-2015), T. C. Çevre ve Orman Bakanlığı, Doğa Koruma ve Milli Parklar Genel Müdürlüğü, Doğa Koruma Dairesi Başkanlığı Sulak Alanlar Şube Müdürlüğü, 2011,Ankara

Polat, S.; Kargı, S.; Güney, Y. (2012). Gümüşsu (Homa) Şelalesi (Çivril-Denizli), Selçuk Üniversitesi Sosyal Bilimler Enstitüsü Dergisi,27/2012, Konya.

Rahemtulla, Y.G.; Wellstead, A.M. (2001). Ecotourism: Understanding Competing Expert and Academic Definitions. Infor. Report Nor- X-380 Canada.

WTO (1994). National and Regional Tourism Planning Methodologies and Case Studies. London UK.

Yağcı, Ö.İ. (2000). Ekoturizm, Gazi Üniv. Fen Bilimleri Enst. , 2000, Yüksek Lisans Tezi, 2000, Ankara.

http://www.pamukkale.gov.tr/tr/content.asp?id=805, Denizli İl Kültür ve Turizm Müdürlüğü, 04.03.2012, sa. 17.00

http://www.webnaturel.com/index.asp?alt_cat_id=61&cat_id=7&ayrintiid=1512Webnaturel, 06.2012, sa.16.00

http://www.botanikbahcesi.com/botanikbahcesi.html, Botanik Bahçesi, 08.06.2012, sa. 17.00
http://www.akademiorganizasyon.com.tr/?/doga-yuruyusu, Akademi Organizasyon 14.07.2012, sa.15.00
http://www.yesilcivril.com/index.php/civril-tarihi-bilgiler/1269-betcesultan-yeryuzuene-ckarlyor,18.07.2012, sa.19.00
http://www.yesilcivril.com/index.php/civril-tarihi-ilgiler/1405-eumenia-,Yeşil Çivril, 18.07.2012, sa. 18.00
http://mgm.gov.tr, Meteoroloji Genel Müdürlüğü, 20.07.2012, sa. 17.00
http://tr.wikipedia.org/wiki/%C3%87ivril, Wikipedia, 19.07.2012, sa. 16.30
http://www.pamukkale.gov.tr/tr/content.asp?id=698, Denizli İl Kültür ve Turizm Müdürlüğü, 20.07.2012, sa.17.00
http://www.pamukkale.gov.tr/TR/content.asp?id=661, Denizli İl Kültür Turizm Müdürlüğü, 11.04.2012, sa. 12.00

CHAPTER NINETEEN

PERCEPTIONS OF THE LOCAL COMMUNITY TOWARDS ECOTOURISM IN EĞIRDIR (ISPARTA)

OSMAN YILMAZ AND HILMI DEMIRKAYA

Introduction

Turkey has alternative tourism sources with historical, cultural and natural beauty in addition to mass tourism alongside the coasts with clean, sunny and sandy beaches. In this context, eco-friendly tourism, or ecotourism, needs to be given priority by locals and has become important more recently (Orhan & Karahan 2010).

Research studies on tourism show that a lot of work has been done on the local people's perception of tourism and their attitudes towards tourists (Yılmaz & Demirkaya, 2011; Jurowski et el., 1997; Yoon et al., 2001; Doganay 2001; Usal 1990). The findings of these studies emphasize the importance of tourism investors and local administrators in following tourism actions by saving local culture and not damaging the ecological environment. Brunt & Courtney (1999) express the necessity of supporting local tourism in the form of sustainable tourism by encouraging the positive contributions and minimizing the negative effects. They emphasize the importance of diagnosing local people's perceptions of tourism and tourists and their attitudes towards visitors for successful tourism activity. Dogan (2004) asserts that it is important that local society receives a share to maintain sustainable tourism. Butler (1980) states that there is a positive relationship between tourism development and local people's attitudes towards it.

As tourism has social effects, changes in the local people's social values, beliefs and culture is inevitable. This effect is related to the number of tourists visiting, their cultural values and duration of stay (Soykan 2004).

The socio-cultural effect of tourism appears as a result of the relationship between local society and tourists (Smith 1995). In the frame of cultural effect, changes in the daily lives of local people are remarkable. Some of those are cultural viewpoints, values, identities and experiencing different cultures (Glasson et al. 1995). Socio-cultural interaction may be positive or negative, and this effect may be seen on tourists as well as local people (Cooper et al. 1998; Efe et al. 2008; Soykan et al. 2012).

If tourists stay in a destination for a long time, local people start to learn their language, intercultural marriages are seen between the young and the local culture starts to loose its distinctiveness (Ryan 1991). The objective of this research is to explore the perspectives of the local people living in Eğirdir, an important tourism center, towards tourism and tourists. The central question was what are the thoughts of local society regarding the socio-cultural effects of tourism together with its perceptions on tourism and ecotourism?

Methods and Material

This study was designed according to the survey method, and both quantitive and qualititive methods were used. The survey included variables regarding the local people, and their psychological and socio-psychological behaviours such as gender, socio-economic level, education level, age, occupation, geographical region, thoughts, perceptions, attitudes and interests. The semi-structured interview form with open questions was used to find how local people see tourism and ecotourism in the qualitative part of the research. In addition, some observations were made in Egirdir and its vicinity.

Sampling: A total of 40 people living in Eğirdir, Isparta, participated in the survey, comprising 24 females and 16 males. Fourteen were in the age range 21–30, 8 in 31–40, 12 41–50, 2 51–60, and 4 61–70. Of these, 28 were born in Eğirdir and grew up there, while 12 had come to Eğirdir for work. Regarding education level, 4 graduated from primary school, 14 from high school, 6 from two-year college, and 16 from four-year college.

When we examine the occupations of the attendants we see that 8 are tradesmen, 6 are salesmen, 16 are teachers, 2 are retired teachers, 4 are repairmen, 2 are secretaries, and 2 are homemakers.

Instruments: The scale comprised 14 items with the aim of determining the socio-cultural effect of tourism with a Likert-type scale. Validity and reliability studies of the scale were done by Demirkaya & Çetin (2010).

In each question, one of the following options was marked by the participants:

(1) Strongly disagree

(2) Disagree
(3) Neutral
(4) Agree
(5) Strongly agree

Semi-structured interview form: there were open questions in this interview form.

Data analysis: the data obtained was evaluated in the SPSS 20.0 program, and the frequency and mean ranges are presented in Table 19.1 below.

In the evaluation of open questions, qualitative context analysis was applied. In this analysis, the answers of all participants were read and turned into categories, reported by interpreting.

Observations and Discussion

First of all the opinions of local people on the socio-cultural impacts of tourism were reviewed, following which the perception of tourism and ecotourism was analyzed from the qualitative dimension.

(1) What are the opinions of local people about the socio-cultural effects of tourism?

A perusal of Table 19.1 reveals that the expression getting the highest arithmetic average ($\bar{x}$= 3.80) from the participants is "Tourism has created jobs for our community"—14 participants replied "totally agree" and 14 gave the answer "agree."

The lowest arithmetic average ($\bar{x}$= 1.80) was for the 13th item expressing "Tourism has resulted in unpleasantly overcrowded beaches, hiking trails, parks and other outdoor places in our community"—14 participants said "strongly disagree," 22 said "disagree," 2 replied "neutral" and 2 of them said "agree."

(2) What are the tourism perceptions of local people?

An examination of the tourism perceptions of local people shows that there are four different tourism perceptions. The first is wandering, seeing new places and different cultures. The second is that tourism is an economic sector developing the economy of the region. The third category emphasizes the indirect effect of tourism and the necessity to protect cultural and natural places from tourism. The fourth category emphasizes cultural interaction. Some of the opinions of the attendants are as follows:

Table 19.1. Frequency and arithmetic mean values of participants' perceptions on the socio-cultural impact of tourism

Tourism Development Impact Items	Mean	Variants (Individual)*				
		5	4	3	2	1
(1) Tourism has created jobs for our community	3.80	14	14	4	6	2
(2) Tourism has attracted investment to our community	3.40	6	14	14	2	4
(3) Our standard of living has increased considerably because of tourism	3.05	4	18	-	12	6
(4) Tourism has given economic benefits to local people and businesses	3.65	10	16	8	2	4
(5) High-spending tourists have negatively affected our way of life	1.90	-	2	-	30	8
(6) Tourism has changed our traditional culture	2.65	2	8	6	22	2
(7) Local residents have suffered from living in a tourism destination area	1.90	2	2	-	22	14
(8) Tourism has encouraged a variety of cultural activities by the local residents	3.40	4	22	2	10	2
(9) Tourism has resulted in more cultural exchange between tourists and residents	3.50	4	20	8	8	-
(10) Tourism has resulted in positive impacts on the cultural identity of our community	3.60	4	24	6	4	2
(11) Tourism has resulted in traffic congestion, noise and pollution	1.85	-	4	-	22	14
(12) Construction of hotels & tourist facilities have destroyed the natural environment	2.05	-	4	4	22	10
(13) Tourism has resulted in unpleasantly overcrowded beaches, hiking trails, parks and other outdoor places in our community	1.80	-	2	2	22	14
(14) Tourism provides more parks and other recreational areas for local residents	3.30	-	24	6	8	2
Overal Mean		2.85				

*1= strongly disagree, 2= disagree, 3= neutral, 4= agree, 5= strongly agree

Category 1:

(Respondent 1) It is an economic and social activity to wander, to see new places and new lifestyles.

(Respondent 7) It is a trip aimed at domestic or overseas travelling.

(Respondent 9) It is a way to see the places you haven't seen and to learn the facts of that new place.

(Respondent 14) It is a domestic or overseas trip to wander, entertain and relax.

(Respondent 20) It means people are moving around and wandering.

Category 2:

(Respondent 2) It is commerce.

(Respondent 10) It is economic contribution.

(Respondent 12) It is economic development.

(Respondent 18) It brings foreign currency to the country.

(Respondent 19) It means future for the country.

Category 3:

(Respondent 13) It is an income resource for the country to protect the history and the nature.

Category 4:

(Respondent 3) It is commerce and cultural exchange.

(Respondent 16) It is to wander, do shopping, be included in cultural exchange and learn about history and nature.

(Respondent 6) It is the unity of different cultures and information exchange.

(Respondent 4) It is wandering, commerce, cultural exchange and relaxing.

(3) What are the ecotourism perceptions of local people?

When the ecotourism perceptions of the participants are examined, it is found that there are two defining categories. The first of these is that it should be a sustainable tourism activity without damaging natural beauty and structures. The economic contribution of tourism was emphasized probably because of the misunderstanding of the affix "eco," in the second one. Some of the opinions of the participants in these categories are as follows.

Category 1:

(Respondent 6) It is development of tourism activities without damaging natural beauty and cultural structures.

(Respondent 9) It is travel made more conscious by wandering around natural areas.

(Respondent 13) It is to open the gates of our country to the visitors by protecting our identity and territory and not damaging our natural resources, historical artifacts and culture.

(Respondent 16) It is a tourism activity done by not polluting and damaging the environment.

(Respondent 20) It is to sustain tourism by not damaging natural beauties.

Category 2:

(Respondent 7) It is the economic contribution of tourism.

(Respondent 10) It is the economic contribution to the country.

(Respondent 12) It is the increase of prosperity and the improvement of the economic welfare of local people.

Korkmaz & Başkalkan (2011) asserted that it is doubtful whether it is possible to protect a natural area allocated for tourism without damaging it. It has been observed that the natural properties of a local area have been damaged because of ecotourism activities conducted without taking legal and precautions and creating a necessary infrastructure.

(4) What is the touristic attractiveness of Eğirdir as a destination centre?

The opinions of local people about Eğirdir's values as a tourism centre have been evaluated in three categories: (1) natural beauty and the lake, (2) historical artifacts, and (3) its being situated on an important road. Some of the opinions of the attendants in these categories are given below.

Category 1:

(Respondent 5) The chance to interact with nature.

(Respondent 6) The lake and natural beauty.

(Respondent 7) Its natural beauty and the lake.

(Respondent 14) To relax and interact with nature.

Category 2:

(Respondent 8) To see historical places.

(Respondent 9) To see natural beauty and historical places.

(Respondent 10) Lake, historical artifacts, geological structure and climate.

(Respondent 18) Historical places and taking pictures.

Category 3:

(Respondent 1) To stop and eat while going somewhere else and to see a few places.

(Respondent 16) To wander and relax. Since they come for just a day, there is no contribution to the economy.

(Respondent 19) Its natural beauty and being situated on an important road.

(5) What do local people do for tourism and tourists?

It is understood that civil servants and homemakers generally show places to the tourists and behave kindly, while teachers try to make students more conscious about tourism and tourists. On the other hand, tradesmen and pension owners emphasize the commercial aspects of tourism. Some attendants state that they do not take part in any activity. Some of the opinions about this issue are:

(Respondent 1) To trade and inform tourists about a place.

(Respondent 2) We give a demonstration.

(Respondent 3) We sell some goods, and we try to introduce the neighbourhood.

(Respondent 4) Trade and demonstration.

(Respondent 5) Nothing.

(Respondent 6) We try to help if someone asks for help about anything.

(Respondent 7) Nothing.

(Respondent 9) I try to make my students conscious.

(Respondent 10) I behave kindly to the tourists, and I try to protect nature.

(Respondent 13) We do not do anything.

(Respondent 14) I do not do any activity for tourism and tourists.

(Respondent 16) We try to learn language and behave kindly.

(Respondent 19) We have been working for accommodation services and nature sports for 27 years.

(6) What cultural activities do local people do for tourism and tourists?

While about half of the respondents consider that no cultural activities are conducted for tourism and tourists in Eğirdir, others state that some activities, such as a fair, a triathlon, sailing and upland festivals, are done. Some of the expressions of the local people about this issue are:

(Respondent 1) No activities for tourists are conducted in Eğirdir.

(Respondent 2) Fair and triathlon.

(Respondent 3) Fair, triathlon and carnival.
(Respondent 6) Festival and sailing.
(Respondent 7) Sports activities in summer.
(Respondent 12) Triatlon and upland festivals.
(Respondent 14) I do not do any activity for tourism and tourists.
(Respondent 20) Festival and water sports.

(7) What are the projects of local people for introducing the culture of the region in the future?

The respondants have no projects for introducing the culture of the region in future. However, a few people mention some projects that are not clear, and they expect these projects as being done by others, not themselves. Some expressions on this issue are:

(Respondent 5) To show Eğirdir's tourism attractions via television, etc.

(Respondent 12) Festivals and carnivals can be organized.

(Respondent 13) If ecological agriculture will help tourism, I want to do it.

(Respondent 14) Its historical and natural beauties can be introduced.

(8) What are the opinions of local people on the necessity of tourism for the economy of the region?

While 32 of the respondants state that tourism is necessary for the economy of the region, 8 consider it unnecessary.

(Respondent 1) Development of Eğirdir, organization of social activities, and developing projects will help to make tourists stay, which is necessary.

(Respondent 9) Yes, a good income may be gained if the region is evaluated well.

(Respondent 13) Yes, as long as it does not damage the nature.

(Respondent 16) Yes, we are aware of that, there is a factory without a chimney.

In the report of the Ministry of Culture and Tourism Investment and Managements General Directorate (2010), it is stated that ecotourism villages, agrotourism areas and tourism development areas will be increased in the region. It is expected that there will be an important development in the economy of the region if these activities are conducted in a planned way.

(9) What are the opinions of local people on the future of tourism in Eğirdir?

(Respondent 1) It is necessary to erase the negative thoughts of people about tourists in order to keep tourists here.

(Respondent 2) Activities should be done, and hotels and beaches should be opened.

(Respondent 5) New activities should be increased.

(Respondent 6) It will develop only with investments on tourism, and different activities and new activities introducing the town need to be done.

(Respondent 7) I do not think that the current situation will go much further.

(Respondent 9) I do not expect a good future if it goes like this. Cleaning and visual activities are absent.

(Respondent 10) It may improve if necessary facilities are funded.

(Respondent 14) I do not think it will be much. I think it will remain as a place to which tourists come just for relaxing.

(Respondent 16) Natural tourism and mountain climbing can be developed.

(Respondent 19) I am less hopeful because local people are not ready for it.

(Respondent 20) I think it will develop gradually.

The waste water of the neighbourhoods around Lake Eğirdir flows to the lake. It can be said that the facilities for tourism can increase pollution in the region if they are opened without necessary infrastructure (Alkan et al. 2009).

The Eğirdir township of Isparta Province is very important in terms of tourism. The most important resource value of the region is Lake Eğirdir, which is the fourth largest lake in Turkey with an area of 487 km^2, lying at an altitude of 917 m, with an average depth of 10–12 m. The deepest point of the lake is 16.5 m. The south-north length of the lake is 50 km, and the east-west length varies between 3–15 km. While Hoyran, which is a relatively small part, is located in the north, the large part in the south is called Eğirdir. Lake Eğirdir, which is the second biggest fresh water lake in Turkey, supplies some of the drinking water need in Isparta (Korkmaz & Başkalkan 2011).

There are two small islands in the lake called Can Ada and Yeşil (Nis) Ada. The former has no buildings and is used as a picnic area, while pensions, hotels and fish restaurants are found in Yeşil Ada. Lake Eğirdir has various functions, such as providing drinking water, fishing, recreation and landscaping, in addition to sports tourism activities such as swimming, sailing and surfing.

Tourists coming to Eğirdir for just one day go for eating, picnics, fishing, sightseeing, observing the animals and plants, taking pictures and visiting historical and cultural sites.

Some of the obstacles for tourism in Eğirdir are the opening of marble quarries, high prices given to tourists, infrastructure problems, pollution of the lake and unconscious hunting.

Eğirdir is a settlement located 918 m above sea level with a Mediterranean climate and continental climate. The natural beauty of the lake and the region attract many native and foreign tourists to the town: 78,000 tourists were accommodated and 97,338 tourists came for one-day visits to Eğirdir in 2010, and 33,880 of them were foreigners. The average length of stay is three days (BAKA 2011, 7).

The accommodation capacity and quality of the town is increasing constantly. The total number of beds is 824, with 9 hotels with 459 beds and 20 pensions with 365 beds. There are also 20 tourism-certificated and municipality-instructed restaurants in the town. The town was included to the Blue Flag Project in 1993 for the first time, and the Altınkum Beach and Camping Facilities gained a blue flag in 1999 (BAKA 2011, 8).

Eğirdir has a geographical structure allowing for activities such as mountain climbing, jogging, walking, cycling, canoeing, skiing, water sports and air sports, and international triathlon races and paragliding festivals are organized every year, providing liveliness to the town (BAKA 2011, 10).

There are a lot of historical and natural places for tourists to visit in Eğirdir, such as Yeşilada, on which Ayastefanos Church is located, Canada, Altınkum Beach, Bedre Cove, Kovada Lake National Park, Yazılı Canyon, Davraz Ski Center, Pisidia Antiocheia, Adada, Eğirdir Fortress, Eğirdir Caravanserai, Ağa Mosque, Hızırbey Mosque, Yılanlıoğlu Mosque, Ada Mosque and Dündar Bey Madrasah.

Suggestions

According to the results of this study, the opinions of local people living in Eğirdir on the socio-economic effects of tourism are positive. They are aware of the fact that employment and infrastructure services will increase thanks to tourism. The respondants do not consider that tourism will bring about negative effects.

Tourism and ecotourism perceptions of the participants are positive. Income level and cultural interaction were emphasized in the perception of local community. For the local community, the importance of tourism is based on its presenting job opportunities. Local people want both tourists' coming to the town and the protection of Eğirdir's historical and natural structure. Besides, it has been observed that local people do not spend any effort on ecotourism.

Below are some suggestions for applying ecological-based tourism activities around Eğirdir:

Studies on marketing should be increased.

Studies on organic agriculture and agricultural tourism should be made.

The local community's earning can be increased by encouraging traditional handcrafts in the region.

Accommodation facilities compatible with local architecture can be opened in the region.

In addition, old houses can be used for tourism after restoration.

Circuits for faith tourism can be determined around the lake.

Nature sports activities can be increased around the lake.

Wildlife observation stations can be opened.

References

Alkan, H.; Korkmaz, M. and Altuntaş, S. (2009). Interactions between local people and lakes: An example from Turkey. Journal of Environmental Engineering and Landscape Management, 17(3),

BAKA (2011). Eğirdir İlçesi Turizm Potansiyeli. Batı Akdeniz Kalkınma Ajansı, s.7.

Brunt, P. & Courtney, P. (1999). Host perceptions of socio-cultural impacts. Annals of Tourism Research, 26(3), 5-12.

Butler, R. W. (1980). The concept of a tourism area cycle of evolution: Implications for the management of resources. Canadian Geographer, 24, 5–12.

Cooper, C., J. Fletcher, S. Wanhill, D. Gilbert, and R. Shepherd (1998), Tourism Principles and Practice, Essex: Pearson Education.

Demirkaya, H. and Çetin, T. (2010). Residents' perception on the social and cultural impacts of tourism in Alanya (Antalya-Turkey). Ekev Akademi Dergisi, 14(42), 383-392.

Efe, R., Sönmez, S., Cürebal, İ., Soykan A. (2008). Balıkesir'in Geoturizm Yöreleri: Marmara Geoturizm Yöresi Örneği. Üçüncü Ulusal Balıkesir Turizm Kongresi. 17-19 Nisan 2008. Bildiriler Kitabı, s. 323-330, Balıkesir.

Glasson, J.; Godfrey, K. and Goodey, B. (1995). Towards visitor impact management: Visitor impacts, carrying capacity and management responses in europe's historic towns and cities. England: Avebury.

Jurowski, C.; Uysal, M. and Williams, D.R. (1997). A theoretical analysis of host community resident reactions to tourism. Journal of Travel Research, 36(2), 3-11.

Korkmaz, M. and Başkalkan, S.N. (2011). Assessment on sustainability of tourism development in Lake Eğirdir region. SDU Faculty of Forestry Journal, 12, 62-69.

Orhan, T. and Karahan, F. (2010). Evaluation f ecotourism potential of Uzundere district and its environs. Artvin Çoruh Unıversity Faculty of Forestry Journal, 11(1), 27-42.

Ryan, C. (1991). Recreational tourism: A social science perspective. London: Routledge.

Smith, S. L.J. (1995). Tourism analysis: A handbook. Harlow: Longman.

Soykan, F. (2004). Bir turizm coğrafyası araştırması: Kırsal alanların turizm potansiyelinin saptanması ve Şirince Köyü'ne (İzmir) uygulanması. İzmir: Ege Üniversitesi Edebiyat Fakültesi Yayınları, No: 122.

Soykan, A.; Sönmez, S., Efe, R.; Cürebal, İ. (2012). Türkiye Turizm Politikalarına ve Ayvalık Adaları Tabiat Parkı Uygulamalarına Eleştirisel Bakış. Kazdağı ve Madra Dağı Belediyeler Birliği, Madra Dağı Ulusal Çalıştayı. 12-14 Ekim 2012, Ayvalık-Balıkesir, Bildiriler Kitabı s. 82-90.

Usal, A. (1990). Turizmin sosyo-kültürel etkileri ve Ege Bölgesi örneği. I Ulusal Turizm Kongresi Bildirileri, DEÜ: Aydın Turizm.

Yılmaz, O. and Demirkaya, H. (2011). Opinions of Local People in Turgutköy (Marmaris) on Ecosystem-Based Tourism. Archives of Applied Science Research, 3(2):465-471.

Yoon, Y.; Gürsoy, D. and Chen, J. (2001). Validating a tourism development theory with structural equation modeling. Tourism Management, 22(4), 363-372.

CHAPTER TWENTY

EVALUATION OF SOCIAL MEDIA USERS' COMMENTS TOWARDS ECO-FRIENDLY ACCOMMODATION ESTABLISHMENTS

SERKAN BERTAN, MURAT BAYRAM
AND BURÇIN KIRLAR

Introduction

Environmental problems are many and varied, but it is necessary to pay attention to some specific problems (Kuvan & Akan 2005). The positive effects of tourism on the environment include providing more park areas for local people, increasing awareness of and the value of the environment, enhancing and protecting the image of the region, developing infrastructure opportunities, improving public activities and protecting touristic values. Negative effects are noise and traffic jams, touristic activities' damage to natural environment, increasing environmental pollution, negative impact on natural resources and damaging touristic values (Yoon et al. 2001; Ko & Stewart 2002; Teye et al. 2002; Kuvan & Akan 2005; Pappas 2008). To protect the environment, to enhance environmental consciousness, and to incite and encourage accommodation establishments for eco-friendly structuring and business features, the "Eco-Friendly Establishment Certificate" (Pine Symbol) has been given to the Tourism Operation Certificated Accommodation Establishments that have the necessary qualifications since 1993, and the classification of eco-friendly accommodation establishments has been developed since 2008 (ktbyatirimisletmeler.gov.tr). Showing the classification of accommodation establishments, the star symbols on the plates arranged in green and the "Eco-Friendly Establishment Certificate" have been given to the establishments that have eco-friendly establishment inscription on their plates (Ibid.).

Table 20.1 shows a list of "Eco-Friendly Accommodation Establishments" in Turkey.

Water saving, energy efficiency, reducing the consumption of harmful substances in the environment and decreasing the amount of waste materials, promoting the use of renewable energy sources, planning the accommodation establishments as eco-friendly and ergonomical, landscaping arrangements and activities, ecological architecture, raising awareness about environmental conciousness, providing education, and cooperation with institutions and organizations are among the important issues on environmental sensivity (Ibid.).

Accepted as a new media and adopted by worldwide internet users, the usage of social media sites is increasing rapidly (Bayram & Bayram, 2012). Until December 2012, the number of active facebook users reached up to one billion per month (newsroom.fb.com). On twitter, whose active users are more than 500 million, an average of 340 million tweets are written per day (blog.twitter.com). Every month 800 million users visit Youtube, on which videos are watched for more than four billion hours per month. Trip Advisor has more than sixty million visits per month and hosts one hundred million user comments and evaluations. As is clear from this data, new internet-oriented technological developments are progressing rapidly. As Buhalis (1998) and Poon (1993) mentioned, consumer behaviour in tourism has always been influenced by the developments in information and communication technologies. In parallel with this development it is obvious that while planning their travels the rising tourist generation prefers to use the internet as a priority, and 83% of leisure travelers and 76% of business travelers plan their travels this way (Google 2012). While it is not yet possiblle to claim a concensus among academics and researchers regarding the concept of social media, of which users and usage rates increase constantly, Kaplan & Haenlein's (2010) description as "a group of internet-based applications enabling to create and change user-centric content and structured on Web 2.0 idea and technological base" is one of the most comprehensive explanations. Allocating six subgroups, Kaplan & Haenlein (2010) conceptualize social media under two main dimensions: social presence/media richness and self-presentation/self-expression. The subgroups are: blogs, social network sites (e.g. Facebook), virtual social worlds (e.g. Second Life), collaborative applications (e.g. Wikipedia), content communities (e.g. Youtube) and virtual game worlds (e.g. World of Warcraft).

Social media, and in particular the review and evaluation sites giving place to other user views, have begun to become consumers' primary information source on the travel planning stage (Chung & Buhalis 2008;

Yoo et al. 2011), which has taken an increasingly important role on travellers' access to information and decision-making behaviour (Yoo et al. 2011). The relationship between social media and tourism started to enter into the field of researchers, and some studies have been done in the international area. Consumer blogs containing reviews of product and service have become one of the fields of social media and tourism (Pan et al. 2007; Pühringer & Taylor 2008; Waldhör & Rind 2008). In terms of marketing, and the contribution and creation and sharing of experiences written by people with similar experiences, these studies show the importance of blogs (Pudliner 2007). Gretzel et al. (2007) reveal that online comments help consumers in the decision-making process, reduce risks and give confidence to consumers. In their study about the influence of user comments on the selection of accommodation establishments, Vermeulen & Seegers (2009) indicate that comments have either a positive or negative effect on the travel planning process. Cox et al. (2009) state that social media is preferred especially on the pre-travel process by users. White (2010) remarks that travel-related photos on Facebook attract user attention and influence the travel plans of people seeing those photos. Fotis et al. (2012) express that social media is used at every stage of the travel planning process for different purposes. Xiang & Gretzel (2010) note that social media plays an important role in tourists' sharing their experiences among themselves. O'Connor (2010) states that review sites like Trip Advisor are important in terms of providing social communication among tourists. Anderson (2012) implicates that comment and review sites have become a prominent state of the consumers' hotel choices. Callarisa et al. (2012) clarify that reviews and evaluations at the online stage are more secure than hastily filled survey forms at hotels. In addition to this, with the progress of the internet, Delarocas (2003) and Gretzel (2007) qualify social media as a revolution which occured in traditional word-of-mouth marketing.

When the effects of social media on consumer choice and establishments' branding efforts are examined from the view of the accommodation sector (Anderson 2012), hospitality sector professionals and executives state seeing a lot of facts proving the influence of social media on visitors' purchasing behavior (effects on prices and occupancy rates). Branding is an important differentiation factor, especially among the similarly characterized destinations and hotel establishments in the conditions of increasing competition (Fesenmaier 2007; Pike 2005). By featuring user-centric content and comments, social media sites can be effective in building brands, communication processes and on the formation of brand loyalty, perceived quality and brand associations (Wang et al. 2011; Xiang

& Gretzel 2010). Hotels can also take advantage of social media to promote their establishments and to get fast and efficient feedback about their services (Bayram & Bayram 2012).

For the purpose of this study, the following are important:

- Perceptions of guests staying in eco-friendly accommodation establishments
- General evaluation of guests staying in eco-friendly accommodation establishments
- Social media usage levels of eco-friendly accommodation establishments
- Information status of eco-friendly accommodation establishments on their websites about their certificate of eco-friendly establishment and about this situation
- Symbols used by accommodation establishments about environmental awareness.

Method

This study is based on the data obtained during March 15–31, 2013 from tripadvisor.com, in which user comments/reviews belonging to the accommodation establishments (see Table 20.1 below) are written and from the official websites of these establishments. The basic criterion of the selection of accommodation establishments on Table 20.1 is their having a certificate of eco-friendly accommodation establishments. For this reason, the study of tourism operation certificated establishments periodically issued by the Culture and Tourism Ministry was used (ktbyatirimisletmeler.gov.tr). According to this study, a total of 29 accommodation establishments are appropriate regarding the stated criterion. In the direction of the main objectives of this study, the official web pages of these establishments were visited to ascertain whether or not establishments have social media accounts, and if establishments have information about their eco-friendly statuses on their websites. On the final stage, user comments related to the establishments on Trip Advisor are shown on Table 20.1 below and evaluated regarding the satisfaction level of users. User comments belonging to the eco-friendly accommodation establishments on Trip Advisor, which is a pioneer web site in terms of online user reviews in the field of tourism and travel, are studied in this research. The reason for selecting Trip Advisor for this study is mainly because it was the first and most prominent website in the field, and remains a popular travel and tourist comment sharing website for millions of internet users. Another reason is that Trip Advisor contains the largest number of comments and gives importance to both security and reality, and is one of the top priority comment sharing websites preferred by consumers. Through its main purpose of delivering unshaped and unmodified content of travel suggestions operating in 30 countries in 29 languages, including Turkish, Trip Advisor is the world's most popular and largest travel community with more than sixty million monthly visits and one hundred million comments and views.

Within the scope of this research there are 7,101 user comments and evaluations related to eco-friendly accommodation establishments on Trip Advisor. While examining all these evaluations (7,101), because of time constraints and actuality of the data, concerning the six main sections (value, location, sleeping quality, rooms, cleanliness and service quality), the most recent one hundred evaluations made by users were included in this study. In this manner, a total of 2,096 user comments were analyzed.

Results

As a part of the study, 2,096 user comments belonging to 29 establishments (see Table 20.1 below) with the eco-friendly accommodation establishment certificate were evaluated. The total user evaluations related to these establishments are 7,011 pcs; thus, 30% of all comments and evaluations has been taken into consideration.

Under six main dimensions, the evaluations made by people staying in eco-friendly accommodation establishments are given in Table 20.2 below. According to the arithmetic means, people mostly appreciate cleanliness of the establishments, followed by rooms, sleeping quality, location, service quality and finally value. From the arithmetic means, it can be said that people have high rates of appreciation of the establishments. Besides the ranking of the establishments' properties, the most appreciated property is cleanliness and the least is value.

According to Fig. 20.1 below, based on the data in Table 20.2, by evaluating the accommodation establishments guests rated the value dimension as 48.7% excellent, 29.2% very good, 14.2% average, 4.5% bad and 3.4% awful. Regarding the "location" dimension, ratings are 53.5% excellent, 25% very good, 15.1% average, 4.2% bad and 2.2 % awful. Another dimension, "sleeping quality," is rated by consumers as 54.4% excellent, 26.6% very good, 12% average, 3.6% bad and 3.4% awful. Evaluating the "rooms" dimension, consumers stated the ratings as 55.1% excellent, 27.1% very good, 12.8% average, 3.1% bad and only 1.9% awful. The evaluations of the "cleanliness" dimension are 64.8% excellent, 23.8% very good, 8% average, 2.2%bad and only 1.2% awful.

Table 20.1. Eco-Friendly Accommodation Establishments.

Name	City	Provınce	Class	Type
Martı Myra	Antalya	Kemer	Hotel	5 Star
Crystal Hotels De Luxe Resort &Spa	Antalya	Kemer	Hotel	5 Star
Amara Dolce Vita	Antalya	Kemer	Hotel	5 Star
Club Hotel Rıxos Tekirova	Antalya	Kemer	Hotel	5 Star
Kemer Holiday Club	Antalya	Kemer	Holiday Village	5 Star
Pırate's Beach Club	Antalya	Kemer	Holiday Village	5 Star
Crystal Hotels Flora Beach	Antalya	Kemer	Holiday Village	5 Star
Crystal Sunrise Queen Luxury Resort &Spa	Antalya	Manavgat	Hotel	5 Star
Otium Eco Club Side	Antalya	Manavgat	Hotel	5 Star
Club Nena	Antalya	Manavgat	Holiday Village	5 Star
Concorde De Luxe Resort	Antalya	Center	Hotel	5 Star
Calısta Luxury Resort Hotel	Antalya	Serik	Hotel	5 Star
Crystal Palace Famıly Resort	Antalya	Serik	Hotel	5 Star
Xanadu Resort Hotel	Antalya	Serik	Hotel	5 Star
Crystal Paraiso Verde Resort & Spa	Antalya	Serik	Hotel	5 Star
Adam & Eve Hotel	Antalya	Serik	Hotel	5 Star
Doğa Residence	Ankara	Çankaya	Hotel	4 Star
Tuğcan Hotel	Gaziantep	Şahinbey	Hotel	5 Star
Antakya Ottoman Palace Thermal Resort	Hatay	Center	Thermal Hotel	5 Star
Four Seaons Hotel	İstanbul	Eminönü	Prıvate Est.	
The President Hotel	İstanbul	Fatih	Hotel	4 Star
Holiday Inn İstanbul City	İstanbul	Fatih	Hotel	5 Star
Crowne Plaza İstanbul Asia Hotel	İstanbul	Pendik	Hotel	5 Star
The Sofa Hotel	İstanbul	Şişli	Prıvate Est.	
Dedeman Konya Hotel	Konya	Selçuklu	Hotel	5 Star
Ersan Holiday Village	Muğla	Bodrum	Holiday Village	5 Star
Hilton Dalaman Golf Resort&Spa	Muğla	Ortaca	Hotel + Golf	5 Star
Double Tree By Hilton Avanos Cappadocıa	Nevşehir	Avanos	Hotel	5 Star
Şanlıurfa Dedeman Hotel	Şanlıurfa	Center	Hotel	5 Star

Resource: ktbyatirimisletmeler.gov.tr/TR,9579/turizm-tesisleri-islemleri.html

In terms of "service quality," the ratings are as 56.1% excellent, 23.6% very good, 11.4% average, 5.1% bad and 3.8% awful.

Table 20.2. Perceptions of people who stayed in eco-friendly accommodation establishments

	Excellent (5)	Very good (4)	Average (3)	Bad (2)	Awful (1)	Arithmetic Mean	Standard Deviation
Value	48,7	29,2	14,2	4,5	3,4	4,1529	1,04589
Location	53,5	25	15,1	4,2	2,2	4,2319	1,00115
Sleeping Quality	54,4	26,6	11,9	3,6	3,4	4,2501	1,02632
Rooms	55,1	27,1	12,7	3,1	1,9	4,3027	0,94126
Cleanliness	64,8	23,8	8	2,1	1,2	4,4885	0,82852
Service Quality	56,1	23,6	11,4	5,1	3,8	4,2310	1,08168

In Table 20.3 below, the general evaluations of guests on accommodation establishments are presented. According to this the evaluations by users were: 1= awful and 5= excellent, and the general appreciation rate of eco-friendly accommodation establishments (N=29) was 4.1. Consequently, this shows that accommodation establishments appreciated in general and the services (especially value, location, sleeping quality, rooms, cleanliness and service quality) offered by them are perceived positively by guests.

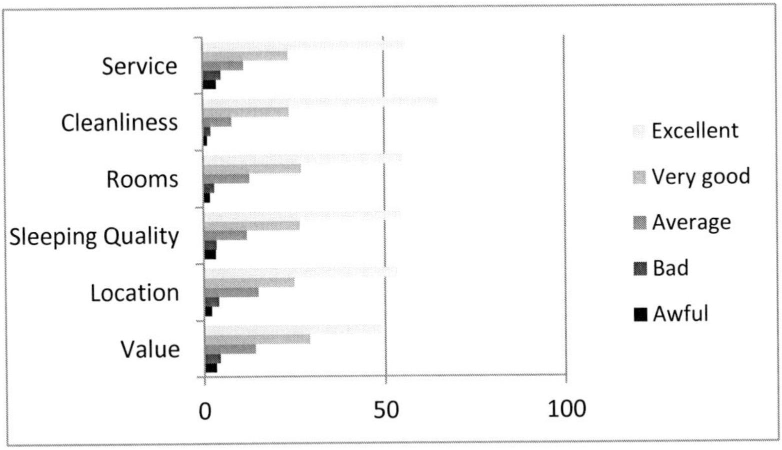

Fig. 20.1. Perceptions of people staying in eco-friendly accommodation establishments

Table 20.3 below shows a distribution of total comments (7,101) on the accommodation establishments on Trip Advisor. According to these assessments made for the eco-friendly certificated accommodation establishments, 58.52% (4.156) are "excellent." It was also observed that assessments of "very good" form 25.68% (1824) of the overall evaluations, "average" 9.51% (676), "bad" 3.7% (263) and awful 2.5% (182).

Table 20.3. General evaluations of guests on eco-friendly accommodation establishments

Eco-Friendly Accommodation Establishments	Total number of Comments	General Evaluation*	Excellent		Very good		Average		Bad		Awful	
			f	%	f	%	f	%	f	%	f	%
A	2	3,5	0	47,92	1	50	1	50	0	0	0	0
B	169	4	81	46,46	45	26,62	26	15,38	9	5,32	8	4,73
C	99	4	46	76,19	33	33,33	13	13,13	5	5,05	2	2,02
Ç	210	4,5	160	54,83	26	12,38	12	5,714	9	4,28	3	1,42
D	124	4	68	22,85	33	26,61	13	10,48	3	2,41	7	5,64
E	35	3,5	8	40	12	34,28	3	8,57	8	22,85	4	11,42
F	15	4	6	32,14	5	33,33	2	13,33	1	6,66	1	6,66
G	28	4	9	68,71	12	42,85	4	14,28	3	10,71	0	0
Ğ	390	4,5	268	62,30	72	18,46	38	9,74	9	2,30	3	0,76
H	252	4,5	157	31,42	63	25	23	9,12	7	2,77	2	0,79
I	35	4	11	62,84	18	51,42	2	5,71	2	5,71	2	5,71
İ	549	4,5	345	75,52	122	22,22	52	9,47	20	3,64	10	1,82
J	331	4,5	250	47,36	52	15,70	18	5,43	6	1,81	5	1,51
K	114	4	54	66,84	36	31,57	11	9,64	11	9,64	2	1,75
L	371	4,5	248	13,46	80	21,56	22	5,92	12	3,23	9	2,42
M	52	3,5	7	53,30	25	48,07	12	23,07	4	7,69	4	7,69
N	257	4	137	39,13	61	23,73	30	11,67	13	5,05	16	6,22
O	23	4	9	27,27	9	39,13	3	13,04	0	0	2	8,69
Ö	22	3	6	85,82	5	22,72	3	13,63	3	13,63	5	22,72
P	515	5	442	42,56	55	10,67	15	2,91	1	0,19	2	0,38
R	780	4,5	332	21,39	363	46,53	64	8,20	13	1,66	8	1,02
S	215	3,5	46	47,88	80	37,20	58	26,97	19	8,83	12	5,58
Ş	71	4,5	34	53,46	26	36,61	10	14,08	0	0	1	1,40
T	260	4	139	59,57	69	26,53	28	10,76	14	5,38	10	3,84
U	141	4,5	84	53,59	42	29,78	9	6,38	5	3,54	1	0,70
Ü	1209	4	648	69,68	294	24,31	145	11,99	72	5,95	50	4,13
V	762	4,5	531	47,36	157	20,60	51	6,69	13	1,70	10	1,31
Y	57	4,5	27	23,07	25	43,85	3	5,26	1	1,75	1	1,75
Z	13	3,5	3	47,92	3	23,07	5	38,46	0	0	2	15,38
Total	7101	Mean	4156		1824		676		263		182	
Percent	% 100	4,1	% 58,52		% 25,68		% 9,51		% 3,70		% 2,56	

* minimum=1, maximum=5

The social media tools used by eco-friendly certificated accommodation establishments are shown in Table 20.4 below. The accommodation establishments represent themselves by creating a profile in a social network, 82.75% (24) of which are on Facebook, 75.86% (22) on Twitter, 13.79% on Trip Advisor, 13.79% on Youtube and 10.34% on Foursquare. In connection with this it is seen that establishments mainly prefer to use Facebook and Twitter.

When the websites of eco-friendly certificated accommodation establishments are analyzed on informing their users about their certificate, it was identified that only 3 of the 29 have no notifications about their being an eco-friendly establishment. Besides this, only 8 establishments give a place to this status on the main pages of their websites, and 2 give coverage of this feature in detail on both the main pages and the subpages of their websites. Eighteen establishments give place to this status on the subpages of their websites only. It is seen that in spite of having the certificate, eco-friendly accommodation establishments do not promote this distinctive feature well.

Table 20.4. Social media usage levels of eco-friendly accommodation establishments

Social Media Tools	Yes		No		Total	
	f	%	f	%	f	%
Facebook	24	82,75	5	17,24	29	100
Twitter	22	75,86	7	24,13	29	100
Tripadvisor	4	13,79	25	86,20	29	100
Youtube	4	13,79	25	86,20	29	100
Foursquare	3	10,34	26	89,65	29	100
Flickr	2	6,89	29	93,10	29	100

Fig. 20.2. The symbols used by accommodation establishments about environmental awareness.

The symbols used by eco-friendly certificated accommodation establishments to express their statuses are shown in Table 20.5. The Culture and Tourism Ministry gives a certificate to the establishments which deserve the title of eco-friendly accommodation establishment, but there is no national symbol or logo designed for this title. Hence, establishments that have this certificate use the symbols they develop themselves, such as a green star surrounding a pine tree, stars under the name of the establishment and green star symbols, as seen in Table 20.5.

Within the scope of this study, 2,096 user comments (30%) belonging to 29 establishments with eco-friendly accommodation establishment certificates were evaluated. Upon examining the comments of people staying in the eco-friendly accommodation establishments, the reviews were evaluated under six main sections and according to the arithmetic mean. People appreciated the cleanliness of the hotels first, followed by rooms, sleeping quality, location of the hotel, service quality and value. Considering the arithmetic means, it can be easily said that people have high rates of appreciation about these establishments. Besides the ranking of establishment properties, the most appreciated section is cleanliness and the least is value. Therefore, it is understood that guests did not see an equivalence between the services they received and the payments they made. By evaluating the accommodation establishments, guests rated "value" as 48.7% excellent, 29.2% very good, 14.2% average, 4.5% bad and 3.4% awful. When the evaluations related to "location" are surveyed, the ratings are 53.5% excellent, 25% very good, 15.1% average, 4.2% bad and 2.2% awful. With regards to "sleeping quality," the ratings are 54.4% excellent, 26.6% very good, 12% average, 3.6% bad and 3.4% awful. Considering "rooms," consumers rated these 55.1% excellent, 27.1% very good, 12.8% average, 3.1% bad and 1.9% awful. With regards to "cleanliness," these are determined as 64.8% excellent, 23.8% very good, 8% average, 2.2% bad and 1.2% awful. In terms of "service quality," the ratings are 56.1% excellent, 23.6% very good, 11.4% average, 5.1% bad and 3.8% awful.

According to the results of the general evaluations by users, on a scale of 1=awful and 5=excellent, the general appreciation rate of eco-friendly accommodation establishments (N=29) is 4.1. Consequently, this shows that accommodation establishments are appreciated in general and the services (especially value, location, sleeping quality, rooms, cleanliness and service quality) offered by them are positively accepted by guests. When the distribution of the total (7,101) comments belonging to eco-friendly certificated accommodation establishments in Trip Advisor are examined, 58.52% (4,156) of the evaluations are stated as "excellent." It is also observed that assessments of "very good" form 25.68 % (1,824) and "average" is 9.51 % (676), "bad" is 3.7% (263) and awful is 2.5% (182). When social media tools used by the eco-friendly accommodation establishments are inspected, it is seen that they represent themselves in proportions of 82.75% (24) on Facebook, 75.86% (22) on Twitter, 13.79% on Trip Advisor, 13.79% on Youtube and 10.34% on Foursquare. In connection with this it is seen that establishments prefer to use Facebook and Twitter. When the websites of eco-friendly certificated accommodation establishments are analyzed regarding informing users about their certificate, only 3 out of 29 establishments have no notifications about their certificate. Besides this, only 8 feature this status on the main pages of their websites and two establishments give coverage to this feature in detail on both the main pages and the subpages of their websites, while 18 feasture it on the subpages of their websites only. It is seen that in spite of having the certificate, eco-friendly accommodation establishments do not properly promote this distinctive feature. When the symbols used by eco-friendly certificated accommodation establishments to express their status are examined, it is understood that the Culture and Tourism Ministry gives a certificate to the establishments but that there is not a national symbol or logo designed for this title. Hence, establishments that have this certificate develop their own symbols, such as a green star surrounding a pine tree, stars under the name of the establishment and green star symbols.

References

Anderson, C.K. (2012). The Impact of Social Media on Lodging Performance. Cornell Hospitality Report, 12(15): 4-11.

Bayram, M. and Bayram, Ü. (2012). Otel İşletmelerinin Sosyal Medya Kullanım Düzeylerinin Değerlendirilmesi. 13. Ulusal Turizm Kongresi. (pp. 1213-1226), 07-09 Aralık, Antalya.

Buhalis, D. (1998). Strategic Use of Information Technologies in the Tourism Industry. Tourism Management, 19(5): 409-421.

Callarisa, L.; García, J.S.; Cardiff, J. and Roshchina, A. (2012). Harnessing Social Media Platforms to Measure Customer-Based Hotel Brand Equity. Tourism Management Perspectives, 4: 73-79.

Chung, Y. and Buhalis, D. (2008). Information Needs in Online Social Networks. Information Technology&Tourism, 10(4): 267-281.

Cox, C.; Burgess, S.; Sellitto, C. ve Buultjens, J. (2009). Role of User-Generated Content in Tourists' Travel Planning Behaviour. Journal of Hospitality Marketing&Management, 18(8): 743-764.

Dellarocas, C. (2003). The Digitization of Word of Mouth: Promise and Challenges of Online Feedback Mechanisms. Management Science, 49(10): 1407-1424.

Facebook Inc., (2013). Key Facts. http://newsroom.fb.com/content/default.aspx?NewsAreaId=22

Fotis, J.; Buhalis, D. and Rossides, N. (2012). Social Media Use and Impact During the Holiday Travel Planning Process. In: Fuchs, M., Ricci, F. and Cantoni, L. (Ed.) Information and Communication Technologies in Tourism 2012 (pp. 13-24). Vienna: Springer-Verlag.

Fesenmaier, D.R. (2007). Introduction: Challenging Destination Promotion. Journal of Travel Research, 46(1): 3-4.
Gretzel, U.; Yoo, K. and Purifoy, M. (2007). Online Travel Review Study: The Role and Impact of Online Travel Reviews. Laboratory for Intelligent Systems in Tourism, (pp:1-70), Texas A&M University.
Google, Inc. (2012). The 2012 Traveler. https://docs.google.com/gview?url=http://www.thinkwithgoogle.com/insights/uploads/940817.pdf&chrome=true
Kaplan A.M. and Haenlein, M. (2010). Users of the World, Unite! The Challenges and Opportunities of Social Media. Business Horizons, 53(1): 59-68.
Ko, D.W. ve Stewart W.P. (2002). A Structural Equation Model of Residents' Attitudes for Tourism Development, Tourism Management, 23: 521-530.
Kuvan, Y. ve Akan, P. (2005). Residents' Attitudes Toward General and Forest-Related Impacts of Tourism: The Case of Belek, Antalya, Tourism Management, 26: 691-706.
Kültür ve Turizm Bakanlığı (2013). http://www.ktbyatirimisletmeler.gov.tr/TR,11596/cevreye-duyarlilik-kampanyasi-yesil-yildiz.html.
Pappas, N. V. (2008). City of Rhodes: Residents' Attitudes Toward Tourism Impacts and Development, Anatolia: An International Journal of Tourism and Hospitality Research, 13 (1): 51-70.
Pan, B.; MacLaurin, T.; Crotts, J.C. (2007). Travel Blogs and Their Implications for Destination Marketing. Journal of Travel Research, 46(1): 35-45.
Poon, A. (1993). Tourism, Technology, Competitive Strategies. Wallingfrod: CAB International.
Pike, S. (2005). Tourism Destination Branding Complexity. Journal of Product and Brand Management, 14 (4): 258-259.
Pudliner, B.A. (2007). Alternative Literature and Tourist Experience: Travel and Tourist Weblogs. Journal of Tourism and Cultural Change, 5(1): 46-59.
Pühringer, S.; Taylor, A. (2008). A Practitioner's Report on Blogs as a Potential Source of Destination Marketing Intelligence. Journal of Vacation Marketing, 14(2): 177-187.
O'Connor, P. (2010). Managing a Hotel's Image on TripAdvisor. Journal of Hospitality Marketing&Management, 19(7): 754-772.
Teye, V.; Sönmez, S.F.; Sırakaya, E. (2002). Residents' Attitudes Toward Tourism Development, Annals of Tourism Research, 29 (3): 668-688.
TripAdvisor (2013). About TripAdvisor. http://www.tripadvisor.com/pages/about_us.html www.tripadvisor.com/pdfs/OnlineTravelReviewReport.pdf. [Accessed the 22th November 2009, 14:55).
Twitter, Inc. (2012). Twitter turns six. http://blog.twitter.com/2012/03/twitter-turns-six.html
Vermeulen, I.E. and Seegers, D. (2009). Tried and Tested: The Impact of Online Hotel Reviews on Consumer Consideration. Tourism Management, 30(1): 123-127.
Waldhör, K. and Rind, A. (2008). Blog Analysis -Mining Virtual Communities Using Statistical and Linguistic Methods For Quality Control in Tourism. In P.

O'Connor, W. Höpken and U. Gretzel (Ed.), Information and Communication Technologies in Tourism 2008 (pp. 453–462). New York: Springer.

Wang, Y.; Hsu, K.; Hsu, S. and Hsieh, P. (2011). Constructing an Index for Brand Equity: A Hospital Example. The Service Industries Journal, 31(2): 311–322.

White, L. (2010). Facebook, Friends and Photos: A Snapshot into Social Networking for Generating Travel Ideas. In: Tourism Informatics: Visual Travel Recommender Systems, Social Communities and User Interface Design. Information Science Reference, Hershey, Pennsylvania (pp. 115-129).

Yoo, K.; Gretzel, U. and Zach, F. (2011). Travel Opinion Leaders and Seekers. In R. Law, M. Fuchs and F. Ricci (Ed.) Information and Communication Technologies in Tourism 2011 (pp. 525-535). New York: Springer.

Youtube Inc. (2012). Statistics. http://www.youtube.com/t/press_statistics

Yoon, Y.; Gursoy, D. ve Chen, J.S. (2001). Validating a Tourism Development Theory With Structural Equation Modelling, Tourism Management, 22 :363-372.

Xiang, Z., and Gretzel, U. (2010). Role of Social Media in Online Travel Information Search. Tourism Management, 31(2): 179-188.

CHAPTER TWENTY-ONE

THE TOURISM POTENTIAL OF THE ANAMUR DISTRICT

TAHSIN TAPUR

Introduction

Tourism is a recreation activity in which people make journeys for rest, fun, curiosity, sports, health, culture, experience, visiting family, attending congresses and seminars, worshiping away from their homes, working and meeting their routine needs (Özgüç 2003).

Tourism is an integrated system made up of subsystems and the field of geography has main share in this from the point of view of hosting tourists and providing for their cultural needs through different geographical sites. Therefore, countries, regions, vicinities and cities in which the tourism demand emerges are each part of geographic sites. In these sites, natural and cultural attractiveness, facilities, services and tourism organizations constitue the tourism elements (Soykan 2000). The fact that natural and cultural assets form a source for tourism existing in different countries, vicinities, regions and localities embody geography and tourism (Emekli 2006, Efe et al. 2008). On the other hand, historixcal legacies and resources develop cultural tourism. Geography is a notion joining cultural and natural legacies together. For example, ancient cities on the delta and travertine areas exist one within the other, such as Hierapolis near Pamukkale travertine, Milet on Büyük Menderes delta, and Efes on Küçük Menderes delta as examples of a synthesis of natural and cultural resources (Doğaner 2001).

This study aims at investigating the tourism potential of the Anamur district and the possibilities to use this potential. Therefore, first of all the literature was reviewed, followed by field surveys and recording observations at the places with touristic values. Along with the data obtained from the literature and field surveys supported by photographs, attempts will be made to enlighten the tourism potential of the Anamur district.

The Anamur district is located within the borders of the Mersin province with an area of 1348 km^2. The city of Gazipaşa (Antalya) lies on the west, Ermenek (Karaman) to the north, Bozyazı (Mersin) to the east and Gülnar (Mersin) to the northeast of the district. In the south it has a coast of 45 kilometres on the Mediterranean Sea (see Fig. 21.1 below). It is located on the Mersin-Antalya state route at a distance of 230 km from Mersin, 265 km from Antalya, 265 km from Karaman and 40 76 km from the Island of Cyprus. Nearly 60% of the district is covered by forests, 20% by meadows and pasture lands, 11% is farming lands and 9% comes under other areas (like bare rocks, sand dune areas and settlement areas).

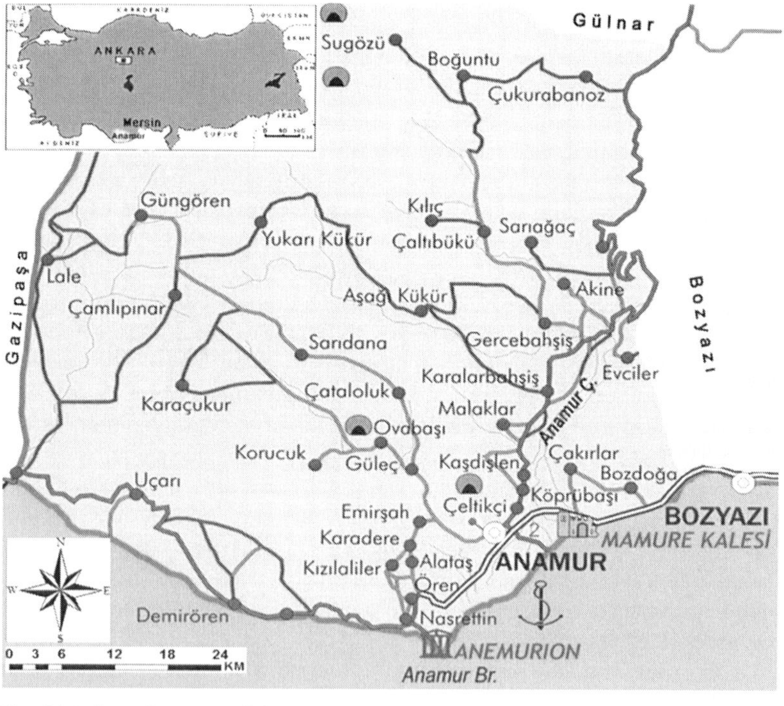

Fig. 21.1. Location map of the Anamur district

The name Anamur comes from the ancient Romance name "Anemurium-wind cape.". At different times in histry, Anamur was under the rule of Luwians, Cilicians, Phoenicians, Hittites, Assyrians, Persians, Romans and then Byzantines. It was captured by the Seljukians during the reign of Alâeddin Keykubat (1228). It was subsequently ruled by the Karamanids but the Ottomans took control in 1859, and it was made a district governorate in 1869 (Anamur 2000). At present, Anamur consists of 1 district, 2 towns and 37 villages. The address-based population registration datas of 2012 show that 62,702 people live in Anamur, of which 34,227 live in the centre of the district and 28,475 live in the villages and towns.

Anamur has very important tourism potential arising from both its physical geography and history. The district abounds in bays with natural beaches, rest and picnic areas, forests, caves, historical settlements, churchs, chapels, public houses, historical castles, historical remains and plateaus, which are regarded as tourism destinations.

Although having the same potential as in other Western Mediterranean districts and those on the Aegean coast, its tourism level has not come to the desired level mainly because of transportation difficulties. The D-400 state route between Silifke-Anamur-Gazipaşa is very narrow and curved, preventing transportation and hindering tourism. The tourism around Anamur is more dependent on domestic tourism via cottages than foreign tourism.

Elements with Tourism Values

The main elements with touristic values in the Anamur district are natural beaches, resting and picnic areas in forests, caves, historical settlements, churches, chapels, public houses, historical castles, historical remains and plateaus.

Sea tourism: Anamur's coast is approximately 45 km from the Bozyazı border to Anıtlı (Kaledran). The mountains run perpendicular to the sea around the Anamur coast, and the sea is deep at many places. However, there are beach areas which can be accepted as important for sea tourism, formed from shingle and sand, where two to three metres from the coast the sea begins to deepen suddenly. However, the beaches at Ören, Cerenler and Taşlık are shallower. The longest of these beaches is 17 kilometres begining from the rest area in the forest in Pullu to the east of Anamur Bay (Fig. 21.2). The most important elements leading to this natural beach are the Anamur (Kocaçay) and Sultansuyu rivers. Besides, there are beach areas in Anıtlı, Ucarı and Melleç.

As sea water temperature is suitable from April until October on natural beaches and bays with good views, sea tourism continues up to the end of the autumn season, the long season adding to the tourism potential of the area. However, this potential is not currently utilised, although it has been improved especially via cottages towards domestic rather than foreign tourism. The distance to the tourism centers visited by foreigners and the limited opportunities for foreign tourism are effective in this situation; 80% of those who take part in domestic tourism activities are those with secondary houses, mainly coming from the Konya and Ankara provinces, but some are also coing from the Malatya, Gaziantep, Kayseri and Kahramanmaraş provinces.

Fig. 21.2. Beach and sand dune areas from the Anamur coast

It is one of the cleanest coasts in Turkey because there are no factories or facilities adding waste to the sea. The littoral current off the coast of Ören contributes in the clean up of the sea. Beach services and facilities are available on the coast at certain points (such as around İskele, Ören, Mamure Castle and Pullu).

There are many summer houses constructed for sea tourism in the Anamur district on the Ören, Bozdoğan, Anıtlı, Ucarı and Melleç coasts. The momentum of summer house constructions on the edge of sands is increasing year by year. The sands are under threat from both housing schemes and waste water disposal into the sea. The main objectives of local governors are to improve the quality of beaches, get them cleaned, raise the awareness of the public and provide shower facilities. One of the factors threatening the natural beach areas is the banana and strawberry greenhouses, which are economically valuable and established on the edges of these beaches.

Picnic areas with touristic value: There are picnic and layover areas, such as in Pullu and Dikilitaş in the forest, Dibek Köprüsü, DSI watering facility, Ala Köprü, Pınarlar fall, Melleç, and Dragon Park.

Pullu Picnic Area: this is an important area located on the Anamur-Silifke route. There is a layover area in the forest for daily use, and the forest reaches the sea. The park has space for 90 caravans or tents, a reception centre, a WC, a buffet, shower and dressing rooms, a country bar, a medical complex, a drinking fountain, and tables and beach facilities among Turkish pines. It is mostly used between May and September.

Dikilitaş Picnic Area: this is located on Anamur-Silifke route, covered with Turkish pines. There are facilities for 200 tents or caravans, an administration building, houses of park officers, a WC, a buffet, a medical complex, a drinking fountain, tables and beach facilities.

Caves with touristic value

Köşekbükü Cave: this is located 9 kilometers northwest of Anamur, reached by asphalt road with a natural view of the forests. The cave is in the Ovabaşı village over an area of 500 m^2. The cave consists of three parts, huzur, şifa and dilek, and features thousands of stalactites, stalagmites and columns. The average compression of the cave is 761 mm, the humidity is 80%, and the temperature is 18^0C. The inner lighting and walking terraces were organised by the provincial special administration. The cave is said to be good for asthma, and asthma sufferers stay in shacks for about 15–20 days (Tapur 2003). However, the cave is presently closed following a forest fire in 2012, which damaged the facilities around the cave and its inside lighting (Fig. 21.3).

Bicikli Cave: this is a 700 metere cave consisting of seven parts in the Abanoz plateau. The name Bicikli comes from its stalactites. Potable water of the plateau is supplied from the spring water in the cave. Although the road to it is smooth and asphalt, there has been no organisation of a tourism substructure.

Çukurpınar Cave: this was found on the Çukurpınar plateau (at 1,880 meters) in the northwest of the Anamur district in 1990. It is the twelfth deepest cave in the world and the deepest in Turkey, formed from katavothre which has shaped nautical Miocene limestones. There are many lakes, falls, potholes and roadways in the cave. It is possible to climb down to 1,450 meters. No tourism substructure exists, including transportation facilities (Tapur 2003).

Üğü Cave: this is located in Güneybahşiş village, 12 km from Anamur. Its total length is 450 meters and the roof height is between 1 and 20 meters. The entrance is dark, with a spring water side. There are five passable parts in the cave. Six lakes have been found in the area, some of which are very deep. The cave has not yet been opened to visitors, but efforts have been made towards this.

Fig. 21.3. Inside and outside views of the Köşebükü Cave

Tourism elements with historical value

Anamur Museum: this was founded in 1992 to display archaeological and cultural objects from the immediate vicinity, mainly the ancient city of Anemurium. The museum carried out excavations in the Bozyazı (Nagidos) necropolis, Aydıncık (Kelenderis) and the recovery of objects from the ancient city remains in Mamure Castle in 1985 and 1986. Today, the number of pieces in the Anamur museum has reached 7,000. Most of these are Anamur pieces brought from Alanya and Silifke and elsewhere, and purchased from the Erdemli museum closed by the ministry. The people of Anamur brought these with full devotion. On the top floor of the museum there are managerial rooms, a library, a photo shop, and a conference hall, and on the ground floor there is a cafeteria, ethnographic folkloric items used in the historical process from nomadic life until permanent settlement, archaeological departments, storerooms for old pieces and a laboratory.

Old Anamur (Anemurium): this is located 6 km in the southwest of the Anamur district, reached via a 2 kilometre road off the Anamur-Gazipaşa route, and is one of the best protected cities of mountainous Cilicia (see Fig. 21.4 below). No exact information is availavble as to when the city was founded. The name was first referred to in the historical records dated from the fourth century BC. Anemurium, which was used as a station during the Roman era because of its closeness to Cyprus, was also a city connected with Germanikopolis (Ermenek), one of the most important Ancient Roman cities around the Taurus Mountains via a land route. It was an important trading city in which natural resources in the area were exported. Anemurium was captured by the Sassanid Empire in 260 and was damaged by pirates in fourth and fifth centuries. The city was charged by Arabians in 650 but abondoned soon after. After this, the Seljuk Sultanate of Rum captured the Mamure Castle, and the province was under the rule of Turks (Mersin İli Turizm Mastır Planı 2010, 74).

The most attractive buildings of the Anemurium city are walls, three baths, a theatre, an odeon (concert hall) and a palaestra. Except for the aqueducts providing spring water to the city, there are church remains belonging to the early Christian era.

Fig. 21.4. West scene of Anemurium ancient city

City Walls: The walls which surround Anemurium are 1.5 m long and 8 m tall, an made of blue limestone from the area. Observation towers on the walls are dated to the Hellenistic era and are made of white limestone.

Odeon: this was built in the second century AD, and is in the form of a rectangle with two entrances. Perches are in the form of hemicycle and have a capacity of 900. The platform is covered with inlay.

Theatre: there are no remains of the perches of the theatre which is located to the north of Odeon (Fig. 21.4). Presently, part of the surrounding walls can be seen.

Public Bath: the building located to the west of the theatre is one of the biggest buildings of the city and is two-storied (Fig. 21.5). It has three warming lobbies and two pools. In the building, a horizontal soil pipe for water distrubtion systems can be seen. There are few remains of the geometric patterned floor inlay.

Fig. 21.5. Public bath and theatre buildings in Anemurium Ancient city

Necropolis Area: this is one of the best protected necropolis sites in the area. There are approximately 350 tombs, dated to the first and fourth centuries AD, designed with barrel vaults, two-storied, one or two-roomed, and surrounded with walls. They were generally built from refined or non-refined rubble stones, and generally in the form of detached sepulchres standing on gradual platforms. Entrance to these sepulchres is provided by very low gates. A front room has been added to some sepulchres. The lobbies of these sepulchers are decorated with frescos and the floor is decorated with inlay (Fig. 21.6).

Aqueducts: There are two rows of aqueducts divided east-west to the north of the city.

Kalınören (Titiopolis): Kalınören is an ancient city to the west of Anamur, located on hills overlooking the sea. There are ruins from Hellenistic, Roman and Byzantium eras, with cisterns, living spaces, church, walls, baths, halls and tomb ruins from the first century AD.

Buildings like a bouleuterion, a ninfeum (fountain), an odeon, and a theatre, which can't be seen in the ancient city at present, are probably under the village settlement. The walls surrounding the city were built from large and small, roughly sculpted polyangular stones. While climbing the hills to the north of the city, the first building encountered is the ruins of the vault-covered bath with three different locations. There is a three-plated-basilica to the west of the bath with a distinct narthex. The floors of the buildings are geoherbally decorated with grey- and white-coloured inlays. There is a necropolis area to the east of the religous structures in which gorgeous tombs exist. However, the ancient city ruins have been largely demolished by the people and transformed into settlements and farming fields (anamur.gen.tr).

Fig. 21.6. Rock tombs in the necropolis area in the ancient city of Anemurium

Mamure Castle: is on the mediterranean coastline, approximately 7 km southeast of Anamur within the borders of Bozdoğan village. Who constructed the castle and when are not known. The castle was built in the third or fourth century AD and extended during the Byzantium and Crusader eras, and is surrounded with a 10 metre wide ditch filled with water for defence. The present castle was built in return for the one demolished during its conquest by Alâeddin Keykubat in 1221.

The castle consists of three parts, comprising the inner court to the east separated with high walls, the outer castle to the west and the inner castle built on rocks to the south. There are two pieces of walls, one within the other, and a road on the walls between bastions, providing passage from one side to another. On this road there are 39 towers, 4 of which are huge (see Fig. 21.7 below). At the outer castle, there are storerooms, cisterns, buildings that were probably habitation areas of soldiers, being centrally planned, having one dome mosque and a fountain. It has seen numerous restorations thanks to which it has managed to survive (Tapur 2003).

According to the historian Shikari, following the conquest and demolition of Anamur and Taşeli, Karamanoğlu Mehmet Bey (1300–1308) defeated his enemies, captured the castle, pulled the churches down and built mosques, and the castle was repaired and named Mamuriye. There are also documents showing that the castle was restored and new parts were added by the mid sixteenth centrury and by the end of the eighteenth century. Lastly, it was restored by the Directorate General of Foundations in 1960 (mersinkulturturizm.gov.tr).

Mamure castle was included in the UNESCO World Heritage Temporer List in April 2012. Within the context of international cooperation agreements for securing the world cultural and natural heritages, the contracting states pay attention to architectural, historical, esthetical, cultural, economical, social and philosophical features besides the circumstances if the assets meet the criterias determined by the World Heritage Committee while prepearing the temporary lists. There is a mosque with one minaret in the castle and a demolished bath in front of the castle to the north of the main road (mersinkulturturizm.gov.tr).

Mamure Castle Mosque: the first built mosque had the classical elements of Ottoman architecture of the sixteenth century dating back to the Karamanids era. The mosque is entered via a stone door with a flat arch (Fig. 21.9). In the construction with the central dome, the passage to the octagonal pulley is provided via pendantive sconces from the Seljuk triangle. In the mosque, brick and stone lines on the walls were bonded harmonizingly and attractiveness was created with nail garnishes from bricks on the passage to the dome and roof trees. The cistern in front of the mosque is decorated with little arches and has a barrel vault.

Fig. 21.7. View of Mamure Castle from various angles

Mamure Bath: the entrance of the bath located to the north of Mamure castle was demolished, and the frigidarium, caldarium and tepidarium parts have survived until now. The small-sized contruction was built with rubbish stone supported with wooden girders. As the construction fell into ruins, repair flasters ruined the wall paintings. It is thought that the bath was built by the Karamanids during the repair of the castle. During the recovery diggings in 1988, sites thought to be baths and houses with inlay floor were built with Khorasan mortar of rubbish stone. Some main marks around the bath show that this place was a settlement integrated with Mamure castle (anamur.gov.tr).

Çoban Castle: is located in Çamlıpınaralanı village on the Anamur-Antalya route. The castle was built on a forested hill overlooking the coast. Today, most of it has been destroyed. The construction is entered from a 1.70 m wide stone door. The large courtryard in the middle of the castle is covered with many rooms with variable functions. Inner surfaces of the sites are plastered and they have loopholes for defence purposes. It is thought that the castle was owned by an Ottoman Bey who built it in the sixteenth and seventeenth centuries (Anamur 2000).

Boncuklu Castle: this consists of ruins in Çeltikçi village, located on the Anamur-Ermenek route, of an ancient settlement with 200 or 300 houses on the slope of a hill. On top of the hill, oviformed walls were built for defence. The cistern in the middle of the castle is vaulted and has thick walls. The chapel in the castle is to the east.

Melleç Han (Tol Kervansaray): this is located on the coast of Demirören village on the Anamur-Antalya route. The construction built with hard red and yellow rubbish stones is two-plated on the north and south sides and covered with a round vault. There is no inscription stating its date of construction, but it must have been built in the fourteenth and fifteenth centuries. The Han which has been uncovered is progressively ruined, day by day, and should be under protection and restored immediately.

Altı Kapı Han: is located on the Anamur-Antalya route to the west of Anıtlı, built with black and yellow stones belonging to the area and with khorasan mortar. The Han opens to six long, round vaulted rooms adjacent to each other with six round, arched entrance gates in an east-west direction. In front of these sites there is a porch with round vaulted shores. The hanhas been used and protected by private property owners for a long time and has been added to current tourism heritage lists. However, any arrangements or protection efforts have not been made. The Han should immediately be restored and taken under protection.

Akarca Han: The building located in the Akarca neighbourhood in front of the Ak mosque was built of rubbish and slate stones and is a one-plated and vault covered construction. While there are no particular documents about its construction date, it is thought to have been built in the eras of the Seljuks or the Karamandis. The Han was used as repair stores and has recently been restored by the Directorate General of Foundations (Fig. 21.8).

Alaköprü: is located on the Anamur (Dragon) river on the Anamur-Ermenek route, built in 1230 by the Karamanids. The bridge consists of one section of 19.65 m on the main stream and a smaller relief section on the east side for flood water (Fig. 21.8). The structure of the main arch has itself a kind of value with important labour and a steady travertine. Its groundwork was built on steady rock floor, and a carrier and reinforcement arch are the most valuable and steady elements of the bridge. The relief section has fallen into ruins over time but has been strengthened via repairs. The whole structure of the bridge is crossbred meshed, except for the arch.

Fig. 21.8. Alaköprü and Akarca Han

Tourism elements with religious value

Akcami: this mosque with a grooved minaret was built in 1326 during the Karamanid era, in an attractive site. There is a historical Akarca Han and a historical bridge in front of it. It has a centrally planned dome made of totally face stone in the Akarca neighbourhood (Fig. 21.9). The mosque is entered through a low arched stone door from the western side. There is a stone mihrab just in front of the entrance and a wooden minbar on the right, but it is not original. In the building, the passage to the dome is provided through a round inner core embedded into the wall on the intervals with relieving beaked arches at corners and side walls, something rarely seen in traditional Turkish architecture. The jambs of the relieving archs are supported with spandrel supports. It has a minaret with one grooved balcony.

Kızıl Church: this consists of ruins of a church in a ruined area known as "külsavuran beleni" in Kızılaliler village, 8 km northwest of Anamur. Although the narthex of the three-plated church found in ruined area has mostly been demolished, its side wall has totally survived. Three windows are found on each side of the building, and the side plates must have been two-storied on the left and right walls open to the outside. Its side rooms to the right and left of abscissa are apparent. The construction looks like a Isaura structure of the fourth or fifth century (mersinkulturturizm.gov.tr).

Otak Village Chapel Building: this is found on the edge of the Kaledran river 10 km to the North of Anıtlı village on the Anamur-Gazipaşa route. The building dates back to the Early Byzantium era. It is a small building with cultic rooms on both sides of a one-plated abscissa made of rubbish and face stone. Presently, it is used by peasants and is mostly unprotected. Some of it has fallen into ruins (Fig. 21.10).

Anıtlı Observation Tower (Köşk): the building in the Kaledran (Anıtlı) village on Anamur-Gazipaşa route, 900 metres north of the highway on the west side of the Kaledran river is known as Köşk among the native population. It was made of face stone on extremely thick ground. The entrance door to the south front is marble, with two-storied 12 stair ladders 15 m high, providing a way to get upstairs. There is a small site on the ground floor and loophole windows are also found here. There are two separate sites upstairs. The north side of the building has been totally demolished. Both floors are covered with round arched vaults. The cover coat of the building has a saddle roof. The observation tower is dated to the fourth or fifth century AD and thought to be an important outpost building of the era (Anamur 2000). It is among the farming fields and is currently unprotected (Fig. 21.10).

Fig. 21.9. Mamure Castle Mosque and Akcami

Fig. 21.10. Anıtlı Köşk ve Otak chapel

Historical Anamur houses

There are examples of civil architecture peculiar to those in the vicinity of the Anamur district and Ören town. The traditional structures of houses are two-storied with barns at the ground floor. The upper living rooms are generally in cruciform shape with doors to rooms at corners from middle sofas and plates vertically crossing one other. In the vicinity, extremely authentic architecture buildings named "kiosks" can be seen, built as a balconied second floor on the barn and one-roomed. Drawing the most attention among the historical Anamur houses are Nazifoğlu Konağı, Karamanoğlu Konağı, Hacı Alibey Konağı, Eski Hükümet Konağı, Hakkı Efendi Konağı, Kısakahyaoğlu Konağı, Şavkı Efendi Konağı, Şevket Erdem Konağı, Hasan Seymen Evi and Mustafa Rıfat (mersinkulturturizm .gov.tr).

Transhumance in Anamur

There are plateaus to the north of Anamur where both locals and people from other provinces lodge. Among them, the Kaş, Abanoz, Akpınar, Halkalı and Barcın plateaus draw attention in terms of tourism potential. These plateaus are important places in terms of substructure and various natural features. In summer, they are utilized by both landlords and lodgers from other provinces.

Kaş Plateau: this is located on the Anamur-Ermenek route and at 1,350 meters altitude. It is possible to see traditional mountain houses alongside modern buildings. The plateau is covered with fir, cedar, juniperus, syrian juniper and buxus trees, and fruit trees are grown in the gardens of houses. Transportation is provided by vans from nearby settlements in the summer. There are picnic areas like Deliktaş and Sıtmalı alongside historical castle ruins around the plateau. In summer, there are workplaces like cafes, barbers, grocers, butchers and bakers to meet the daily needs of the public (anamur.gov.tr).

Abanoz Plateau: this is located on the Anamur-Ermenek route, founded in the polje base between two reciprocal mountains. As the katavothre draining the water of polje is very little in winter, ponding occurs at the polje base. The plateau and its environment are covered with fir, cedar, juniperus, syrian juniper, and buxus trees. Fruits, vegetables and ornamental plants are generally grown in front of the houses which are utilised as cottages in the summer. Traditional plateau houses displaced with modern buildings and villas are also found on the plateau.

Transportation is provided by vans from Anamur and Bozyazı. It is a place where those working in Anamur can make round trips. On the plateau, many social and commercial foundations exist which meet the needs of the population (cafes, mosques, barbers, grocers, markets, butchers and bakers). These are sufficient and there are no substructure problems like electricity, transportation, water or telephone. On Fridays, an open market is set up. The population of the plateau is 50 people on average in winter, but rises to 7,500 in summers. The increase of population on the plateau causes problems like irregular settlement, concrete houses which are not in line with the nature of plateaus and environmental pollution (Tapur 2009).

Akpınar Plateau: this is located on the Anamur-Ermenek route. It is one of the plateaus where past administrators and aghas of Anamur have lived. It is possible to meet all kinds of daily needs on the plateau. Fruit and vegetable farming has improved and during summer an open market happens every week. As the unearthed limestones can be cut with a knife before being exposed to sunlight, locals use them for producing toys and sculptures.

Accommodation facilities

To meet the accommodation needs of tourists visiting Anamur there are 33 facilities, 3 of wich have certificates from the Ministry of Tourism (Luna Piena Otel-2 stars, Yan Otel-1 star, Anamuryum Aquapark), and 30 are municipality certificated. 15 of these facilities are hotels, 4 are motels, 3 apartment hotels, 5 rest houses, 2 mocamps and 2 campsites. The total capacity of accommodation is 443 rooms, 2,052 beds, 290 tents and 12 bungalows (Mersin Kültür & Turizm Müdürlüğü 2012). Additionally, on the Dragon river side the Pullu and Dikilitaş picnic areas in the forest are good places to pitch a tent and camp.

Conclusions

Anamur district has an important tourism potential with its camping and picnic areas inside the forest, caves, historical settlements, religious tourism elements, hans, historical castles, historical ruins, plateaus and especially coastal tourism. However, this potential is not utilised completely for various reasons. Although Anamur has same tourism values with Antalya, Muğla and Izmir, transportation problems, tourism substructures and introduction of the vicinity are not to the desired level.

It is easy to improve the vicinity with the wealth potential from tourism activities and to make it a well-known tourism area. For this purpose, the places mentioned here should be fully determined. The tourism substructures should be completed by designating both historical areas and the many beaches and sandy areas as tourism areas.

In order to carry out a planned improvement in tourism centres, physical plans should be made and alternative tourism ways planned on an advanced level. Some ideas and studies exist on this topic. The most important of these is that an area of 30 ha was declared as the Melleç Tourism Center 25 km west of Anamur, announced on April 19, 1989. Of this area, 20,144.9 ha are forest, 4 ha are 2B fields, 3 ha are public property, 9 ha are private property and 5 ha are roads and streams. There are two daily services running on the existing main road. The most important problem of the Melleç Tourism Center is the property, as there is no investment in this tourism center. Additionally, Anamur is among the new "tourism cities" in Turkey as per the Tourism Strategies 2023 Report. However, no action has been instigated in this regard.

Existing facilities towards tourism take place on the coastline. The facilities to be built up from now on should be planned systematically and close to the existing facilities. Additionally, the improvement of the river tourism sector at the Çarıklar Dragon river and the sea and beach tourism sector at the Anamur coast, the Mamure Castle environment, Ören, Melleç and Anıtlı should be supported.

The efforts for the improvement of transportation should be accelerated. Double highway and tunnel constructions on the Gazipaşa-Silifke part of the Antalya-Mersin D-400 route which began in 2010 are still going on. Also, the fact that the Gazipaşa airport was put into service in 2011 is considered to contribute to Anamur's improvement of tourism.

Considering the slope and ridge circumstances, the most important aspect of Anamur is probably the sea. The improvement of both commercial services and tourism activities in both Anamur and its surroundings depends on port services. So, the efficiency of the seaport should increase and efforts for improving sea transportation will be sufficient for this. After the seaport begins service, besides attracting tourists from foreign countries it will be easier for ferry voyages between Anamur and Cyprus as it is very close to the island, and Anamur-centred Mediterranean tours can be organised.

In Anamur, tourism and farming activities are conducted one within the other. In many areas, while buildings for tourism are built on first class farming lands, greenhouses near beaches are constructed. This is not a good utilisation of land and should be stopped, and separate areas for tourism and farming should be determined.

Plateaus such as Abanoz, Kaş, and Akpınar on the Taşeli steppe have been settled with the aim of making them summer places. On the plateaus, modern buildings have begun to spoil the natural ecosystem. The necessary construction plans should be applied in this connection. Although these are sufficient for the current tourism potential, commercial zones serving tourism are highly limited. Commercial zones towards tourism (restaurants, entertainment venues, souvenir shops, rent a car services, shopping centers etc.) should be improved and increased in number.

References

Anamur (2000). Anamur İlçe Turizm Müdürlüğü. Anamur.

Doğaner, S. (2001). Türkiye Turizm Coğrafyası. İstanbul: Çantay Kitabevi

Efe, R., Sönmez, S., Cürebal, İ., Soykan A. (2008). Balıkesir'in Geoturizm Yöreleri: Marmara Geoturizm Yöresi Örneği. Üçüncü Ulusal Balıkesir Turizm Kongresi. 17-19 Nisan 2008. Bildiriler Kitabı, s. 323-330, Balıkesir.

Emekli, G. (2006). Coğrafya, Kültür ve Turizm: Kültürel Turizm. Ege Coğrafya Dergisi 15: 51-59. İzmir.

Mersin İl Turizm Müdürlüğü (2012). Turizm İstatistik Verileri, Mersin.

Mersin İli Turizm Master Planı (2010). T.C. Mersin İl Özel İdaresi, Mersin.

Özgüç, N. (2003). Turizm Coğrafyası Özellikler-Bölgeler. İstanbul: Çantay Kitabevi.

Soykan, F. (2000). Turizm Coğrafyası ve Turizm Planlaması, Ege Coğrafya Dergisi 11: 39-55. İzmir.

Tapur, T. (2003). Anamur-Silifke Arası Kıyı Bölgesinin Coğrafi Etüdü, Sel. Ün. Sosyal Bilimler Enstitüsü Yayınlanmamış Doktora Tezi, Konya.

—. (2009). Abanoz Yaylası (Mersin), Sel. Ün. Sosyal Bilimler Enstitüsü Dergisi Sayı 21, Sayfa: 471-485, Konya.

www.anamur.gen.tr/erişim06.12.2012.

www.anamur.gov.tr/erişim06.12.2012.

www.mersinkulturturizm.gov.tr/belge/1-57504/anamur.html/erişim06.12.2012.

Chapter Twenty-Two

Dipsiz Lake and Dipsiz Lake Waterfall (Doğanşar-Sivas)

Vedat Karadeniz

Introduction

The economic activities which are not harmful for nature stand out with the concept of sustainable development. Among these, the tourism sector is considerably related to environmental quality. This sector has gained importance due especially to economical input since the second half of the twentieth century (Selimoğlu 2004,1). However, an unconscious plundering of nature for economic interests causes lot of destruction to our nature, ending in irreversible damage. The importance of the natural environment is now better understood as we leap forward because the concept of tourism in which the environment factor is not considered is producing degradative impacts. As a result, people have started to prefer hidden natural places instead of crowded places in the sense of tourism. The desire of the people to be alone with nature, away from the hustle and bustle of the city, has led to the use of such wilderness for tourism.

In recent years, nature tourism, regarded as tourism which is relatively less harmful for our nature, plays an important role in the evaluation of rural areas as touristic places (Özgüç 1998, 153). One of the important resources that must be considered within nature tourism is lakes and waterfalls.

Most lakes and waterfalls are thought of as natural touristic resources since they are natural wonders and potential centres for tourism, important as in terms of health, with their fresh air and aquatic ecosystems, as well as mental rest (Doğanay 2001, 344–366). However, these resources aren't evaluated in terms of tourism in Turkey. Whereas most of them can be used within tourism by removing some infrastructure problems and advertising, they can contribute to the national economy at both national and areas of this study, possessing important values, and are waiting for investment and advertisement within nature tourism. Our survey shows that no work has been done on the aforesaid natural sites; we find only a white paper with a few sentences and some news on the internet about Dipsiz Lake and Dipsiz Lake Waterfall in a book named “Atlas of Waterfalls of Turkey”, and in a study conducted by Kahraman (2008).

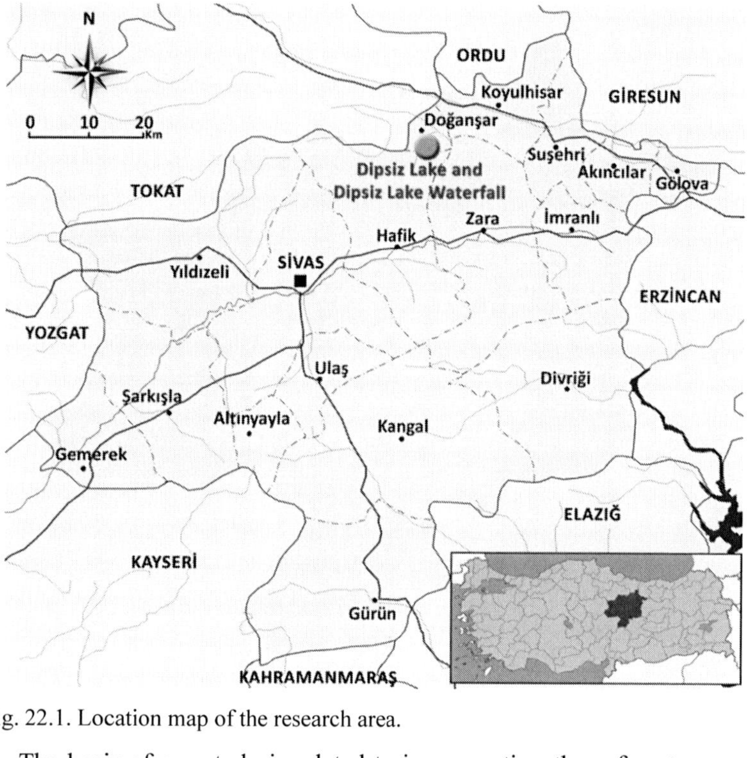

Fig. 22.1. Location map of the research area.

The basis of our study is related to incorporating these forest-covered natural sites into the national economy and look to the contributions expected for the region, where geographic conditions limit economic activities. The site was visited in June 2012 and April 2013, and observation and measurements were made.

Study Site

Dipsiz Lake and Dipsiz Lake Waterfall are in the Çatpınar village, 82 km from Sivas and 20 km from Doğanşar (Fig. 22.1), between the coordinates of 40° 05′ north latitude and 37° 33′ east longitude. The site is within the administrative boundaries of Doğanşar, Sivas.

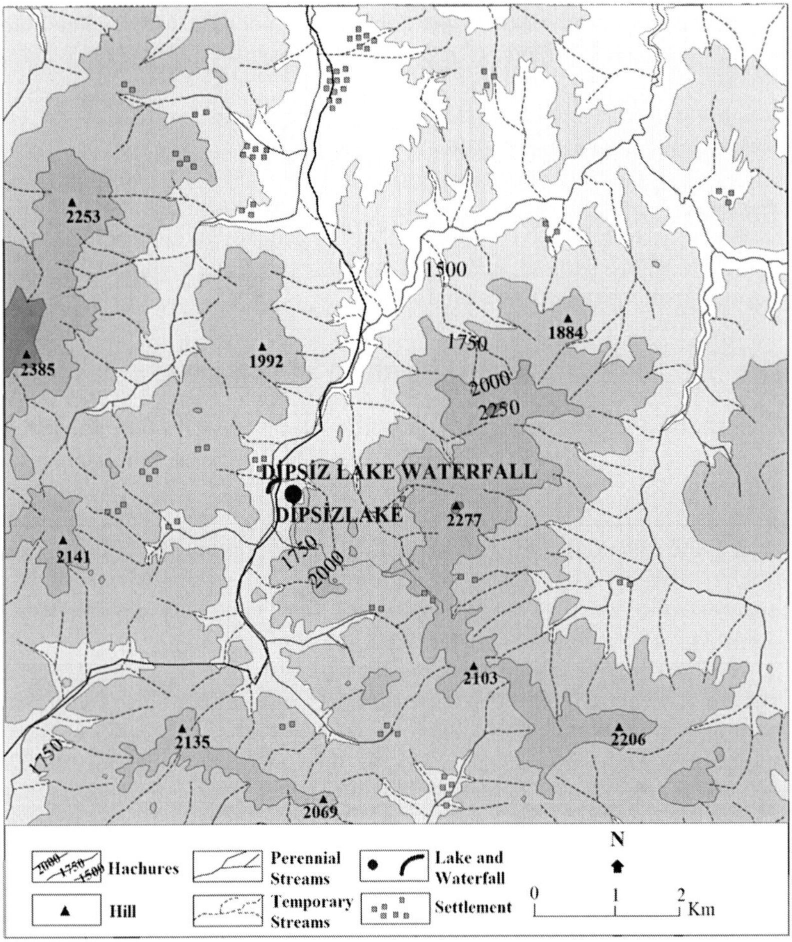

Fig. 22.2. The topographical map of the research area.

The former is located 250 m to the east of the Hafik-Doğanşar highway and to the south of the turn off to the Dikenli and Gökseki plateaus, and the latter is located on the east slope of the Dipsiz river, just on the west side of the main road.

One can reach this site from Sivas after turning off on the Sivas-Erzincan highway just 1 km from the centre of Hafik, following which is a

42 km road to the destination. The Hafik-Doğanşar road is paved except for some damaged parts; because of the topography of the area, the road is mostly winding and rough, since the land is sloping. It is an important advantage for accessibility that the lake and waterfall are on the edge of the main road. Those who have private cars can reach the area by using this route. For those who don't, two private minibuses run from the bus station in the district and village.

The altitude of Dipsiz Lake is 1,620 m above sea level and the valley bottom of Dipsiz River is 1,514 m. The excess water of the lake passes under the road of the Gökseki and Dikenli plateaus through the sinkhole and rises to the surface again 200 m later, constituting the waterfall by pouring over the travertine after 100 m.

Natural Environmental Features

The base of the study area consists of Mesozoic ophiolitic series, covered by the Doğanşar formation, in which Eocene old flyshes are common. The formation is presented with Eocene, including argillaceous limestone and sandy limestone, which starts with pebble stone and goes on with sandstone and clay stone alternations (Yılmaz 1982–1983, 7–9; Yıldız 1990, 33–37). The youngest formation in the area where the lake and waterfall exist is the travertines of the Quaternary period, which constitute these two sites of natural beauty.

The area is based tectonically on the Alpine-Himalayan orogenic system. The materials in geosynclinal areas rose in that period and gave way to folded mountains in Turkey, which got exposed to the orogenic movements of Alpine (Atalay 1987, 47–54). Büyük Kızıldağ Hill (2,277 m), Dokuztekne Hill (2,103 m), Toroluk Hill (1,992 m), Susuzkır Hill (2,135 m), Tütüyen Hill (2,141 m) and Karabacak Hill (2,385 m) can be counted among the important hills of this system around the waterfall between the North Anatolian Mountains and the Taurus Mountains. The mountainous areas of tectonic origin have been deeply incised by streams forming young valleys.

Since Dipsiz Lake and Dipsiz Lake Waterfall are located in the North Anatolian Fault Zone, seismic movements which occasionally happen also have important effects on shaping the area. In this zone, crushed and shredded rocks, hot and cold water resources, travertine formations, ponds, and young volcanic cones have come into being (Ketin 1969, 1–27). Hence, since the stream bed of the rivers merges into Yeşilırmak from the north, andesite and basalt rocks are often encountered in ponds around Gölçiftliği, Kuzgölcük, Güneygölcük, Cıbıl Hill, Tekeli Mountain, Gühertaş,

Dilekpınarı, Yıkılgan, Aşıçalı, Çal Mountain, Kabak Çalı and the Keşiş Mountains. Even the cavities on the top of the Aşıçalı Mountain, named Karasak and Karlık by local people, are typical volcanic shapes.

Fig. 22.3. A view of Dipsiz Valley over the travertine terrace

The data of the Doğanşar Meteorological Station (1987–1994) was used for the evaluation of climatic features. According to this, the average annual temperature lies around 7.5 °C, the lowest temperature in January (-3.9 °C), whereas the highest temperature in August is 16.5 °C. The annual temperature range is over 20 °C. Similar climatic characteristics are observed at about 400–500 m between the lake and waterfall in Doğanşar. The average annual precipitation is 575.1 mm, with 36.5% falling in spring, 12.6% in summer, 27.1% in autumn, and 23.8% in winter. The maximum precipitation occurs in May with 91.1 mm. The months of July and August are arid. Accordingly, in the area located in a transition region between the Central Anatolia and the Black Sea regions, a transitional climate predominates lying between continental and Black Sea climates.

The region shows great vegetation diversity, the precipitation over 500 mm allowing for dry forests, the vegetation of which has been widely destroyed for use as firewood and building. In the part where the lake and waterfall exist, black pine (Pinus nigra), scotch pine (Pinus sylvestris), hornbeam (Carpinus sp.), oak (Quercus sp.) and juniper (Juniperus sp.) constitute important plant species (Atalay&Efe 2010, Atalay&Efe 2012). In the areas where forest vegetation has been removed, astragalus (Astragalus sp.), mullein (Verbascum sp.) and speedwell (Veronica sp.) are dominant. At higher elevations, alpine meadows cover a large area. The land of the district is mostly high and mountainous, which has resulted in the expansion of meadow vegetation. This situation is quite important in the sense of livestock activities in particular.

Fig. 22.4. Views from Dipsiz Lake

Formation of the Lake and Waterfall

Dipsiz Lake is a typical karstic lake formed on the travertine structure of the quaternary period. In the area where the lake exists, the groundwater has carbon dioxide with calcium carbonate ($CaCO_3$) around it due to melting under pressure. The calcium carbonate ($CaCO_3$), which has melted in the water, constitutes the travertine structure by precipitating in the form of very thin layers on rocks as a result of the evaporation of carbon dioxide. A similar situation is observed in karst areas where the water rises to the surface and the depressed area (Atalay et al. 2008). The excess water of the lake maintains its flow over the travertine structure. But in time, the water of the lake is outcropped again by a sinkhole on its base about 200 m later by passing under the road to the Gökseki and Dikenli plateaus. Today, the dry valley between the lake and where the water again rises to the surface confirms this formation.

Fig. 22.5. Views from Dipsiz Lake Waterfall constituted by the water arising from Dipsiz Lake.

Though the lake was spread over a large area when it formed for the first time, the sinkhole on its base has emptied most of the water out. its surface area is about 400 m^2 and average depth is 16 m.

After the water of Dipsiz Lake rises to the surface, it passes under the road of Sivas-Doğanşar and forms the waterfall by falling over the eastern slope of the Dipsiz River 100 m away (Fig. 22.5). The karstic processes have affected the formation of the waterfall as in the lake. The travertine step is constituted by calcium carbonate ($CaCO_3$) sediments in the composition of the karstic source water, which pours out the water of Dipsiz Lake, precipitating in the part in which water falls into the Dipsiz River. Thus, the waterfall consists of the water pouring over the travertine step. The formation of the waterfall is similar to other waterfalls in Turkey.

The height of Dipsiz Lake Waterfall is about 46 m but changes in some parts. A 46 m part of the elevation difference of about 106 m between the lake surface and the valley bottom of Dipsiz River, where the waterfall pours, constitutes the waterfall. The western slope of the travertine step, which has accumulated for millions of years and which actually continues to accumulate in the part where the waterfall has formed, is very steep and its inclination is about 85°. In this section, the Dipsiz River valley is just like a canyon. The source water forming the waterfall spreads on a slope of about 30 m wide and merges with the Dipsiz River. Most of the water falls over two points in the central section in the form of drizzle in other sections. Especially in the summer months, the water falls only over two main points based on the decrease in the source of water flow. This water merges with the river in three steps in the central section of the fall surface. Of these, the first step is 16 m high, the second is 24 m high and the last, constituted by the water just before it merges with the river, is 6 m high. The water that falls rapidly hits the blocks below separated from the travertine step and merges with the river. Because of this, there is no pothole under the waterfall. The surface of the travertine step is quite smooth and covered with bushes, which are not longer than 1 m. The small caves under the waterfall also add beauty to the formation.

The Touristic Potential of Dipsiz Lake and Dipsiz Lake Waterfall

Doğanşar town, has important natural attractions. Among those mentioned are Karasak and Karlık Cavities on the top of the Aşı Çalı Mountain, the Karakaya Canyon on Tekeliiçi Rill, volcanic cones on the top of the Aşı Çalı, Çaldağı and Keşiş mountains, the caves and more than

30 plateaus on the Aşı Çalı Mountain. The rich forest vegetation around the lake and waterfall is important in particular regard to mountain tourism. Thus, those who come here also have the chance to see those alternative destinations.

There are no difficulties in terms of transportation. These two sites can be approached by the road of Sivas-Doğanşar, which has not been evaluated enough in the sense of tourism. Some physical arrangements have been made to meet the needs of visitors, the most important of which consists of the four fountains to get drinking water. There are a few picnic tables by the lake side which can be used for resting for picnics (Fig. 22.6). In spite of these little arrangements made around the lake, no arrangements are seen around the waterfall, which is just below the main road. It is highly risky to see the waterfall from below by walking down to the Dipsiz River valley, where the slopes are extremely steep and there is no pathway. Therefore, a difficult journey is required to see the waterfall closely, and it is therefore easier to come down over the Dipsiz River by following the valley.

Conclusions and Recommendations

The number of people who want to be far away from the depressing environment of cities and see relatively unspoiled natural places is gradually increasing. One of the important choices is nature tourism to provide these people a chance to get close to nature and to diversify the rural economy, to create a new field of employment for rural populations, and to reduce migration from rural areas to cities. Aquatic environments, which are one of the touristic values, should be evaluated within the scope of nature tourism in Turkey. In this connection, one of the wonders of nature which expects investment is Dipsiz Lake and Dipsiz Lake Waterfall in Doğanşar town, where the economic activities are limited due to geographical conditions, and the use of nature for tourism will contribute to the local economy. However, some arrangements must be made around the lake and waterfall, which are given below.

The first thing is advertising activities, because the media today is very effective . An effective advertisement in both local and national media will increase the interest of people. Putting a signboard in Doğanşar at the turn off of the Sivas-Erzincan road is very important in this connection.

Different environmental arrangements around the lake and waterfall are needed without harming the natural surroundings. Visitors have difficulties especially in getting to the lower parts of the waterfall. Going down the valley bottom from the top of the waterfall is possible only over the bare travertine rocks. Therefore, stairs must be built on the right side of the waterfall. Viewing areas must be prepared under the waterfall and suitable sitting places installed. A part of the stabilized road used to reach the lake from the main road must be covered with asphalt and improved. Besides this, a WC, a car park and baking stones must be built to meet the needs of the visitors, and the number of picnic tables must be raised.

Fig. 22.6. The fountains and picnic tables around Dipsizgöl for visitors

The rich flora of the area can be evaluated for tourist purposes, especially in terms of camping. The grasslands around the Dipsiz Lake and forest areas on the slopes have an important potential for camping, but as these areas are private it is difficult to invest in them in terms of tourism. The field can be publicized and the entrepreneurs provided with information about the building facilities. The need for eating and drinking can be met in some gardens or restaurants which should be built in this part. The oberj type of accommodation facility can also be built for those who want a holiday for a few days with lake and waterfall views and clean air.

Doğanşar has a rugged topography with a large high plateau areas, canyons and valleys, interesting volcanic formations, and caves among the leading landforms. These natural landscape elements can be offered in a range of integrated tourism to the visitors together with the lakes and waterfalls. The plateaus especially, which have gradually been losing their economic importance, can be rearranged with recreative purposes, as in most parts of the country.

In Doğanşar, where land conditions obstruct the basic economic activities, migration is inevitable, and nature-based tourism activities can be a possible alternative for local people. Migration can be decreased as those who live in rural areas around there can be pushed into tourism. For this purpose, first of all, the awareness of tourism of local residents must be constituted and an entrepreneurial spirit must be actuated.

References

Atalay, I. (1987). Introduction to Geomorphology of Turkey. Ege Univ. Publ. No:9, İzmir.

Atalay, I.; Efe, R.; Soykan, A., Cürebal, İ., Sarı, C. (2008). Formation of Antalya Travertine Deposits and Karstic Ground Water Systems. In: Efe, Cravins, Ozturk, Atalay (Eds). Environment and Culture in the Mediterranean Region. Part I, Chapter Six, pp. 93-108. Cambridge Scholars Publishing. ISBN (13): 9781847186584, 2008, Newcastle, UK.

Atalay, I.; Efe, R. (2010). Anadolu Karaçamı (Pinus nigra Arnold subsp. pallasiana (Lamb.) Holmboe) nın Ekolojisi ve Tohum Nakli Açsından Bölgelere Ayrılması. Ecology of the Anatolian Black Pine (Pinus nigra Arnold subsp. Pallasiana (Lamb.) Holmboe) and its Dividing into Regions in Terms of Seed Transfer. Çevre ve Orman Bakanlığı, Orman Ağaçları ve Tohumları Islah Araştırma Müdürlüğü. Bakanlık Yayınları no:424, Müdürlük Yayınları no:37. Ankara. ISBN 978-605-393-066-2.

Atalay, I.; Efe, R. (2012). Sarıçam (Pinus sylvestris var. sylvestris) Ormanlarının Ekolojisi ve Tohum Nakli Açsından Bölgelere Ayrılması. Ecology of Scots Pine ((Pinus sylvestris var. sylvestris) Forests and their Dividing into Regions in Terms of Seed transfer (Extended English Abstact). Orman ve Su işleri Bakanlığı, Orman Ağaçları ve Tohumları Islah Araştırma Müdürlüğü. Müdürlük Yayınları no:45. Ankara. ISBN 975-605-4610-11-2.

Doğanay, H. (2001). Turkey Tourism Geography. Çizgi Publishing House, Konya.

Kahraman, F. (2008). The Topography of Doğanşar District and Its Effects on the Lifestyle of People, Proceedings of Sivas Symposium During the Republican Period (27th-31st October, 2008), Sivas.

Özgüç, N. (1998). Tourism Geography. Çantay Publishing, İstanbul.

Selimoğlu, Ö. (2004). Ecotourism in the World and Turkey, İstanbul Chamber of Commerce Survey and Research Branch, Istanbul.

Yıldız, H. (1990). Geologic and Tectonic Characteristics of the North Anatolian Fault Zone Between Koyulhisar (Sivas)-Reşadiye (Tokat). Cumhuriyet University Graduate School of Science and Technology, Unpublished M.Sc.Thesis, Sivas.

Yılmaz, A. (1982-1983). Basic Geological Characteristics and Ophiolitic Melange Position of Tokat (Dumanlıdağ) and Sivas (Çeltekdağı) vicinities. MTA Rewiev, No:99-100, 1-18.

CHAPTER TWENTY-THREE

THE USE OF YÖRÜK (NOMAD) MIGRATION ROUTES IN THE WEST OF ANTALYA FOR TOURISM PURPOSES

CEMALI SARI AND HILMI DEMIRKAYA

Introduction

It is known that there are a large number of road systems belonging to different civilizations and various historical periods in southern Turkey around Antalya. Today, most of these roads are identified and registered. Therefore, they have also somehow been subject to culture tourism, although not at the desired level.

The routes of some roads in the study area are as follows:

- The road starting from Alanya in the east of the region and extending toward the west, heading from the site of Şarapsa Han to the north, going through Güzelbağ, passing over Alara River through Kemer Bridge near Gündoğmuş district and reaching the Bozkır district of Konya through the vicinity of the Söbüçimen plateau.
- The road starting from Side and extending through Şıhlar and Sırtköy, reaching Konya via Ürünlü.
- The road following Köprüçay valley to the west of the river via Oluk Bridge situated in the north of Beşkonak Village, reaching Selge (Altınkaya village) by merging with another route from the west, heading from here to the west and reaching the north of Perge via Hasgebe.
- The road starting from Perge and reaching the east of Lake Eğirdir and Yalvaç via Karabavlu village located in the north of the Aksu river valley.
- The road to Kepezüstü through the vicinity of Varsak village, located in the very northeast of Antalya and reaching Pisidia from here via the road to Döşeme Boğazı.

- The road from the southwest, Lycia, to Olbia (Hurma village) and reaching Antalya (Ercenk 1992, 2).

The last of these roads, named Lycia Road, has gained a great reputation in the last ten years in Turkey and the rest of the world. The Lycia road starts from Fethiye Ovacık and ends in Hisarçandır, and is 502 km long. Having been used by the local people from the Lycian period up to the end of the Ottoman Empire, this road lost its importance with the development of the highways in the Republican period. After being marked by Kate Clow and the volunteers under the sponsorship of the Garanti Bank and the Ministry of Culture in 1998–99, it was put into the service of hikers in 2000 (Ercenk 1992).

However, research regarding the movement of yoruks, who have been constituting an important part of the population of the research field recently and the subject of a great number of studies with their social, cultural and economic activities, between the summer pasture and the winter quarters namely migration routes, has not yet been done. The active lives of yoruks continued decreasingly in the field until 30 to 40 years ago, and today this lifestyle has ceased completely (Efe et al. 2012). While some of the roads the yoruks used during their seasonal migrations between the coast, the fall range and the plateau are also used actively today, while others are no longer used. Although there is a lot of information concerning the yoruk lifestyle making up a significant part of the cultural wealth regarding the recent history of the field, their migration routes between the summer and winter quarters have not yet been able to take an important place in literature. This study aims to map the yoruk migration routes in the field (on a 1:25,000 scale topographic map), share the acquired findings with the local and national tourism shareholders, and therefore fill the gaps in this area.

Material and Method

This research has been designed in accordance with the interview and document analysis of the qualitative research methods. A semi-structured interview form was prepared by the researchers. The researchers went to the villages and conducted face-to-face interviews with the individuals who took part in the yoruk migration in the past. While one of the researchers asked questions of the participants, the other recorded the statements of the participant. Afterwards, the acquired data was analyzed via text content analysis in the office. Moreover, the statements of the participants were supported through the literature review concerning the subject.

The migration routes and the areas of plateaus, winter quarters and fall ranges situated along these routes were marked on the 1:25,000 scale topographic map. The photos taken by the authors were also used to contribute to the visual quality of the research.

Western Taurus Plateaus

The Taurus Mountains, which provide ample opportunities in terms of transhumance in Turkey, belong to the south part of the mountain system of the Alps, and are spread over a large area. The part of this mountain range in the north of the Gulf of Antalya forms the Western Taurus Mountains. A large number of plateaus exist on different levels of elevations of the Western Taurus Mountains.

The Feslikan (1,850 m), Karçukuru (2,000 m), Morver (1,600 m), Saklıkent (1,800 m), Üçoluk (1,100 m), Söğütcuması (1,350 m), Ördübek (1,450 m), Av1an (1,000 m), Karagöl (1,400 m), Yazır (1,000 m), Beydağı (2,000 m), Göllü (1,200 m), Kartal (2,050 m), İmecik (1,900 m), Altıyaka (1,050 m), Çakmak (1,600 m), Çukuryayla (1,500 m), and Söbüce (2,050 m) plateaus are situated on this mountainous terrain. Plateaus are generally above 1,000 m here (Sarı 2010, 214).

It can be said that transhumance depends on two historical functions in the Western Taurus. The first of these is transhumance with the purpose of stockbreeding, which has been practiced for a long period, and the second is for resting and cooling down. High temperatures with accompanying high humidities experienced on the coastline in the summer months have made the Western Taurus rising behind the coast an ideal place for transhumance (Sarı 2013).

Today, the Western Taurus plateaus are the best-known and most desired ones by the residents of the coastal region, especially in the hot season. The researchers, doing a great number of researches concerning the plateaus in recent years, and the organization of many festivals in the plateaus by the local governments every year have enhanced the recognition of Western Taurus plateaus by the people. However, the ways of transportation to the plateau have not been dealt with with the same meticulous care, and the symbolized yoruk migration has been tried to animate the asphalt avenues and streets of the cities.

It is also important to state that not all the roads used for the transportation to the plateaus in the Western Taurus have cultural and touristic value. Nevertheless, the identification and registration of migration routes can be important for the study of a culture and its transference to the next generations. Thus, by doing a detailed study of the field, identifying and gathering the necessary inventories, the question of which of these may be used in touristic product diversification can be understood more easily. Consequently, yoruk migration routes will also have been included in the road systems of Antalya.

Summer Pastures and Winter Quarters of Western Taurus Yoruks

Today, the Taurus mountains, especially the Western Taurus, spring to mind when "the yoruk" is mentioned, because the plains in the south of the Western Taurus and the plateaus with an elevation starting from 1,000 m and going up to 3,000 m are the places yoruks most desire.

Although the coastal areas in the Antalya region provide great opportunities for pasture until May, these pastures begin to dry up at the beginning of the summer. However, the grass of high plateaus starts to grow at the same time. The reverse of this process takes place in the autumn. While winter comes to the plateaus, the weather on the coast gets milder. The yoruk have based production on this dilemma of nature. On the other hand, this phenomenon means a two-phase life in terms of both time and place—"winter quarter" and "summer pasture." Therefore, the yoruk has to move in certain periods of the year, called Yoruk migration (Efe 1998; Kılınç 2010, 23).

The yoruks used to spend the period from November until the beginning of May in the winter quarter. The winter quarters are generally the ample plains of the coastline, particularly Finike, Kumluca, Kemer and the vicinity. These plains are the places between the valleys and the nooks with a temperate climate. They were not agricultural lands but large meadows, most of which were rented from villagers. The main feature required for the winter quarter was the grass, security and ease of heating.

As for the principle thing for the yoruks, it was the plateau. They used to set off for the plateaus of Kızılcadağ and Beydağı at the beginning of May every year, and stay there until the middle of September. The continuity of the subsistence used to depend on the fruitfulness of the plateau. For the yoruk, the plateau meant the mountain weather, immense pasture, security, freedom, abundance and cold water.

The yoruks used to go down to the lower parts, the fall ranges (Avlan, Göltarla, Söğütcuması, Altınyaka), when the plateau started to get colder, staying for about one month. At the beginning of November they used to set off for the winter quater (Mavikent, Beykonak, Çavuşköy, Çaltı, Çayiçi, Yenikışla, Güzören, Belen and Yazır).

Yoruk Migration in the Western Taurus

The yoruks are nomadic people who make their living from animals alone and the products derived from them. The basis of the livelihoods of Yoruks is stockbreeding. Their lives are generally spent dealing with stockbreeding in the summer pastures and the winter quarters. In fact, they have to go to the plateaus when the grass dries up on the coasts in the summer and go down to the green plains of the coasts when it snows in the winter so that they can feed the animals (Çimrin 2008, 2012).

The shortest migration route between the summer pasture and the winter quarter takes a period of one week. This period can also be fifteen days and one month. The yoruk migration involves glory, colourfulness and liveliness (Seyirci 2000, 6).

According to the interviews conducted in the research field, the yoruks have a certain route to their winter quarters, summer pastures and fall ranges. No migration is done with a different route. Having spent the winter on the coast, the yoruks start to prepare for migration at the beginning of May. The migration date is announced by the lord three or four days in advance. The migrations are generally done on Thursday. On the migration day, they hang the bells on the flocks of sheep and goats and set off at the break of dawn. After sunrise, the camels are loaded and embroidered rugs are laid on them. Big bells are hung on the camel saddles, and small bells are hung on their bridles, and they are bound one after the other. A donkey goes in front, and five or six camels constitute a convoy. The new bride or the young daughter of the family pulls the convoys of camels, and the nomad lord does not go beyond the young bride, and is respectful to the women. As for the head of the family, he stands near the migration convoy on his horse. If the bride has another thing to do sometime, the old men do it. However, they do this by getting on the donkey. So, from that moment, the migration has started. When different convoys line up one after the other, a magnificent yoruk migration comes out. The most glorious migration is the plateau migration. The young men liven up the migration by riding horses and shooting. It is an exact splendor for the yoruk, and is the product of rich experience and culture.

The animals and people walk unceasingly until they arrive at the place they will stay for the night. When they get to the accommodation destination during the migration, everyone settles in their own habitation. Only the camels take a break of ten minutes in this journey. The places that are used to spend the night have been determined thanks to the experiences gained throughout centuries. The loads of the camels are untied during the stay and the tents are not set up, except for in rainy weather (they do not set off until it stops raining, either). Water is certainly present in the accommodation place. Even if there are no streams, cisterns exist there. This journey lasts at least a month. They never cook during the journey, eating bread, cheese and cottage cheese.

In this way, they arrive at the plateau by stopping over, and almost the same route is followed on the way back to the winter quarter. While the accommodation places are within short distances in the summer migration, the accommodation areas are at longer distances during the fall migration, namely the migration from the summer pasture to the winter quarter. Shopping is done in the bazaars and villages on the way.

The practice of migrating and settling among the yoruks has changed a lot in the last forty years. In recent years, concrete buildings have been raised on the plateaus by keeping up with the changes. The number of the black tents has decreased to such an extent that it is almost nonexistent. Private cars have taken the place of horses. Tractors and trucks that complete the migration in one day have substituted the camels carrying the households. The majority of the yoruks in the Antalya region that consisted of up to tens of thousands of people until the 1960s has started to adopt a semi-migrant settler life by building houses close to the coastal villages in recent years. Every year, a number of nomad groups have settled permanently in villages and towns (Zaman 2007).

In recent years, yoruk festivals have been organized by the local governments and non-governmental organizations so that the yoruk traditions and customs are not forgotten.

The Yoruk Migration Routes of The Western Taurus Plateaus

It is known that there were many yoruk tribes moving between the coastline and the mountainous area in the west of Antalya until fifty or sixty years ago. Almost all of these tribes adopted a sedentary life, and the settlements of most of these people are known. It is also known that a large number of yoruk tribes exist in the field thanks to the publications on Antalya yoruks, and the observations and researches done in many cities, towns and villages in the Western Taurus. The most important for these tribes as big groups are as follows.

While the majority of them live in the regions of Serik, Manavgat and Döşemealtı, there is also a small Karakoyunlu tribe that spends the winter in the west of Antalya. This tribe spends the winters in the Mavikent (Yenice) village of the Kumluca district while spending the summers in the Yapraklı (Göğü), Salur and Küçük Söğle villages of the Elmalı district and the Çıyırcık plateau of the Geçmen village, and some of them go up to the Kofu mountain (see Fig. 23.1 below).

A group smaller than the Karakoyunlu tribe go to the Belen plateau of the Kemer district. As for their winter quarter, it is in the vicinity of the Çamyuva (Ağva) village (Seyirci 2000, 34).

Töngüşlü yoruks, who spend the summer on the Kızılcadağ (Patlıca) plateau of the Elmalı district, have settled in the Mavikent Karaöz quarter and the Çavuşköy (Adrasan) Gökçealan quarter of the Kumluca district (Seyirci 2000, 24). Having spent the summer in Kızılcadağ, Töngüşlü yoruks go down to the plain in the vicinity of Elmalı Yalnızdam after August 15. They return to the winter quarter after October 15 (Seyirci 2000, 24, Fig. 23.1 below).

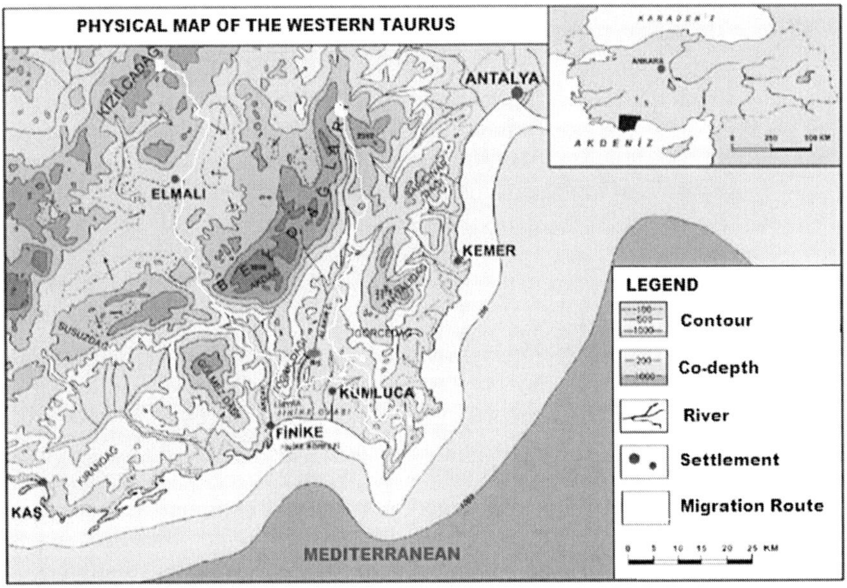

Fig. 23.1. Physical map of the Western Taurus and its surroundings

Karahacılı yoruks stay in Çavuşköy (Adrasan) in the winter, and go up to the Hacıyusuflar plateau of the Elmalı district in the spring.

Yeniosmanlı yoruks spend the winter in Çavuşköy, and go to Elmalı Dereköy (Ortaköy) in the spring.

Fig. 23.2. The Yoruk migration routes from Gölcük to Kuzca villages

The yoruks in Kumluca and Kemer used to go up to the site of Karaçukuru between the Bey Mountains (Kılınç 2010).

All of these clans in the vicinity of Elmalı come to the vicinity of Karagöl after harvest time so that they can graze their animals and sell their sheep and goats (Ekiz 2001).

Another part of these yoruk tribes who spend the winter in the Güzören and Beykonak villages of the Kumluca district set off for the north throughout the Alakır valley in the spring and go to Altınkaya (to Güzle, which is one of its quarters now). Then, they reach the site called Ayı mağarı near the Söğütcuması plateau of Kuzca village. As for the summer, they spend it on the Beydağı plateau.

The Use of Yoruk Migration Routes in The Western Taurus for Tourism Purposes

The Western Taurus constitute a potential for the implementation of sustainable tourism plans just next to the South Antalya Tourism Development Area, and for the formation of many alternative types of tourism with its unspoilt and clean natural structure that has not been affected by the unplanned tourist developments, one of which are the Yoruk migration routes. There are a large number of plateaus and migration routes used for the transportation to these plateaus in the field. The route starting from Kumluca and extending toward the Beydağ plateau through the natural and cultural beauties of the Alakır valley is the most suitable (Fig. 23.1). As for the Kumluca, Finike and Elmalı Kızılcadağ Yoruk migration routes, they can be organized for bicycle tours with its length of approximately 150 km, because this route runs parallel with the modern highways, to a large extent.

The natural vegetation, the Alakır valley and the Alakır stream constitute the natural touristic resources of the Kumluca Beydağı yoruk migration route.

Alakır valley extends between the bifurcated Bey Mountains, and the Beydağı plateau constitutes the head of the valley. The valley opens into the Gulf of Finike by getting larger. Kumluca is located in this area.

The fact that the Alakır valley extends in a north-south direction has paved the way for the effects of the Mediterranean climate to reach the north, and due to this a great variety of tasty fruits are grown in the villages of Karacaören, Karaağaç, Altınkaya, Çaltı, Gölcük, Kuzca and Büyükalan due.

The rich vegetation has an important place in the power of the valley as a tourist attraction. The plant geography is quite generous to the interested people in terms of the richness and the variety of plants. The natural vegetation consists of perennial plants due to the convenience of the topography and the elevation.

Out of the perennial plants, the Mediterranean maquis community is usually found at altitudes up to 600 m above sea level. The most prevalent members of the maquis community are the cermes oak (*Quercus coccifera*), holm oak (*Quercus ilex*), mastic (*Pistacia lentiscus*), bay laurel (*Laurus nobilis*), carob (*Ceratonia siliqua*), phllyrea (*Phillyrea media*) and sandalwood (*Arbutus andrachne*) (Efe et al. 2008; Atalay et al. 2014).

Among the perennial plants, forest trees and shrubs form communities according to the altitudes.

The Turkish pine (*Pinus brutia*), which is the most important forest tree, occurs at altitudes up to 1,200 m, occuring in the site of the Akdağ Kozağacı, Çatlı, Sarıcasu, Altınyaka, Gölcük, Kuzca and Büyükalan villages.

The cedar (*Cedrus libani*) and Turkish pine constitute the basis at altitudes between 1,000 m and 1,400 m, some of which have a diameter at a breast height of 180 cm and are 1,000 years old. The black pine also has a widespread occurrence at an altitude of 1,400 m (Büyükalan). The black pine and cedar can be found dispersed among juniper forests (in the site of Aladağ of Kuzca village) at altitudes as high as 1,100 m in some cases. Cedar extends widely across the Akdağ Dibek forests in the villages of Kuzca and Karacaören in the area (Atalay&Efe 2010, 2011).

The fir and the juniper are present at elevations between 1,400–1,700 m. Firs, black pines, cedars and junipers occur together. Likewise, junipers are present along with black pines (in Gölcük village, Çayır site, Büyükalan, Kuzca village Akdağ site), and its most prevalent type is found at elevations above 1,500 m. Three of its most widespread types are Juniperus feotidissima, Juniperus oxycedrus and Juniperus exelca.

Furthermore, trees such as willow, poplar, seabuckthorn, wild pear, elm, maple and plane tree are commonly encountered in the Alakırçayı valley and lower lands.

The Alakır valley is quite rich in terms of the variety of wildlife as well as natural vegetation, mainly fallow deer, wild goat, fox, wolf, bear and the rarely encountered wild boar.

The Alakır stream, which has given its name to the valley, also has an important part in the attractiveness of the valley as well as the natural vegetation and the wildlife. The stream originates in the Bey mountains in the north of Kumluca and forms a basin through the creeks in the high mountain range surrounding its north, west and south sides, reaches the Mediterranean Sea covering a distance of 80 km through the big valley between the coastal mountain range and the Asıl Bey mountains in the west. It starts from the south of Dereköy in the North, passes by the villages of Büyükalan, Kuzca, Çatlı, Karacaören, Çayiçi and Kavak in a north-south direction and flows into the Mediterranean Sea. The basin of the Alakır stream, which is not a very long river, turns into a quite powerful river especially after the rains in the spring months, and changes into a simple stream in the summer months.

As for Akdağ and Katrandağı, they are extremely appropriate for trekking with their fresh air and unspoilt nature. The cedar forests are quite interesting, and visitors can see the flora and fauna in their own habitats.

Regarding the Kumluca Beydağı yoruk migration route, the villages and ancient cities situated along the route constitute the cultural touristic resources of this route.

The field has been an irreplaceable location for a great variety of human activities throughout the years. Besides the rich natural features, its cultural features are also a part of its important tourist attractions. The villages provide rich opportunities for those who want to meet and integrate with the local culture. The traditions and customs of the people can be seen by staying in rural settlements, and the regional cuisine can be tried.

The ancient cities of Rhodapolis, Idebessos, Akalissos and Kitanaura, situated among the pine forests along the Alakır valley, shed light on the history of the beautiful geography of this area. The presence of these historical cities shows that the field has been inhabited since ancient times. Both the cultural values in which the past and the present are integrated and the magnificent scenery resulting from the rich natural vegetation that consists of Turkish pine, cedar, fir, juniper and oak species, turning the Kumluca Beydağı yoruk migration route into a potential tourism area.

Fig. 23.3. Beydağı plateau

Agriculture practiced without chemicals and artificial fertilizers, sheep and goat farming, cattle farming, blacksmithing (especially knives, the fame of which do not extend beyond Antalya, although they are liked by a lot by people), gourd violin making (the fixed musical instrument of the weddings of the Teke region which is made in Kuzca and played by most of the villagers), and wood embroidery (wooden spoons and various ornaments) constitute great potential for tourism (Sarı 2007, 284).

Conclusion and Suggestions

It is known that a great number of road systems belonging to diverse civilizations and a variety of historical periods exist in Antalya. One of the most important of these road systems are Yoruk migration routes, since they reflect the recent period culture of the field.

In recent years, the many researches on the plateaus and the organization of festivals by the local government every year have enabled the Western Taurus to be more widely known. However, the ways of transportation to the plateau have not been handled with the same attention, and the symbolic yoruk migration has occurred on the asphalt avenues and streets of the cities.

In the Western Taurus, the Kumluca-Finike-Elmalı-Kızılcadağ and Kumluca Beydağı Yoruk migration routes can be introduced into Antalya's tourism through convenient planning and carrying out the necessary works. These migration routes are many kilometres long, passing by villages, towns and cities while going from winter quarters to fall ranges and summer pastures. It is possible for these routes to be organized as trekking routes for day-tours, several day trips and even weekly trips and ten-day trips in accordance with the determined objective. As a matter of fact, it is virtually impossible to determine an unlimited route in the field due to the construction of new highways and their growing prevalence over time, along with the construction of forest roads, enlargement of the boundaries of the settlements and the formation of new ones. Therefore, some parts of the Kumluca Beydağı Yoruk migration route, which can be identified, can be used for weekend trekking.

The Yoruk migration routes have come out as the cultural experience of a practical, secure, systematic and rapid life. The apparent difference of these roads from the others is that there are no human structures_on it. However, it should also be kept in mind that the ancient cities in the field create an integrity with their immediate surroundings (such as the Kumluca-Finike-Elmalı-Kızılcadağ and Kumluca BeydağıYoruk migration routes). As for the situation that makes the Kumluca Beydağı Yoruk migration route attractive, it is its located in the mountainous field in which the human effect is still at a minimum today. Thus, the natural scenery along the route will be a magnificent vision for the visitors to these routes.

The tourists trekking and cycling can cycle with rawhide sandals on the Western Taurus Yoruk migration routes, and cycling tourism can also be introduced into the migration routes of the field.

The researches on cycle tracks should be given importance by the local governments and highway officials so that the convenient natural and cultural geographic features of the research field can be introduced into mountain and road cycling.

In the case that the infrastructure of this sport is completed in a short period, the Western Taurus can be an alternative to the prominent countries of the world for mountain and road cycling with its geographical features.

References

Atalay, I.; Efe, R. (2010). Anadolu Karaçamı (Pinus nigra Arnold subsp. pallasiana (Lamb.) Holmboe) nın Ekolojisi ve Tohum Nakli Açsından Bölgelere Ayrılması. Ecology of the Anatolian Black Pine (Pinus nigra Arnold subsp. Pallasiana (Lamb.) Holmboe) and its Dividing into Regions in Terms of Seed Transfer. Çevre ve Orman Bakanlığı, Orman Ağaçları ve Tohumları Islah Araştırma Müdürlüğü. Bakanlık Yayınları no:424, Müdürlük Yayınları no:37. Ankara. ISBN 978-605-393-066-2.

—. (2011). Ecological Attributes and Distribution of Anatolian Black Pine [*Pinus nigra* Arnold. subsp. pallasiana Lamb.Holmboe)] in Turkey. Journal of Environmental Biology 31, 61-70.

Atalay, I., Efe, R.; Öztürk, M. (2014). Effects of Topography and Climate on the Ecology of Taurus Mountains in the Mediterranean Region of Turkey. Elsevier, Procedia - Social and Behavioral Sciences. Volume 120, Pages 142–156

Çimrin, H. (2008). Antalya'da Zaman: Antalya Yörükleri (5), Antalya BuGün Gazetesi, 23.08.2008.

—. (2012). Antalya Yörükleri. http://www.antalyabugun.com/?page=makale&MID=2534 (10.05.1213 tarihinde erişilmiştir).

Efe, R. (1998) Ermenek Çayı Havzası; Doğal Ortam Özellikleri. FA.Ü. yayınları no 1. İstanbul. pp. 210, ISBN 975-303-001-0

Efe, R.; Atalay, I.; Soykan, A. (2008). Mediterranean Ecosystems of Turkey: Ecology of Taurus Mountains. In: Efe, Cravins, Ozturk, Atalay (Eds). Environment and Culture in the Mediterranean Region. Part I, Chapter One p. 3-37, Cambridge Scholars Publishing. ISBN (13): 9781847186584, 2008, Newcastle, UK.

Efe, R., Soykan, A., Cürebal, İ., Sönmez, S. (2012). Madra Dağı'nın Kuzeybatısında Yaylacılık Faaliyetleri. Kazdağı ve Madra Dağı Belediyeler Birliği, Kaz Dağları Ulusal Çalıştayı. 12-14 Ekim 2012, Ayvalık-Balıkesir, Bildiriler Kitabı s. 107-124.

Ekiz, A. (2001). Dünden Bugüne Elmalı. Antalya: Akdeniz Yayınları, Damla Ofset Matbaası.

Ercenk, G. (1992). "Pamfilya Bölgesi ve Çevresi Eski Yol Sistemi, (Belleten, C.),Cilt: LVI S.:216, Antalya.

Kılınç, N. (2010). Antalya Yöresi Yörük Fıkraları Üzerine Bir Araştırma. Süleyman Demirel Üniversitesi Sosyal Bilimler Enstitüsü, Yayınlanmamış Yüksek Lisans Tezi. Isparta.

Sarı, C. (2007), Alternative Tourism and Touristic Product Variation Strategies: TheAlakır (Kumluca) Valley, GeoMed 2007 International Symposium on Geography, Pp. 277-285 June 5-8, 2007, Kemer, Antalya, Turkey.

—. (2010). Türk Yayla ve Yaylacılık Kültürünün Batı Toroslardaki (Beydağları) İzleri, IV. Uluslararası Türk Kültürü ile Sanatları Kongresi/Sanat Etkinlikleri (IV. International Turkish Culture and Arts Congress/Art Activity), s.211-220, 2-7 Kasım 2009, Kahire/Sharm El Sheikh, Mısır.

—. (2013). Batı Toroslarda Yaylaların Fonksiyonel Değişimi Süreci ve Yayla Şenlikleri, Marmara Coğrafya Dergisi, S.: 27, s.: 242-261, İstanbul.
Seyirci, M. (2000). Batı Akdeniz Bölgesi Yörükleri. İstanbul: Der Yayınları.
Zaman, S. (2007). Antalya Beydağları Yaylaları. Erzurum: Eser Ofset Matbaa.

CHAPTER TWENTY-FOUR

AN ESSAY ON VILLAGE WOMEN'S MARKETS IN THE ALAPLI (ZONGULDAK) DISTRICT CENTRE AND THE ROLES OF PEDDLER (STALLHOLDER) WOMEN DURING THE VILLAGE MARKET PROCESS

DÖNDÜ ÜÇEÇAM KARAGEL
AND HULUSI KARAGEL

Introduction

The weekly bazaars set up in many towns and cities in Turkey are public spaces in which the small scale female producers directly involved in agricultural production can continue this activity permitted by their traditions. Women are small scale producers who bring the surplus products which they produce for the consumption of the family, like vegetables, fruit, milk, yoghurt, cheese, dry legumes and dough products, to the nearest local bazaar. This way, the stallholder women earn an income which they can use to purchase items like food that they can't produce themselves, as well as other things they need, and thusly partake in a sort of barter (Aliefendioğlu 2000, 87). Going to the bazaar is a part of the culture and traditional life of our country. The bazaars set up in the urban and rural areas in different times of the week are reflections of cultural-spatial differences. Bazaars allow for the marketing of the surplus products produced in rural areas and create an environment where exchanges and social mingling can take place (Özgüç & Mitchell 2000, 35). The peddling of the women in the study field begins with them putting up the agricultural produce they have grown in the local bazaars, which are set up in nearby city centres and towns, and continues in a way that has not greatly changed over the years. Women can start this activity at almost any age and they regard it as the easiest manner of earning an income. Because of this, there are women who started peddling after the age of fifty because of the economic situation of their families and they feel regretful that they have not tried this before. Peddling allows women to do some of the tasks they accomplish at home, such as cooking and taking care of children, while they are working. Husbands and housemates such as sisters-in-law and mothers-in-law also share in the workload.

Women in the study field have gained their economic independence and thus the possibility to be self sufficient, largely in the bazaars. The bazaars have become functional places where peddling women can leave the rural area, make several social connections and make use of their surplus produce. However, social and family structures have started to change in recent years due to urbanization, and the advancements in information and communication technologies have caused changes in the approach regarding bazaars. In the study field, the peddling activities of the women villagers are traditions mostly passed on from mother-in-law to daughter-in-law, and from mother to daughter. Even though this traditional economic structure is slowly losing its socio-cultural effect, it still resists the newly developing market and supermarket chains. In this respect, the roles of the women villagers carry a great importance in keeping these activities sustainable. Therefore, the purpose of this study is to examine the roles of these women in the private and public fields during production and reproduction processes and to study what these roles help them gain or lose.

Economic instability which has made it difficult for families to earn a living is also effective in leading women in small cities, as well as large cities and metropolises, to become economic providers. Bazaars in the study field provide the closest and easiest readymade opportunity and established order for women to both keep a tradition alive and bring in extra income. The role of women producing their own products is strong in their active roles in the agricultural produce trade. All of the women peddlers personally sell the surplus of what they produce in their gardens themselves. Aliefendioğlu's (2000) statements about Akçakoca show that a collective approval for women's peddling in bazaars is also apparent in Alaplı.

Women are not considered to be economically active for what they produce for the household. However, when women systematically and consistently bring their produce to the local bazaars and sell them, they are considered to be "productive" (O'Connel 1994). Only a small portion of what women do in their active participation in the agricultural production process is considered to be actual work, and most of the other tasks are accepted as part of the women's traditional roles in society, and as activities that must be done. The main signs of modernization are urbanization, increase in literacy, increase in non-agricultural activities and mass media products gaining popularity. One of the most important signs of modernization in the agricultural field is production aimed at bazaars and the connection formed with cities. Modernization in women's lives has caused conflicting results, economically and culturally. Rural transformation in Turkey has led women to work in agriculture and men to work in fields other than agriculture and has increased the workload of women in labour-heavy tasks, while men's public authority has expanded (Ertürk 1996). In Middle Eastern countries, women have taken men's place as the head of the house with the seasonal or continuous migration of men to work in the cities. The migration of men to work in non-agricultural fields has caused women to take over the agricultural work, thus "feminizing the agricultural field." This situation has empowered women and increased their social standing at home and therefore in society (Aliefendioğlu 2000; Gülçubuk & Yasan 2010).The fact that most women prefer to do things without having to ask their husbands' permission shows that the majority of the working women care about their independence (Bolak 1993). In this respect, peddling women consider themselves to be more social, cooperative, active and independent than the women who live in the village and cannot connect with the city. The women say that selling their produce in the bazaars has increased their role in the family, as well as making them feel better compared to sitting at home and expecting money from their husbands.

Topçuoğlu tried to emphasize the point that it is not necessarily a good thing that men go outside the village, province or country in search of an opportunity to save money and increase their social statuses by leaving their wives behind in villages to be the household heads. It is obvious that the man is the head of the house both in traditional and legal terms. Therefore, the cases where women become household heads do not represent good situations, and should not be interpreted as a boost to the woman's social stature because it means there is no husband, older male relative or grown-up son. We cannot ignore the benefits the man provides to his family and wife in our real societal life and deny that the woman becoming the head of the household is a reflection of a certain misfortune (Topçuoğlu 1978).

Studies have shown that gender mainstreaming closely affects the distribution of work. The assumptions and preconceptions about gender roles are important when deciding whether a job is a woman's or a man's. Once gender mainstreaming differentiation crystallizes within a job or career, tasks are classified and defined according to genders and a strong normative pressure is put on the market (Ecevit 1998). All societies have observed the physiological differences between men and women, splitting the society roughly in two and assigning different roles and statures to the groups. The rights, tasks and values attached to these groups change greatly from society to society and men and women are sometimes seen superior to their counterpart. Parallel to this, the tasks of men and women in society have been assigned differently, and men's work has generally been identified as tasks that require greater physical strength. The role of the patriarch for both the families and religions has mostly been assigned to men, and it is considered to be natural for women to be under men's patronage. However, we observe these traits at a higher ratio in societies where simple technology is prevalent and in agricultural societies. In the industrialized Western world where technology has developed and the urbanization process has been completed, the statutes of women and men and the roles, rights and tasks attached to them have changed greatly, and many tasks that were considered to be men's work have been undertaken by women, and vice versa (Saran 1979). Even though women working and bringing in a certain income does not always affect their monetary independence and their roles in the home, the traditional grounds for men's authority are being questioned ideologically, and the responsibilities of men are being defined in a way that encompasses roles within the household (Bolak 1993). In the economic climate of today it is not always possible for a man to provide for the family on his own. On the other hand, women generally take part in every stage of production in societal life. To speak of the absolute leadership of the man in a family where the woman works would be ignoring the conditions of the age and the place of women in society. Women partake in the family income because men cannot earn enough to carry the entire economic weight of the family. Because of this, the husband and wife both have to support the family economically. Needs and living conditions are effective in the management of family. If a woman works she becomes economically independent and contributes to the family income, and the man's superiority in this respect will lessen in the same ratio because he is not superior to his wife as a human being (Fazlur 1987).

Social status is the degree an individual holds in the social hierarchy determined by the place of a person in the social group he or she inhabits, decided in relation to the places of the other members of the group. It is an evaluative degree assigned to an individual from outside sources, and this degree is not formed randomly as there are certain measures and criteria which decide social status in each society. As a result, each person has a place in society, certain roles are expected from them and certain opportunities and degrees of respect are provided according to this place

(Topçuoğlu 1978). The remarks of the peddler women, such as "men can't do this work, they don't know how to do this, we always do these thing, this is our job," are clear indicators of the roles divided as a result of gender discrimination between men and women.

The free labour provided by women as an extension of their biological roles, which is immeasurable in worth, is important in terms of the roles of men and women in society, the work division in the home and economical indicators. This free labour has a balance shifting, critical importance in women being represented in the decision making processes and mechanisms in the family, the society and on regional and global levels. Patriarchal decision-making mechanisms have contributed to men working fewer hours by opening up agricultural developments to men and ignoring women, while women's working hours have remained the same or have increased. Whether or not a woman is employed for money, she spends more time than the man working on tasks for the bazaar that are not valued (Aliefendioğlu 2000). Work divisions based on gender mainstreaming have been present over generations as the oldest forms of work division known to all cultures and have been one of the cultural elements least affected by cultural change, and are regarded as something that is almost accepted as natural. Work division based on gender mainstreaming began with dividing work in the family and went on to become a network that has encompassed the entire society. The signs that women are seen as secondary can be seen everywhere in this work division (Moser 1989, 1800). In gender mainstreaming there are two basic, related dimensions of the tasks women perform. Men and boys do not perform these kinds of tasks, and these are jobs with lower statuses. Though it varies from culture to culture, these are generally tasks done at or around the home or for which there is no income or value. All of these tasks are generally considered to be women's responsibilities. Girls are usually encouraged to undertake these tasks, while boys are generally kept away from this kind of work so that children accept their societal roles (Aliefendioğlu 2000).

Study Area

The study was conducted in the Alaplı district centre in Zonguldak (Fig. 24.1) in Wednesday and Sunday bazaars and in the organic produce market (Fig. 24.2), set up by the Alaplı municipality and which is open four days a week. The aim of the study is to examine the roles the working village women in the marketing process and how these roles affect them socio-economically.

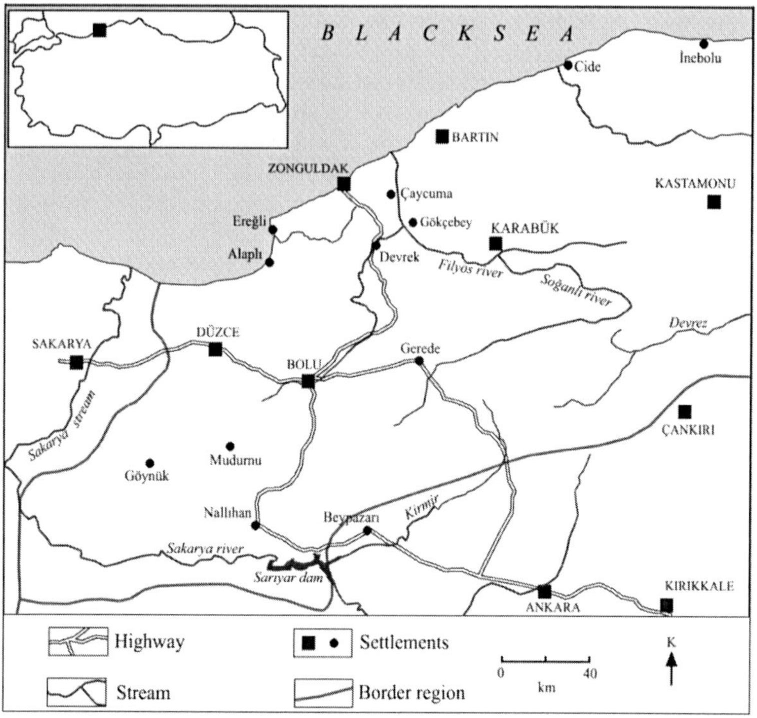

Fig. 24.1. Location of study area

The largest market space in the district centre is the covered market place which belongs to the municipality, set up once a week and opened for service in 2007. The market place has a quite organized and clean visage, is covered around its perimeter and has booths where the peddling tradesmen and the village women have their own stalls (Fig. 24.3). The peddling village women generally sell a large amount of their fresh produce in the stalls they set up by roadsides in the places called "site" or "neighborhood bazaars" in the early morning (Fig 24.3), before moving to the actual market place around nine or ten o'clock at the latest. The women spend the rest of the day at the municipality bazaar until the rest of their produce is sold. On Sundays the market place in the 690 Houses Neighborhood, where a smaller number of peddlers are present, is used. The peddlers are also arranged along the street in the neighborhood here (Fig. 24.2). The peddling tradesmen and the village women have a mixed distribution here.

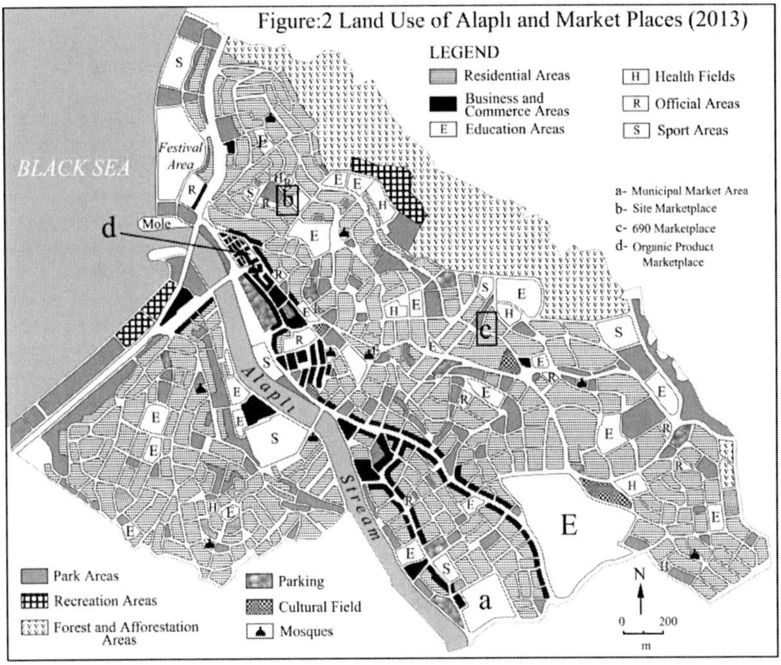

Fig. 24.2. Land use of Alaplı and market places (2013)

The organic produce market opened by the Alaplı Municipality in 2012 provides services four days a week on Mondays, Wednesdays, Fridays and Saturdays. Until 2007, this space was used by the village women on weekly bazaar days. Since the bazaar was set up by the municipality in the city centre on Wednesdays, on these days the organic produce market is only used to sell the dry and moist pastries and the handmade lace and embroidery products that the housewives make. The marketplace is covered for protection against rain and heat and is designed to be a very organized selling place with stalls and seats (Fig. 24.2). Due to the capacity of the market place, only 35–40 peddlers can be active at a time. However, 214 women entrepreneurs have applied to use the marketplace. Most of the applicants are villagers who peddle, mostly from Alaplı and some from Ereğli (Alaplı Municipality 2012). Only those registered with the police are allowed to be active in the market place, which is open four days a week. While the number of peddlers was around 40 when the marketplace first opened, the number has dropped seriously over time because the free conditions expected from the area did not materialize. Sometimes, the number of peddlers drops to as low as 3–5 peddlers.

Fig. 24.3. Images of village women in the bazaar in the Alaplı city centre

The market places set up in the study field are not only market places where organic produce is sold, but also a space where peddling tradesmen and village women can make sales together in the "Municipality Covered Market Place" on Wednesdays, the area that is set aside for them. This space is about one fourth of the area which peddling tradesmen use. While the village women's produce consists of the fruit and vegetables they produce in their own gardens, the peddling tradesmen sell clothes, jewelry, glassware, shoes, dry legumes and several varieties of ready foods as well as vegetables and fruit. In the "site bazaars" set up between the streets early on Wednesday mornings, the number of peddling tradesmen is usually not more than 5. The activity here is completely made up by the sales and interest towards the village women's fresh fruits and vegetables, and 100 village women were counted here at times where the bazaar was very crowded. The fact that the oldest woman in the bazaar says that she remembers both her grandmothers making sales in this market place tells us that the market place has been used for a long time. The reason that this market place has not lost its importance over this long a time is that it provides an environment where people who live nearby can buy produce easily. Also, the transfer of the Municipality Bazaar to outside the Alaplı city centre has increased the importance of this market place.

In agricultural societies where technology is not very advanced, women are usually considered as the group who do certain things in the fields, make the foodstuff ready for consumption and somehow use what is left over once the needs of the family have been fulfilled. A large part of these tasks can be carried out next to the home or in a garden or field close by (Saran 1979). The production fields of the peddling women in the study field are also very close to their homes (Saran 1979). Also, most of the women who come to the market places in the Alaplı city centre live in nearby villages. The distance between the city centre and the villages who take part in peddling are 10 km at most, and most of the peddlers come from villages which are 3–5 km away.

Material and Method

While the dates aren't known for certain, bazaars in the study field have been set up by village women for a long time. These market places were places where women could bring their surplus produce and turn this into profit. The trade activities of peddling village women are known and observed not only in the study field but also in many parts of Turkey, however the Black Sea Region is especially known for activities in this field. What Trak said about peddling in Turkey is very important as well:

> Peddling is the most widespread activity in Turkey. It has spread in villages, towns and cities and reaches the whole country and allows exchanges in bazaars between producer villages with little population with consumer towns and cities. Because of this, the bazaar dates which are known to almost everyone, are the place where large movement of population is observed between the villages and cities, and these people enter into a heated commerce. The factor which plays the most important part in this is procuring mutual needs and provision of an income.
>
> (Trak 1950, 110–111)

Using fieldwork in the study field, this study aimed to show how and where the peddling village women produce their products, and how they make a program for their time and tasks. Basic data were collected using the results of surveys, interviews and in-depth interviews conducted on 85 peddling villager women in the July to August periods of 2010, 2011 and 2012 in the four market places set up in the Alaplı city centre. The women were not only monitored in the market places but were also visited in their villages to see the production phases of the produce sold in the bazaar first hand. This way, the activities from production to marketing and the roles undertaken in this process were made more concrete with observations.

There were problems from time to time during one-on-one interviews with the women and collecting enough data. However, the fact that the researcher was a woman created a sense of trust and the problems were minimized. Also, being customers at the bazaar at times outside of the research increased the chances of our being known by the village women which made it easier to communicate. Very personal information such as names and surnames and even telephone numbers to communicate were obtained in the interviews. The purpose of asking for names in the questionnaires was to keep the same people from answering more than once, and also helped us to track down those who had and hadn't been interviewed. Phone numbers were requested so that they could be reached in case things were forgotten, missing or incomplete while preparing the study text. The number of those who didn't accept to be interviewed during the study was very small—these were the people who were afraid that the information they provided would be officially processed and may cause them trouble in the future, and who said that their husbands had warned them about this. However, as the research in the market place progressed we became familiar, and they came to believe that no harm would come to them from us, and most importantly when the other women in the market place told them that what they were doing wasn't right, they agreed to see us. The warm and friendly relationships we developed with the village women in the market places affected the progression of the research positively. There were even those who voluntarily guided the research from time to time. Those who said that they didn't have time for the questionnaire and in-depth interview because they were peddling at the market place greatly supported the research outside of the work place. The fact that we interviewed many women for the study made it possible to gain healthy data. A study report was prepared in the light of the information and findings obtained.

Findings and Discussion

There are certain roles and responsibilities that the village women in the market places of Alaplı have, ranging from agricultural production to marketing. These roles push the women workforce to work harder in production and re-production processes. As Oral (1986) has stated, the women need to fit many roles and images, such as the: "motherhood role, family role, female bird makes the nest function, societal roles, good-women, semi-holy motherhood, sisterhood images and the role of continuing the traditions," which have given them too much responsibility and too many roles (Oral 1986). The rise in opportunities to work abroad since the 1960s as a result of industrialization and inadequate urbanization has caused women to be both family heads and to take over the agricultural work in the wake of the male population migrating for work, mainly in rural areas, and forming "families with female household heads." While this might be seen as advancement, the conditions of underdevelopment and increased responsibilities have completely countered any advancement (Arıkan 1988). Migration of the male population both in country and abroad, on the rise since 1980s in the study field, has left women alone with their gardens and farms.

As of the end of the 1990s, most of the female workforce in Turkey was either in agricultural work or worked as unpaid family workers. The agricultural work in the rural areas is the production field that women have taken a big part in. Women are a fundamental force that keeps the rural economy strong in Turkey and in developing countries. Alongside the responsibilities they have as housewives and mothers, women also join in the agricultural work with their physical efforts (Üçeçam Karagel 2010).

Women, regardless of whether they work for money, carry many responsibilities and jobs given to them by the society. The foremost of these is housework. Looking after kids, cooking, washing dishes, ironing, cleaning etc. are the most widely known tasks women undertake. In rural areas there is also field and garden work, animal care and production of animal products which are heavy labour provided free for the family. According to time management research, when they live together with one or two people, even if they have no kids, women in the family spend two or three times more on production compared to men in similar situations. Besides the work cited above, women also have an active role in marketing. This information brings such questions to mind as: Why do the village women feel the need to undertake such activities? How long have they been involved in this activity and what is the real reason or reasons to continue? It is hard to have a certain idea of how long village women in the study field have been selling surplus products. However, what the oldest village women, who is 86 years old, has to say is important, as she remembers bringing products like eggs, butter and cheese to the other villages with her grandmother, and that this was common in their village. According to the oral information gathered from research in the field this activity has been done in the region for more than a hundred years and is thought to be normal. Due to the land formation, which is mountainous and fragmented by rivers, family farms are also smaller and more fragmented. As Özgüç states, the technique and kind of agriculture have a undeniable effect on the participation ratio of women in the Black Sea Region. For example, women workers are dominant in fields where hoe and tree agriculture are common, while women are the minority in fields where tractor and any kind of plough is common (Özgüç 2006). The problematic topographic structure in the Black Sea Region limits the use of machines and tools in agriculture. This situation brings the work force into focus, showing that women work as much as men. Also, the kind of produce grown in the region affects the role of the female population in production. Products like flax, hemp, corn, nuts and tea, that require care and work for every step of production from planting to harvest, need larger work forces. In addition to this, the fact that men also work in other job fields such as industry, trade, mining, forestry, animal husbandry and fishing, and that these jobs require them be out of the village or to be in another city, makes it necessary for the women to join agricultural work. As such, women living in the rural areas of the Black Sea Region have been involved in the agricultural sector to a great extent (Üçeçam Karagel & Karagel 2009).

Village women have kept their important role in the village markets over the years in the study field. Peddling continues to be a task handed down from mother to daughter, or mostly from mother-in-law to daughter-in-law. It is common to find that the mother doesn't want her daughter to be a peddler, but after she gets married her husband or mother-in-law makes decisions as to whether she will work in bazaars. Of the 85 women interviewed at the marketplace, only two were single and came to the bazaar with their fathers. Sales at the bazaar also belonged solely to women, and fathers helped their daughters by sitting at their sides. In our study group, 59 women became peddlers after getting married and 24 before, the latter saying that before marriage they came here with their mothers who taught them the job, and maintained the work because they were used to it and the pleasure derived from earning money. They mentioned that when they had children, they could not come to the market place for a certain period, but that either their mother-in-law or one of their sisters-in-law kept their stalls running. They said that they had been at the same stall at the marketplace for years and that it is important not to leave the stall empty. Even though the bride takes over the job from the mother-in-law, the mother-in-law comes to marketplace in her place when the bride is busy, sick or has recently given birth.

In this sense, market place activities have important roles in transferring the work to the daughters and brides in the family by teaching them the job. The men do not usually contribute to the roles that the women undertake. Handing the job down from older people to younger people is caused by not wanting to lose the position in the market place since it is an important source of income. The stalls that the families use are theirs traditionally and have been maintained for many years.

The general reasons for the village women to go to the marketplace are related to seeing it as a means to earn extra income for the family, and an independent source of income, meaning thay do not have to ask their husbands for money and do not waste the extra produce. In this sense it is very important due to its contribution to the family budget. The biggest difference between the peddling undertaken by men and by women in bazaars is that the former is considered vital to the family income while the latter is considered to be additional income from selling the extra produce. Due to this reason, women who do market trading are never considered to be the main breadwinners, vital to the family budget.

The marketplace is a very important economic field for the village women, whether bringing in extra income or providing the sole income for the family. Forty-three of the village women at the market said that their husbands were retired, 2 said their fathers were retired and 30 said their husbands had their own jobs, whereas the husbands of 5 women were unemployed. Five of the women were unmarried. Most women said that their husbands' incomes weren't sufficient and that their job, while hard, paid good money and that they couldn't quit. Help came from the husbands in the garden work and moving the products to the marketplace and back. There were many saying that this was their main source of income, and that they would not be able to make ends meet without this work, with everything at their homes being bought with the help of selling goods at the marketplace. As a result of this character, which makes up one of the most important results of the research, the marketplace has been a place of constant activity for the women and will continue to be handed down through the generations.

Peddling is regarded as an activity that requires certain characteristics. One woman said that market trading was a job that anybody could do, and added that "if you can calculate in your head and if the products you sell are clean; you're good." However, the other peddler women said that it wasn't a job that everybody could do, that there were people who quit because they couldn't do it and it required certain qualities (Aliefendioğlu 2000). Learning and teaching about the job are noteworthy in terms of effective sales in the marketplace. There are big differences between peddlers with many years of experience and those who have just started the job. In order to sell the produce at the marketplace, the products first have to be grown and harvested and taken to the market place and organized. It is necessary for the salesperson to know how to use the scale, pay attention to dialogue with the customer, calculate the amount of change, keep customers, keep produce fresh and clean and to cooperate with other peddlers. Almost all women thank their mothers, grandmothers and mothers-in-law after learning the job by saying "God be pleased with her. She taught me this job. We are earning money thanks to her."

Not having to ask for money from their husbands is an important distinction for peddler women compared to other women who do not work in the bazaars. Even if peddling doesn't provide dependable or constant income and security, it is an independent source of income. Most of the women say that the extra income covers the needs of the house that they didn't have the money for, that they spend it on the kids, that they put money aside for hard times and that they could do it without needing to ask their husbands. Some of the women said that the income from the market was very good and that it covered all household expenses, including wedding expenses for the children, insurance for both herself and her husband, and farm costs like tractors. The women mentioned that having a large field, maintaining greenhouses, having someone help them with their work and being able to get products to all of the markets are effective means in earning more money at the marketplace: "If you have a small garden and can only go the market once a week, you can only cover the expenses of the household." It was interesting to note that some of the women who were interviewed bragged about how many greenhouses they had.

In the village market, types of vegetables are widely sold as agricultural products because of the ease in growing and harvesting them. The village women mostly grow and sell vegetables. The area from which the women come is reflected in the variety of the products they sell. Women from sloping villages have smaller fields and smaller vegetable varieties for this reason. However, vegetable variety is richer in villages within valleys with easily watered land. This situation also effects how frequently women can go to the market. Those who live in the Alaplı Stream Valley have wider and flatter fields on which to build greenhouses. As a result, they can go to both the Alaplı and Ereğli markets when they are most crowded (Mondays and Saturdays). The frequency of attending market greatly affects the amount of income. The women said that if they could bring produce to the market every day they were able to make very good money, even using the expression that they made "more money than the engineer." Since the variety of products plays a determinant role in the frequency of going to the marketplace, the number of persons working in areas for agricultural production is also important. Due to the general characteristic of the rural area, the effect of patriarchal family structure is rather high. Forty-two of the village women interviewed in the study field lived with their daughters-in-law and 32 with their mothers-in-law in two or three storey houses, and the income belonged to the whole family. Eleven of the village women interviewed in the study field stated that they did not live with their daughters-in-law or mothers-in-law.

The preparation phase for the products brought to the market is rather difficult and time consuming. Planning and organizing the production process and market trading are all undertaken by women. The role handled by women during market trading is very clear. When others are not available to help out in field or garden work or with animals, it is more difficult to prepare more products for the marketplace and the frequency of going to market decreases, which means a lower income. Therefore, elderly women who have married sons prefer to live with their daughters-in-law in the same house for ease of cooperation. In households where the mother-in-law and daughter-in-law live together, one of them goes to the marketplace and the one who stays at home takes care of housework, childcare and preparation of products. The son who works in the city handles the transportation of goods to the market with his own cars if he has one, otherwise an arrangement is made with the other peddler women to rent a car to get to the market place. If the individual who takes products to the market falls ill or has something else to do, she stays home and the other women of the household go to the marketplace. The process lets women share the work load and gives them ownership of their roles. The fathers-in-law and mothers-in-law whose daughters-in-law do not live with them envy the daughters-in-law of other families who do. They believe that living in separate houses is disrupting the integrity of family unity. When none of the women in the household can take the produce to the market, the role is not undertaken by the men.

The reasons for men not selling products in the market are many and come from different perspectives, and trying to express the situation from only one side may be misleading. Therefore, it will be more correct to explain the case with information gleaned from the village women and their spouses. Regarding the reasons why their spouses cannot sell produce at the market, village women state that their husbands have retirement income or they already work in another job. This job has been done by women for a long time and men would not be able to do it; even if they understood how, they would be ashamed and would not be able to sell anything. The women were observed to not let anyone criticize their husbands in this respect. They stated that their husbands went back to the village during bazaar periods to work in the garden or stayed in cafes until they were finished. Husbands came to pick up their wives during evening hours and they went home together.

While there were women who mentioned that their husbands helped them with the garden work, there were also women who said that their husbands did not help them with anything, that they waited at cafes until evening, and also took the money they had earned. The fact that almost all sellers are women affects the ease of working comfortably. Therefore, women also prefer to work without their husbands at their sides. It is also mentioned that this situation is effective in their going to the market place with more ease. There are also individuals who state that customers do not want to purchase items from male sellers in village markets and it is important to have women peddlers there. Maybe that is the most important reason behind the support provided to women by their spouses in working as peddlers and selling produce at the market.

What village women say about their husbands in terms of their help during the bazaar period is interesting. Men are ashamed to be among women, cannot make sales to customers and have to give reductions in prices when an acquaintance wants to shop, whereas women are not like that and are masters of this profession. However, this is not completely the case. It is possible to see many male peddlers in the village markets set up among neighbourhoods, but peddling is the primary job for these males whereas the husbands of the village women have other jobs as their primary occupation or are retired and do not want to be busy with peddling, thinking that they already have another income. Only five males in the study field worked in this regard, and were retired but were having financial difficulties and wanted to improve their finances by making more sales at the market place by working along with their wives.

After being accustomed to sales in the marketplace, village women prefer to work in the bazaars instead of staying in the village and handling household and garden work. The effects and change caused by being working women are voiced by them in the belief that they are improved in terms of socialization compared to women who stay in the village and who do not work in the village market. They stated that they fought with the hardships of the world all by themselves, and that they knew all the places like the back of their hands, whereas in the past they did not even know where to go in the market; now, they had no fear of anyone. In this sense, village markets are social areas in which women can see relatives and acquaintances, talk about their problems and relax with friends, presenting some cooperative behaviours.

During the weeks in which the produce is not sufficient, the village women try to keep up their presence by bringing the produce grown for household consumption. They also present behaviours that include gentility so as not to lose continuing customers and not feeling embarrassed towards them. In this sense, the importance of the roles that they undertake also increases. Orders for the next market day are accepted as well as providing phone numbers for customers for future orders. That way, the effective position in the market place can be strengthened, personal relationships can be kept warm and income can be increased.

Conclusions

Women have undertaken very important roles for the sustainability and economic income of their families during all stages of life. Free agricultural labour in rural areas still continues today. While it is more common in urban areas for women to be employed for money, women can earn money for daily work by working in the fields of others in rural areas. A commercial activity created by village women is observed in the study area. Marketplaces provide the women with an organized establishment and readymade opportunity to continue the tradition of market trading and bring in extra income. The fact that women are the producers of their own products is highly effective in their active role in agricultural produce commerce. Although the history of this activity is not clearly known, the fact that it has been undertaken for many years has given the women some particular roles. The fact that this activity has been undertaken by women in the study field for long periods has created the belief that "these are done by women in our area; males do not sell goods in village markets," and has caused more pressure on women to handle specific workloads and responsibilities in terms of gender mainstreaming. This situation has resulted in a social structure that accepts it as normal for women to bring and sell their products in marketplaces in addition to being producers in the field of agriculture. The most important role for the women in this social structure is to gain their economic independence and teach and transfer what they know to their daughters and sisters-in-laws who will sustain their positions in the marketplace and keep up the work after them.

References

Aliefendioğlu, H. (2000). Bir Kamusal Alan Olarak Akçakoca Köylü Pazarı ve Pazarcı Kadınlar, Hacettepe Üniversitesi, Sosyal Bilimler Enstitüsü, Doktora Tezi (Basılmamış), Ankara.

Arıkan, G. (1988). Kırsal Kesimde Kadın Olmak", Hacettepe Ünv. Edebiyat Fakültesi Dergisi, C:5, S:2, s.1-16, Ankara.

Bolak, H. (1993). Aile İçi Kadın Erkek İlişkilerinin Çok Boyutlu Kavramlaştırılmasına Yönelik Öneriler, 1980'ler Türkiye'sinde Kadın Bakış Açısından Kadınlar, İletişim Yayınları, s.235-246, İstanbul.

Ecevit, Y. (1998). Türkiye'de Ücretli Kadın Emeğinin Toplumsal Cinsiyet Temelinde Analiz", 75 Yılda Kadınlar ve Erkekler, Tarih Vakfı Yayınları, s.267-284, İstanbul.

Ertürk, Y. (1996). Alternatif Kalkınma Stratejileri, Toplumsal Cinsiyet Kadın ve Eşitlik, ODTÜ Gelişme Dergisi, 23 (3), 341-356.

Fazlur, R. (1987). Ana Konularıyla Kur'an (Çeviren: A. Açıkgenç), Ankara.

Gülçubuk, B.; Yasan, Z. (2010). İşsizlik Yoksulluk ve Göç Kıskacında Kırsal Alanda Kadın Emeği, 21. Yüzyılın Eşiğinde Kadınlar, Değişim ve Güçlenme, Uluslararası Multidisipliner Kadın Kongresi (13-16 Ekim 2009) Bildiri Kitabı C:3, s.90-97, İzmir.

Moser, C. (1989). Gender Planning in the Third World: Meeting Practicaland Strategic Gender Needs, World Development, 17 (11), 1799-1825.

Oral, Z. (1986). Kadın Olmak, Milliyet Yayınları: 54, İstanbul.

Özgüç, N.; Mitchell W. (2000). Şehirlerin Alternatif Alışveriş Mekanları: İstanbul'da Haftalık Pazarlar, Mimar Sinan Üniversitesi, Mimarlık Fakültesi Dergisi, Tasarım-Kuram, C:1, S:2 ,s.35-58, İstanbul.

Özgüç, N. (2006). Türkiye Ziraatında Kadın Gücünün Dağılışı:1955–2000, İnsan ve Mekân (Prof. Dr. Erol Tümertekin'e 80. Yıl Armağanı), Çantay Kitabevi, s.203–213, İstanbul.

Saran, N. (1979). Sosyal Antropolog Gözüyle Toplumumuzda Kadın, İ.Ü. Edebiyat Fakültesi, Sosyal Antropoloji ve Etnoloji Dergisi 3, s.1-12, İstanbul.

Trak, S. (1950). İktisadi ve Ticari Türkiye, Cilt:I, Milli Eğitim Bakanlığı, Yüksek Ekonomi ve Ticaret Okulu Yayınlarından, Sayı:30, Teknik Kitap ve Mecmua Basımevi, İzmir.

Topçuoğlu, H. (1978). Türk Toplumunda Kadının Sosyal Statüsü, Türkiye Kadın Yılı Kongresi, Ayyıldız Matbaası, s.53-110, Ankara

Üçeçam Karagel, D.;. Karagel, H. (2009). Türkiye Kırsalında Kadın İşgücü (1990-2000). e-Journal of New World Sciences Academy Social Sciences, Volume:4, Number:3, 4A0008, 116-136.

Üçeçam Karagel, D. (2010). Türkiye Kırsalında Tarımda Çalışan Kadın Nüfus (1990-2000), The Journal of International Social Research (Uluslararası Sosyal Araştırmalar Dergisi), Kadın Araştırmaları Özel Sayısı, Volume:3, Issue:13, s.254-267.

CHAPTER TWENTY-FIVE

A TEMPORARY SETTLING WITH A CHANGING FUNCTION IN GAZIANTEP: VINEYARD HOUSES

MEHMET EMIN SÖNMEZ

Introduction

Viniculture[1] is carried out in many countries around the world, particularly between latitudes of 34° and 50° north, owing to the fact that in this type of agriculture water demand is not so high and they can grow in areas with deep underground waters (İzbırak 1947, 410). In areas between the above-mentioned parallels, table grapes and wine grapes are produced a lot, while vineyards for dried grapes cannot grow beyond latitude 45° north. Thus, Turkey in general and Gaziantep in particular are among leading areas producing grapes for table, wine and drying. It is a fact that the Mediterranean climate zone covers areas where viniculture is most widespread. According to Tümertekin & Özgüç (2011), quoting Zimolzak & Stansfield (1983), Europe is divided into three zones by dominant drinks in respective areas:

1. "wine-drinking Europe" as an extension of the Ancient Roman Empire
2. "beer-drinking Europe" extending to the northwest of the continent
3. "vodka-drinking Europe" comprising Scandinavia, Russia and Eastern Europe.

Such a division indicates that the wine zone covers the Mediterranean basin. The reason is the close relationship between consumption habits and produce. Considering that drinks are made from produce in their near vicinity to a considerable extent, viniculture is substantial across the Mediterranean zone, also covering Anatolia. This production-consumption

[1] Economic activities performed in vineyards or grape fields are called viniculture (Doğanay & Coşkun 2012, s. 317).

relationship is valid not only for today, but it is estimated that viniculture and vineyard culture in Anatolia dates back to 3500 B.C. The commercial products found in the Uluburun Batığı (merchant ship), which sank around 1300 B.C. in the Aegean town Kaş, reveal the importance of grapes and wine (Pulak 2004, 87). Similarly, the İvriz Relief of the eighth century B.C. documents that viniculture was present in Anatolia in the past. On the relief, Tarhunzas, the Hittite Storm God, offers grapes and wheat ears to the Hittite King Varpalavas (Özyar 2005, 13). These records show that grapes were predominantly produced, used and traded in ancient Anatolia as they are today.

Climatic conditions are really influential on grape production in Anatolia. The hot and arid summer of Turkey is a great advantage for producing grapes because grapes need a long and hot dry season for growing. Especially in areas like Gaziantep, where the winter is relatively cold, more sugar is accumulated in grapes during summer. Arid summers allow for the drying of grapes as needed (Uzun 2003, 16–17). This, in turn, makes the region an ideal area for viniculture.

According to Evliya Çelebiye, there were seventy thousand vineyards in and around Gaziantep during the seventeenth century. Besides, molasses and dried layers of fruit pulp made of grapes had a considerable share in the economy of Gaziantep at that time (Dağlı, et al., 2005, 50). The decisions taken during the *Gaziantep Mıntıka Kongresi*[2] (Gaziantep Regional Congress) reveal that viniculture was an important activity in the area during the early Republic years, and decisions were made at the congress regarding Antep molasses. To improve the quality of molasses produced in the villages of Gaziantep, the slogan "Those who do not make good molasses cannot sell it" was introduced during the *Gaziantep Mıntıka Kongresi* in 1927 (Mortan & Arolat 2009, 25). So, we can infer that viniculture has been going on for a long time before it lost significance in the area.

The fact is that viniculture is one of the main income-generating activities not only around Gaziantep but also all over Turkey. This has resulted in the emerence of vineyard houses as a type of seasonal country cottage peculiar to Gaziantep since early times (Özgür 2011, 32). Such settlements were inhabited during certain periods of the year and are not gathered in one place, being used for agricultural activities (Tanoğlu 1969, 254). In Anatolia, vineyard houses are used for agricultural production especially during the season when grapes mature and are dried (Girgin

[2] Established to implement economic decisions taken at İo impEconomic Congress between February 17 and March 4, 1923.

1991, 153). However, vineyard houses not only have economic but also social meaning (Somuncu 1996, 107). According to İzbırak (1947), nearly seventy years ago vineyard houses in Kayseri were used as summer resorts rather than farms. It has also been reported by Somuncu (1996) that use of vineyard houses as resorts can also be seen in some other cities of Turkey. The vineyard houses in Ankara have constituted living spaces where the production and consumption cycle is experienced in parallel with seasonal movements within the traditional urban life during the late nineteenth and early twentieth centuries (Türkyılmaz 2012, 1).

These units are shaped by people on the basis of social and economic status as well as environmental conditions and their use for various purposes. Therefore, building materials and architectural features of vineyard houses vary from person to person as well as region to region. The vineyard houses in the study area can be thought of as temporary rural settlements in which some or all of the units such as pergolas, huts, homes, servicemen's homes and pools exist in varying parcel sizes allocated for agricultural activities, though used as summer resorts. The use of vineyard houses across Turkey is not so common today, and recent studies (Somuncu 1996; Kadıoğlu 2010) show that vineyard houses started losing their functional and structural characteristics due to changes in social, cultural and economic conditions. In this study, functional and structural changes taking place in vineyard houses around Gaziantep have been investigated along with causes and negative effects.

Materal and Method

This chapter covers studies on the characteristics of historical vineyard houses in Gaziantep, as well as the causes and effects of the changes in their function, together with the causes and environmental effects of the emergence of secondary houses mostly built as vineyard houses in the same region.

A typical vineyard house was selected as a sample to display the characteristics of such houses, and a detailed layout of the sample house was prepared. In addition, a field survey was done to find out the distribution and functional changes in vineyard houses. People above fifty who once lived in vineyard houses were interviewed during the field survey. Moreover, the compatibility between the area of today's vineyard houses and land use was identified by using the neighbourhood's land use ability and maps from1984.

Physical Characteristics of Gaziantep Vineyard Houses

As a part of geographical landscape, a house says a lot about a specific neighbourhood as well as the dominant culture there. Houses reflect residents' social and economic needs and the characteristics of the natural setting where they are located (Tümertekin & Özgüç 2011, s. 364). Actually, only a few historical vineyard houses have remained in Gaziantep; they act as narrators of their natural environment and the social life in the past. During interviews, it was found that the vineyard houses in the study area were mainly used for agricultural purposes, but also for recreation and fun. People went there in early April and did not come back until the end of October. In April, the lower parts of the grapevines were hoed and weeds were removed, and during the same period insecticide was applied to the vines. Weeds were cleared twice in May and June by tractors all over the vineyard to better feed the vines, followed by grape picking in mid-August. Lastly, grapes were left to dry in the first week of September.

In Gaziantep, vineyard houses were built of limestone collected from the surrounding places up to the 1980s. The sizes, pergolas and pools of these houses and quality of the building material and workmanship largely depended on the homeowner's income level. So, the stones used in some of the vineyard houses were well carved, while some others displayed unskilled labour. The vineyard houses in the area, like other parts of Turkey, consisted of houses, pools, pergolas and huts where watchmen or servants stayed (Figs. 25.1 and 25.2). Our sample vineyard house is located near Kadı Değirmeni (west of Gaziantep along the tramway). We tried to find a real example which bears all historical characteristics of Gaziantep vineyard houses without any change so that the study can be built on sound foundations. Apart from that sample, other vineyard houses around the Burç Quarter were visited for reference during the discussion on historical vineyard houses in Gaziantep and its neighbourhood.

The historical vineyard houses in Gaziantep are mostly two-storey (ground and first floor) and 100 m^2 (Fig. 25.3 and 25.4). In the sample vineyard house, there are two rooms on each floor, and one of the rooms on the ground floor is used as kitchen while the other is a living room (see Fig. 25.3 below). Each of the rooms has 6 windows, with 24 windows in the whole house. It shows that vineyard houses are airy and also used for cooling during summer. We can say that vineyard houses in Gaziantep are often built for recreational purposes as in other parts of Turkey. The two rooms upstairs are used as bedrooms (see Fig. 25.4 below). In addition, there is a bathroom and toilet, 3 m^2 each, across the entrance on the ground floor (see Fig. 25.3 below). Like other vineyard houses, they are made from limestone, except for the roof, wood is used for coating the ceiling, and the roof is made using iron sheets and tiles. Thus, the natural environment is visible in historical vineyard houses of Gaziantep (Fig. 25.1 and 25.5)

One of the indispensable elements in these vineyard houses is the pool, covering 4–5 m² on average. Such pools are mostly built for irrigating the surrounding vineyards and gardens (see Fig. 25.4 below). The pool in our sample house is drilled on 3 sides so that outgoing waters can be directed towards canals irrigating gardens and vineyards. When the pool is filled with water, it gradually empties because the water leaks out and reaches the irrigation canals. This proves that the pool was used for irrigation. The pool is fed by the adjacent water tank, but the tank is no longer in use.

Fig. 25.1. Typical vineyard houses of Gaziantep with ancillary units

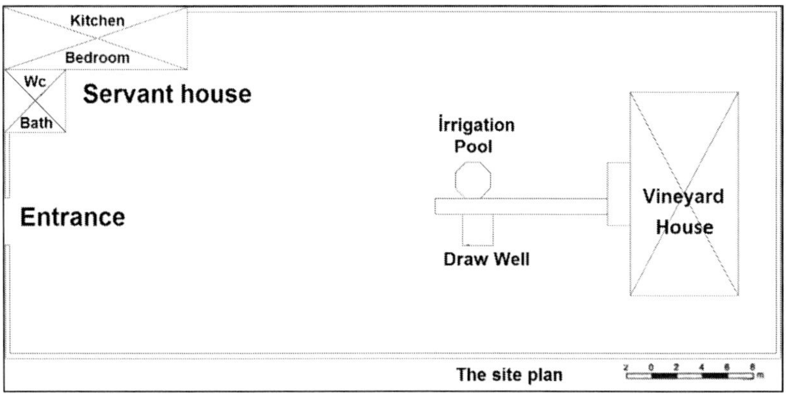

Fig. 25.2. General layout of Gaziantep vineyard house

Another component of the vineyard house is the house in which servants live. This can be found in most Gaziantep vineyard houses, but not in other parts of Turkey. It is understood from the building stones and size of the structure that the unit is not built so neatly, with the stones cut only roughly and with no decorative element. In addition, the structure is only 2–2.5 m heigh. On the other hand, the ceiling of the vineyard house is nearly 5–6 meters high. It is understood that this unit located at the entrance is used by servants who are responsible for cleaning and care of the compound. The structure of the servants' quarters in the study sample consists of three parts. There are two adjacent equivalent rooms, each approximately 30 m². One of them serves as a bedroom, while the other is a kitchen. The third part, which is also adjacent, is 6 m² and is designed as a toilet and a bathroom (Fig. 25.2 and 25.7).

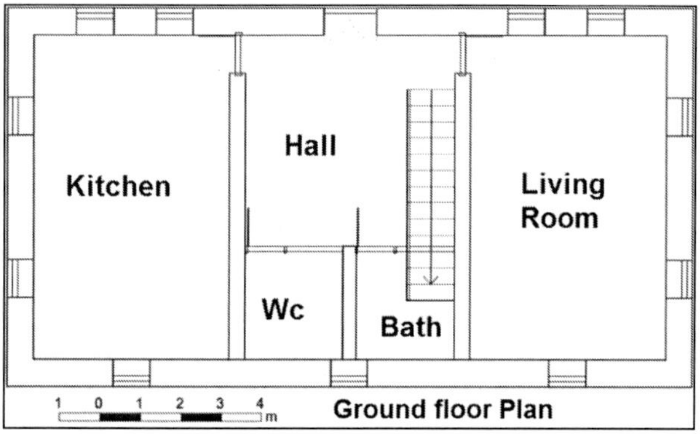

Fig. 25.3. Ground floor of vineyard house

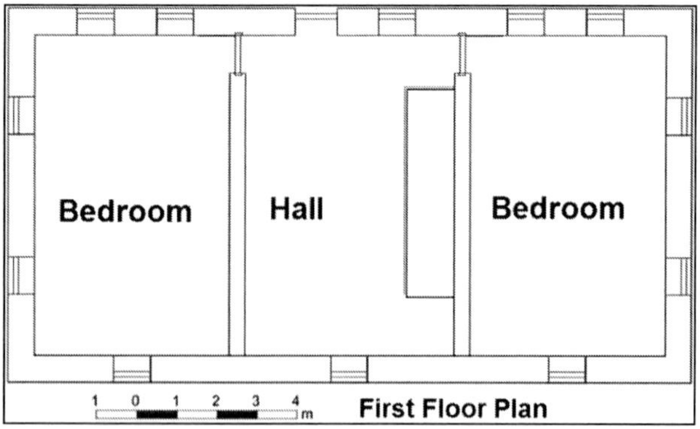

Fig. 25.4. First floor plan of vineyard house

Fig. 25.5. A typical Gaziantep vineyard house or summer resort

Fig. 25.6. The pool situated in the middle of the garden and used for irrigation

Existence of a servants' house or watchman's hut in the majority of the Gaziantep vineyard houses indicates that homeowners in the area were relatively well off. Also, such vineyard houses are mostly used for recreational purposes.

Fig. 25.7. The building allocated for servants or watchmen in vineyard houses.

The exterior wall is another noteworthy component of vineyard houses. The stone wall, nearly 1 m high, encompasses the whole vineyard house also including the pool, pergola, hut and servants' house. The compound is separated from vineyards or neighbouring vineyard houses, and the sample vineyard is surrounded with stones. Generally, the stones are derived from the neighbourhood and not carved finely. The relatively low height of the wall and inattentive arrangement of the stones means that the things inside are too unvaluable to take security measures. This also indicates that such homes are only temporarily inhabited most of the time.

Another aspect of these structures is the large area affected by the vineyard house, though in the immediate area the house and supplementary units are not so big. To exemplify, the total area of the compound on which the sample vineyard house extends is nearly 1,200 m² (Figs. 25.1. and 25.2). However, the homeowner has around 6–7 decars there. It is understood from the interviews that most of the vineyards were built on plots of tens of decars and grapes were produced there in the past.

Taking into consideration the vineyard house, the physical characteristics of which are discussed in detail, it is seen that historical vineyard houses in Gaziantep are shaped by the surrounding natural environments and used for recreation. Also, we found from the interviews that the sample vineyard house was purchased in 1961 together with a farm of 12 decars. The land outside the vineyard is not used at all today. However, interviewees say that those parts were once covered in vine stocks and fruit trees. This plot included in the city of Gaziantep remains idle today because of an inheritance dispute, thanks to which we were able to find one typical vineyard house with characteristics peculiar to the Gaziantep neighbourhood; otherwise, it would have been replaced by high rise blocks.

As seen in the current example, vineyard houses used to be inhabited not only for agricultural production but also recreational purposes in Gaziantep. The 24 windows and showy buildings are also clues about the same.

Secondary Houses or New Vineyard Houses in Gaziantep

As vineyard houses evolved into secondary houses, called new vineyard houses in Gaziantep, their structural characteristics, sizes and functions have largely changed. Still, they show some resemblance in outlook. The secondary houses are also bordered by a wall on a land of several decars. However, the walls are higher than 2–3 m and made from concrete rather than stones. There is usually a wide entrance door facing to the road. On the right or left side of the gate is a house around 80–120 m^2 for watchmen or servants. This unit may not exist in some of the current vineyard houses. There is a car parking area on the opposite side. Vineyard houses are generally situated on an area of a few decars, and the majority is reserved for a garden. There is a modern swimming pool around 40–80 m^2 (Fig. 25.8). The main building in the new version is splendid and mostly multistorey (2 to 5), and covers a much larger area than the historical ones, varying between 200–250 m^2 (Figs. 25.8 and 25.9). The most common building materials are tiles, gas concrete blocks, woods and ytong type concrete. These structures have a relatively weak relation with the natural surroundings compared to the historical ones.

Not only the physical characteristics but also how they are used and in what frequency are different from the historical vineyard houses. In the past, such places were mainly used in late spring and summer, when grapes grew and were picked up, whereas the new versions are being used at weekends. They are also used throughout the whole summer and occassionaly during the week in other seasons. Therefore, these new houses have almost nothing in common with the originals in function. They are distinguished from summer villas as they are visited any time in the year. Departing from that, secondary houses would be a suitable term. They are built as vineyard houses mainly due to the legal issues. Such places are within municipal adjacent areas, and new domiciles can be built on a few decars purchased in those areas (Figs. 25.8 and 25.9).

Fig. 25.8. An example of secondary house with modified functions called Gaziantep vineyard house.

Fig. 25.9. Secondary houses replacing the historical vineyard houses along the Gaziantep-Kilis highway (a) secondary house, (b) swimming pool, (c) servant's house.

Causes and Effects of Changing Physical and Functional Characteristics of Gaziantep Vineyard Houses

Only few good examples of the above-mentioned historical Gaziantep vineyard houses exist today. Up to the 1960s, vineyard houses situated in the Alleben valley neighbourhood were destroyed. The vineyard house in our study is standing in the Batal-Höyük District, being purchased by its current owner in 1961 (Fig. 24.1). The Alleben Valley and its vicinity were huge productive farms for fruit and vegetable production until 1970s. However, as a result of expansion of the city the area is dominated by newly built high rise blocks of houses (Fig. 25.10). The agricultural capacity of the area has been lost and the Gaziantep vineyard houses were replaced by the new versions, secondary houses, because of the rapid urbanisation.

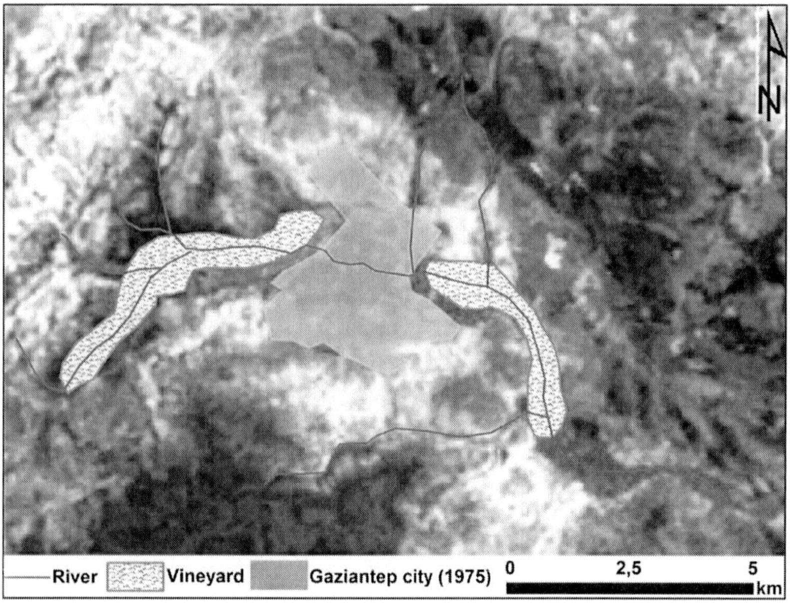

Fig. 25.10. Some parts of Gaziantep with an abundance of vineyard houses until the 1970s (the total area of Gaziantep was almost 9.2 km² in 1975).

The development described above has caused a shifting of new vineyard houses (secondary) towards Burç Quarter in the west. Since that place has a relatively greater height, though situated in the valley, and is green and cool, it heavily attracted the building of secondary houses (Fig. 25.11). The secondary houses started becoming widespread in the 1990s when industrialization and urbanization gained speed in Gaziantep. In the early 1990s, many rich people emerged and the city became crowded while all types of pollution increased. Therefore, the countryside became more attractive, and this aroused interest in using the vineyard houses close the the center of Gaziantep for resting and recreation. As a result, Gaziantep vineyard houses shifted from agricultural to recreational places. In fact, there are more than eighty thousand workers in the organized industrial sites of Gaziantep (Sönmez 2012). The overall population of the city is almost 1.5 million. Such a rapid urbanization and industrialization movement affected all agricultural products negatively in Gaziantep, and vineyards were particularly affected more than others.

According to the Turkish Statistical Institute (TÜİK), around 350 thousand tons of table grapes, wine grapes and grapes for drying were produced from 500 thousand vine stocks in 1991 in Gaziantep; this number decreased by 180 thousand, making the amount of grape production 90 thousand in 2012 (see Table 25.1 below). It is obvious that grape production and number of vine stocks has regressed nearly four times in twenty years. This is closely related to the fact that there has been a shift in the function of vineyard houses because of rapid population growth and industrialization in Gaziantep. Another reason could be the preference for pistachio nuts to grapes on a higher income basis. For example, there were around 13 million pistachio trees and annual production was nearly 20 thousand tons in Gaziantep in 1991, and the number of trees and production reached almost 20 million and 57 thousand tons, respectively, in 2012 (see Table 25.1 below). The number of pistachio trees has increased by almost 60% and production has increased by almost 300% in Gaziantep during the last twenty years. Such a replacement in horticultural pattern in Gaziantep has affected viniculture and vineyard areas negatively. The viniculture is no longer an income generating activity in Gaziantep because of its decreased production after the 1990s.

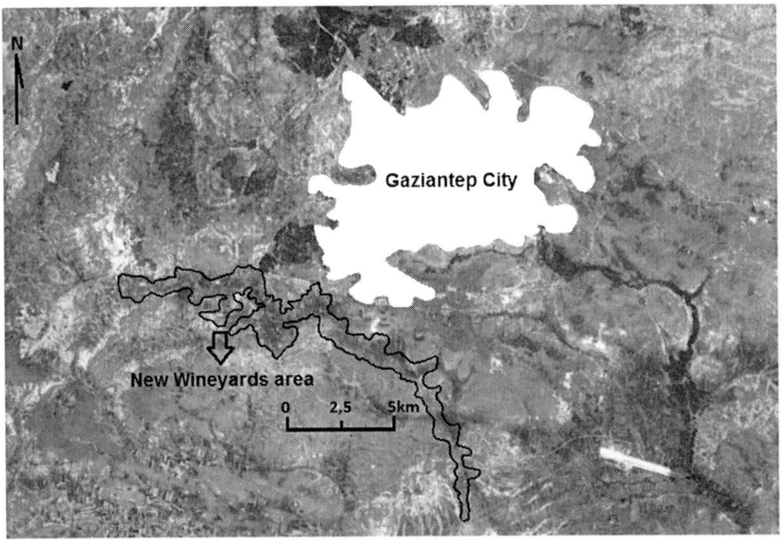

Fig. 25.11. A densely occupied area by so-called vineyard houses in Gaziantep (The area of Gaziantep city in 2013 is nearly 50 km²)

Furthermore, the relationships between rural and urban sites became greater as a result of expansion of the transportation network and fundamental changes in the means of transportation. This, in turn, weakened the relationships between rural areas and the natural environment, making the former less dependent on the latter. As a consequence, the influence of the natural environment on building materials and patterns in rural areas was reduced to a minimum. As an example, the interviewees explained that authentic vineyard houses were no longer built after the 1980s, during which there were close relationships between the rural and urban parts. In addition, the number of vehicles increased drastically in the 1990s thanks to which those engaged in viniculture could make one-day trips from the home to the farm. The development in transportation and convergence of rural and urban parts is one of the reasons why vineyard houses were not built or built as modern houses for different purposes in Gaziantep after the 1980s.

Table 25.1. Changes in the number and production of grapes and pistachio trees in Gaziantep province

	1991		2012	
	The number of trees	**Output (ton)**	**The number of trees**	**Output (ton)**
Grapes	492870	346108	179354	87971
Pistachios	13153500	19355	19436135	56233

The changes that took place in the functions and patterns of Gaziantep vineyard houses give hints about social values and economic needs (Tümertekin & Özgüç 2011, s. 364). It can be inferred from the physical changes in those structures that social, cultural amd economic activities also changed in the neighbourhood. In fact, the physical characteristics of the vineyard houses in the study area go hand in hand with the economic changes taking place in Gaziantep.

The fact is that vineyard houses have lost their function and agricultural lands are being used for other purposes. Farms are split into fragments and agricultural production has decreased, bringing about many problems such as destruction of meadows and the natural environment.

In reality, large vineyards in the study area are split into 2 or 3 decar parcels and then sold (Fig. 25.3 and 25.12). Also, the cost of building plots has increased to almost urban site levels due to the huge demand for these areas. This resembles the splitting of lands continuously due to inheritance issues. As a consequence, agricultural production data shows a constant decrease. However, we found during the interviews that a 6 or 7 decar vineyard is big enough to support a family of 5 people. These places without construction permits can be used for housing only if vineyard houses are built in them. This allows landowners to have secondary houses without facing any legal challenges, following which new infrastructures and expenses come to the front. The vineyards in those areas are removed and other plants with high demands of water are planted. A study suggests that secondary houses all over Turkey, like the new-style vineyard houses in Gaziantep, are just a waste of resources because those houses are only inhabited for around 1.5 months a year (Manisa & Görgülü 2008, 69). On top of this, most of the vineyards are located in places more suitable for agriculture than their surroundings. Housing in those areas leads to the destruction of the scarce fertile agricultural lands. Instead, new vineyards are planted on stony and steeply sloping areas, which rapidly increase the amount of areas with a high erosion risk in the vicinity.

Fig. 25.12. Due to expansion of secondary houses in Gaziantep, orchards and vineyards disappear and the wrong use of lands increases

Discussion

The study area reveals the significant changes that have taken place on vineyard houses from both architectural and functional aspects. The houses built as cottages and vineyard houses are not in compliance with governing laws. Concerning the preservation of agricultural production in places outside the zoning and urban areas, article 63 under Section 6 of the regulations for unplanned lands stipulates:

> In parcels contained outside inhabiting places without an upper scale plan; cottages or vineyard houses for one family, country teahouses, restaurants and their outbuildings can be built provided that construction area ratio does not exceed 5 %, total construction area does not go over (250) m² under any circumstances, valance levels do not exceeed the natural ground by (6.50) m and 2 floors, and they are not closer than (5.00) m to the road and parcel boundaries (www.mevzuat.gov.tr 2013).

Nevertheless, almost none of the examples in the study area are found to be compliant with the regulations. As explained under new vineyard houses, some of the real examples are as high as four to five floors, with a swimming pool, a servants' house and the core house covering nearly 500 m². It is clearly seen that most of the new vineyard houses are built in breach of governing laws from the view of space and height.

In addition, such constructions built as vineyard houses or cottages are against the law of land use. In the Law Number 5403 on Soil Conservation and Land Use issued in 2005, some amendments were made in order to prevent the misuse and wrong use of land and to promote preventive precautions. According to article 13 of the law, farm lands can be used for other purposes only for the common good and under *force majeure*:

> Absolute farm lands, special crop lands, planted farm lands and irrigated farm lands cannot be used for other purposes than farming. However, if there are not alternative areas and it is considered eligible by the Board; the Ministry can permit using those lands for other purposes on condition of respecting soil conservation projects for following needs:
> (a) Strategic needs for defense,
> (b) Need for temporary settlement area arising aftermaths of natural disasters,
> (c) Oil and natural gas exploration and operation activities,
> (d) Mining activities about which a common good decision is taken by the relevant ministry,
> (e) Plans and investments about which a common good decision is taken by ministries (www.mevzuat.gov.tr 2013).

In the Gaziantep neighbourhood, valley plains and karstic depressions are cultivated heavily; therefore, they contain valuable farmlands. Of these, the Alleben stream valley has been largely replaced by new residential areas as a result of the expansion of Gaziantep city. Except for these, Nizip stream valley in the north and Tüzel and Kızılhisar streams in the south and southwest, and Sabun suyu valley in the west are significant farm lands regarding land use. Lands of types I–IV which are good for farming are intense in those areas (Fig. 25.13). Particularly in the Tüzel and Kızılhisar valleys stretching in a southeast-northwest direction in the south and southwest and karstic depressions, soil thickness is high and the underground water level is relatively high; thus, they are still being used for farming. For this reason, these areas are predominantly limestone and are the backbones of the Gaziantep Plateau, on which farming lands are scarce. Nevertheless, these agricultural lands are divided into even smaller plots, eventually turning into secondary housing zones.

This is completely against Law Number 4302 on Land Conservation issued in 2005 for the preservation of farming lands, because absolute farm lands, which are quite limited all over the region, exist here. Considering the land use map for Gaziantep city and its surroundings prepared in 1984 by the Rural Affairs Directorate, the above-mentioned significant farm lands represent the vineyard areas (Fig. 25.14).

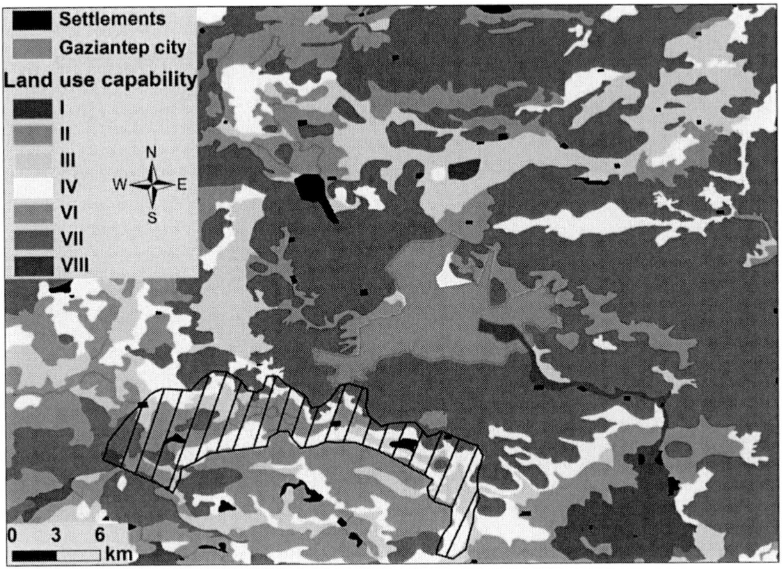

Fig. 25.13. Land use ability in Gaziantep city and surroundings (arrows refer to farm lands threatened by secondary houses)

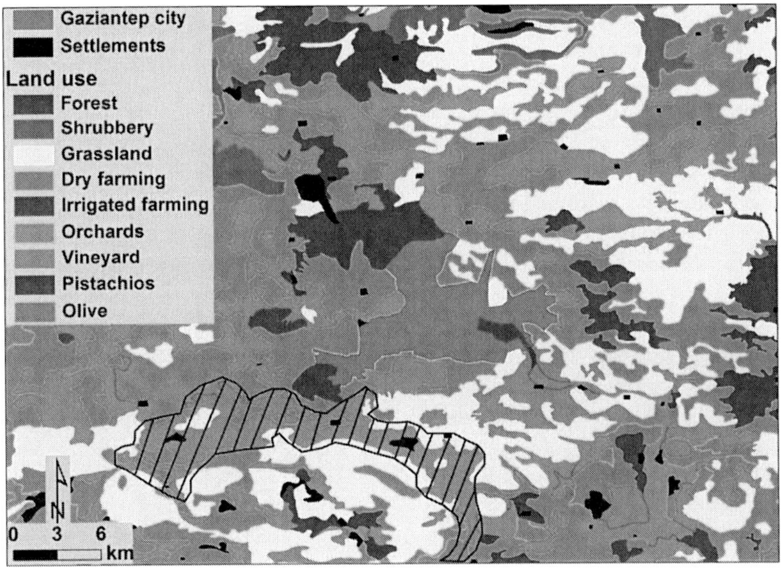

Fig. 25.14. Land use status of Gaziantep city and surroundings in 1984 (arrows refer to vineyard lands threatened by secondary houses)

There are also large areas suitable for farming on these large karstic depressions and valley plains. It is understood from interviews and observations that wide farming lands still exist in those areas. However, the majority of these farming lands are split into small parcels of 2–5 decars. Such sizes are too small for one family to sustain living on a vineyard. On these plots where vineyard houses are settled, vinestocks are removed giving considerable space to fig, peach, apricot, apple and cherry trees. This indicates that the units built as vineyard houses have totally lost their functions.

Conclusion

While rapid growth of the city causes a narrowing of farm lands, the growing population make it almost compulsory for well-off people to obtain second houses. Unfortunately, such economic and social changes in the neighbourhood have affected vineyard houses the most and created the the disappearance of the historical and architectural functions of vineyard houses.

While such changes are taking place the functioning of vineyard houses can be closely related to Gaziantep's social development and economic growth. Recently, Gaziantep has developed in both population and economy, transportation has improved and driving has become more widespread in the neighbourhood. In parallel, the number of vineyard houses built as cottages has drastically increased, and existing ones have gone through significant structural and functional changes. The increased number of so-called vineyard houses in the area is related to legal matters. A considerable amount of farm lands is lost in these areas which do not have a construction plan. We recommend that new construction zones can be opened for secondary houses, which would be a better solution than building on farm lands.

In addition, since valley plains in the neighbourhood are occupied by secondary houses, vineyards are being planted in unsuitable areas. This heightens the risk of erosion on those high and steeply sloping areas. Also, it must be born in mind that many of these new houses are not used for almost eleven months a year, which is an economic waste.

References

Dağlı, Y.; Kahraman, S.A.; Dankoff, R. (2005). Evliya Çelebi Seyahatnamesi IX. Kitap. Yapı Kredi Yayınları, Istanbul

Doğanay, H.; Coşkun, Ö. (2012). Tarım Coğrafyası. Pagem yayıncılık, Ankara.

Girgin, M. (1991). Alaşehir Çayı Vadisinde Bağ Evleri. Atatürk Üniv. Kazım Karabekir Eğitim Fakültesi Dergisi, 2(2).

İzbırak, R. (1947). Kayseri Şehrinin İşleme ve Gelişmesinde Bağcılığın Etkileri. 401-418.

Kadıoğlu, Y. (2010). Türkiye'de Fonksiyonel Özelliğini Kaybetmiş Geçici Yerleşmelere Bir Örnek: Tavas Bağ Evleri. Marmara Coğrafya Dergisi (21), 98-114.

Manisa, K.; Görgülü, T. (2008). İkincil Konutların Turizm Sektöründe Yeniden Kullanılabilmesine İlişkin Bir Model. YTÜ Arch. Fac.E-Journal, 3 (1), 68-78.

Mortan, K.; Arolat, O.S. (2009). Gaziantep Ekonomisine Bakış. İstanbul: Heyamola yayınları.

Özgür, E.M. (2011). Yerleşme Coğrafyasına Giriş. Ankara.

Özyar, A. (2005). Geç Hitit Krallıkları. Arkeoatlas (4), 10-35.

Pulak, C. (2004). İhtiyar Akdeniz. Arkeoatlas (3), 78-89.

Somuncu, M. (1996). Kayseri'de Bağcılık. Ankara Üniversitesi Dil ve Tarih-Coğrafya Fakültesi Coğrafya Araştırmaları Dergisi(12), 107-133.

Sönmez, M.E. (2012). Yerleşme Yeri Seçimi ve Alansal Gelişimi Açısından Gaziantep. Malatya: Özserhat Yayıncılık.

Tanoğlu, A. (1969). Nüfus ve Yerleşme Coğrafyası. İstanbul: İstanbul Üniversitesi Yayınları.

Tümertekin, E.; Özgüç, N. (2011). Beşeri Coğrafya, İnsan, Kültür, Mekân. İstanbul: Çantay Kitabevi.

TÜİK (2013). http://tuikapp.tuik.gov.tr/bitkiselapp/bitkisel.zul

Uzun, İ. (2003). Bağcılık. Antalya.

CHAPTER TWENTY-SIX

TOURISM GEOGRAPHY AND TOURISM PLANNING OF THE HAZIM DAĞLI NATURAL PARK (YAPRAKLI - ÇANKIRI)

NURHAN KOÇAN, ABDURRAHMAN DINÇ AND NEŞE YÜCESOY

Introduction

The people living in the intensive and difficult conditions in the cities have been directed towards tourism activities to be closer to nature. Therefore, the untouched features of rural areas are used for tourism activities.

The sustainability of tourism activities that contribute to the economic, social and cultural values of countries depends on the protection of the natural and cultural values and their improvement. Therefore, tourism performed with sustainability principles in the interesting natural and cultural features of nature provides visitors with scientific, aesthetic, recreational, cultural and instructional new knowledge and experiences, and the socio-economic development of the population living in this area has been became as a resource (Weaver 1999). For the functioning of sustainable tourism, cooperation between the tourism industry and the local community and their involvement in its planning must occur. There is no doubt that environment-based tourism planning provides an environment which is improved in economic, social and physical terms of (Gunn 1994).

In order to protect the nature and land we live on, and for them to benefit from their existing potentials, improving these for the next generation may be possible by planning and providing constant care and supervision (Basal 1998). This can only be accomplished with sustainable physical planning decisions. The planning aims to maintain the environment's resources and take optimum advantage of the resources, improving people's quality of life.

Considering the economic, environmental and social changes to diversify recreational/tourism activities of the natural and cultural resource values in rural areas, providing development and infrastructure arrangements are significant gains. The rational and sustainable use of the resources is a positive effect for the socio-cultural development of society and a more extensive approach is accepted by the public.

Material and Method

In this study, the Hazim Dagli Natural Park and its close surroundings in the Cankiri province of the Yaprakli district have been evaluated. In order to obtain data for this area, projects and brochures were obtained from the Yaprakli Governor's office and the District Directorate of Agriculture. To reveal the natural and cultural structure of the area, land observation has been carried out and photographs have been taken, while previously studies comprised the relevant literature.

The working area has been considered in 1/25,000 and 1/5,000 scales. The former has examined its relationship with the surrounding areas and has recommended including the environmental areas. In the latter, the application has been developed giving priority to the boundaries. Under the main theme of the study with camping and recreation tourism, health and sports tourism and cultural and scientific tourism as sub-headings, three scenarios have been developed. In this study AutoCad 2012 software was used to transfer the boundaries of the study area to computer and Photoshop CS5 software was used to prepare the visual layouts. Proposed scenarios and developed strategies for the area and its surroundings are presented in the form of landscape planning layouts. Physical planning scenarios were formed to develop the tourism potential of the area, taking into account the working scale of the development process cover to be five years for the 1/5,000 scale and twenty-five years for the 1/25,000 scale.

Physical Geography Characteristics of the Area

Geographical location: The Hazim Dagli Natural Park is located within the boundaries of the Yaprakli District of the Cankiri Province in the Central Anatolia Region. The nature park is located in north of the Yaprakli District and the river basin of Central Kizilirmak, surrounded by the Ilgaz and Tosya districts in the north, the Cankiri Central District from the west and south, and the Bayat District from the east. The geographical coordinate of the area is 40° 50'04"–40° 49'20" north latitudes and 33° 45'51"–33° 46'54" east longitudes.

The Hazim Dagli Natural Park was determined in 2004 as the "Yuklukoy Plateau Picnic Place" with the Cankiri Province Environment and Forestry Directorate proposal and approval of the Ministry then expanded it to include a pond made by the State Hydraulic Works 5. The Region Directorate announced it as a "Nature Park" in August 2009.

The area comprises 39.5 hectares wooded areas, 84 hectares open areas, and 2.5 hectares ponds, totalling 126 hectares. The south and east of the area is surrounded by forested areas, while north of the area the Yenipazar Plateau and southwest of the area the Buyuk Plateau are available. The nature park is not only used as a recreation area also as a plateau (such as Kucuk, Satikaya, Yukarioz, Asagioz, Saricicek plateaus).

Geological and geomorphologic features: Ophiolitic series and basalts make up the geological structure of the Hazim Dagli Natural Park. Neogene inner sea sediments are dominant in region which is bounded by the Ilgaz Massif elevation in the north. Oligocene conglomerates show an interesting thickness in the region which is represented by marine gray marls from the Eocene period, thick red-coloured conglomerates from the Oligocene period, Mioseevaporit series and Pliocene conglomerate and sandstone. The region is located in to the north of Anatolide and south of the North Anatolian Fault Zone, and between the Ilgaz Massif and the collapsed zone of Central Anatolia. Large faults have often occurred in the area (Calapkulu & Shadaydeh 1967), and secondary faults in these tectonic axes formed the current topography of the area.

The park area has a gently rolling topography surrounded by different elevations, on a slight slope from the southwest to northeast, drawing attention to the forest area in the centre.

The ridge forms the north and west boundaries of the area which combines the Gokmenpinar and Eyriova Hills. The nature park forms the highest point of the area in the Gokmenpinar Hills located in the south at an altitude of 1,731 m, as are the hills which form its eastern, southern and southwestern borders. The nature park has plateau features with hills and ridges of a non-rough geomorphology which occur as a result of geodynamic movements. The most sloping places of the area are at 10% in the forest areas surrounding it, covering 76.6 hectares or 61% of the area. The open areas in the forest have 5% slopes, covering 49.14 hectares or 39% of the area (Milli Parklar 2010).

Climatic features: The Yaprakli District is located in the Black Sea climate zone of the specific Central Anatolia Region terrestrial transition zone. Yaprakli Meteorological Station data and Cankiri Province Weather Station data were used for the climatic data of the area. The Yaprakli District has cool winters and warm summers, with an annual temperature of 11.1 °C in Cankiri. The highest average temperature is 23.1 °C in July and the lowest is -0.6 °C in January. The annual average humidity is 65%. The average annual rainfall is 392 mm. The average wind speed is 1.6 m/sn and the dominant wind direction is west-northwest in Cankiri. The altitude of the Yaprakli District is 1,225 meters and the annual average rainfall is 538 mm. The average relative humidity changes between 61% and 67% in the nature park region. The strongest wind is 2.1 m/sc. The annual average temperature is 9.1 °C. Depending on the height of the area and the surroundings, the winters are cool and the summers are warm (Meteoroloji Genel Mudurlugu 2011).

Hydrography: Located in the northern part of the Central Kizilirmak Basin, the Hazim Dagli Natural Park has rivers and groundwater streams with regard to the hydrographical features. A pond and a seasonal creek that feeds the pond are available in the natural park. This creek, extending in an east-west direction of the area, is carrying plenty of water during the rainy periods but is dry in summer. In addition, there are freshwater resources in the area, and eight water resources are used for drinking water. These resources are distributed uniformly in the parking area and meet its water requirements.

Soil: The Yaprakli District has forest areas and a rich flora in general. The Hazim Dagli Natural Park includes VI. and VII. class land; 103 hectares (81.7%) are VI. class land and 23 hectares (%18.3) are VII. class land (Milli Parklar 2010).

Due to the gentle slope and forestland, there is no risk of erosion. Considering the soil structure of the area and its surroundings, brown forest soil quality can be illustrated. The drainage is best in this soil (Milli Parklar 2010).

Flora: The rich forest areas in the Hazim Dagli Natural Park are composed of *Pinus sylvestris* and *Pinus nigra* communities. In these regions the wooded areas are medium-closed with small-medium trees. In almost all areas of trees *Juniperus oxycedrus* is intensively at the bottom layer as a living cover. Also, *Alyssum xanthocarpum*, *Anthemis triumfettii*, *Arenaria serpyllifolia*, *Asyneuma rigidum*, *Bunium microcarpum*, *Campanula glomerata*, *Carthamus lanatus*, *Centaurea triumfettii*, *Epilobium hirsutum*, *Erodium cicutarium*, *Globularia trichosantha*, *Lapsana communis*, *Leguosia pentagonia*, *Pilosella piloselloides*, *Rumex* *scutatus*, *Silene vulgaris* and *Tripleurospermum oreades* species are available in these areas (Milli Parklar 2010).

Carex erophila, *Cerastium gracile*, *Cirisum elodes*, *Holosteum umbellatum*, *Juncus inflexus*, *Plantago lanceolata*, *Ranunculus arvensis*, *Salix alba*, *Sisymbrium altissimum*, *Typha shuttlewortii* and *Veronica anagallis-aquatica species* are available in the pond and its surroundings at 1,661 m. *Anchusa leptophylla*, *Ballota nigra*, *Crocus ancyrensis*, *Digitalis lamarckii*, *Scorzonera tomentosa* and *Senecio hypochianeus* species are region endemic plants (Milli Parklar 2010).

Anchusa leptophylla, *Ballota nigra*, *Cirsium alatum*, *Crocus ancyrensis*, *Digitalis lamarckii*, *Helichrysum arenarium*, *Onosma mutabile*, *Scorzonera tomentosa*, *Senecio hypochianeus*, and *Stachys cretica* species are endemic plants of the Hazim Dagli Natural Parks (Milli Parklar 2010).

In order to determine the flora of the region, plant sampling was undertaken in different vegetation periods, and 32 families, 90 genders and 110 taxa have been identified (Dölarslan & Gül 2012).

Fauna: *Rana macrocnemis*, *Testudo graeca*, *Alectoris chukar*, *Merops apiaster*, *Sturnus vulgaris*, *Erinaceus concolor*, *Ophisops elegans*, *Saxicola rubetra*, *Lepus europaeus* and *Crocidura leucodon* species are available in the area. Also, *Tachybaptus ruficollis*, *Egretta garzetta* and *Ardea cinerea* species have used in the pond as housing and feeding. In addition, *Natrix natrix* are found in the area (Milli Parklar 2010).

Transportation: The Hazim Dagli Natural Park is located 44 km from the Cankiri Province and 14 km from the Yaprakli District. Transportation in the area is possible via the Ankara-Cankiri-Kastamonu-Inebolu (D765) way in a north-south direction or upon the Cerkes-Ilgaz (E-80) way in a west-east direction to the Cankiri Province Centre, and from this point on the 30 km highway to the Yaprakli District Centre and the 14 km stabilized forest way. Transport to the area is provided by private vehicles and it currently has no public transport vehicles (Milli Parklar 2010).

Conlusion and Recommendations

The purpose of this study is to provide landscape sustainability in the study area by creating a place for tourism and recreational activities and to contribute to the local economy.

Scenario 1: Camping and Recreation Tourism

Vision: In this scenario, the topographic and geographic features of the area have been evaluated on the basis of the balance of protection and use. The area has been intended for use in daily active/passive camping and recreation activities. With this scenario the future of the area has been observed as a recreation-oriented attraction point by uncovering and developing its existing potential.

The conceptual elements of this scenario are social centre, glass tunnel, observation terrace, accommodation, and camping ground.

Predicted time periods for the realization of this scenario:

Five years for the formation of the concept to cooperate with the municipality and relevant departments and mobilization of the project

Twenty-five years for the formation of camping areas and implementation of the social centre.

In the camping and recreation tourism scenario:

The area is evaluated as an integrated tourism context which has important natural and cultural landscape values, such as Cankiri, Ilgaz, Tosya, and Korgun, and close to the values in terms of geographical features of the Ankara, Kastamonu, Karabuk and Corum provinces. The area has been developed within these settlements as an alternative to camping and recreation areas.

Daily and accommodation camping sites have been planned in the area, and sample designs have been made for the camping types. The proposed camping area is considered in terms of potential of the area and amount of visitors.

An A-type camping area has tents, caravans, motor caravans and bungalow type facilities with high potential for visitors, while B-type camping areas have recreational resources values and visiting potentials, providing opportunities for daily use.

In terms of daily use, a camping zone, a rest house (bungalows) zone, and a social centre and other recreational facilities zone have been determined in the use zone. Picnic units, sports area, a children's playground, parking, s buffet, s fountain, s toilet, an oven and social use units have been designed as examples in the daily use zone.

Caravan camps, motorhome camps, tents, a rural casino, a rural café, and pedestrian sidewalks have been considered for organized camping areas, unlike in the daily use zone. Bungalows and overnight units have been considered in the camping zone.

In this concept a social centre, an administrative centre, food and beverage units, a theater and cinema halls have been proposed. Accommodation units are designed to be al wooden units. Taking into account the geographical location of the different heights of the area, the terraces are important in providing the opportunities for observation to visitors.

Glass tunnel has been considered to be as special design in order to keep it alive in every season. This design aims at winter tourism especially, providing places in which to spend more time with glass to observe the different features of the area.

Scenario 2: Health and sports tourism

Vision: In this scenario, the area has been evaluated to reveal its potential for health and fitness. The area and its surroundings have strengths as alternatives to sporting events. However, the existing potential is not considered to be sufficient. Various sports activities are planned for this concept. Within the scope of the scenario about the future of the area, developing the potential for sportive activities is proposed, so this center is designed with escape from the unhealthy urban environments to natural areas in mind.

The concept elements of this scenario are mountain biking, life courses, mini golf, observation terraces, and fishing.

Predicted time periods for the realization of this scenario:

Five years for the formation of the concept to cooperate with the municipality and relevant departments and mobilization of the project

Twenty-five years for the creation of accommodation units in the area by cooperating with tour companies, in addition to making it a touring route and moving sports clubs there.

In the health and sports tourism scenario:

The moving topographical structure of the area and the presence of water and forests makes it an alternative to the other district of Cankiri and the surrounding provinces for sports. The area within the tour route has shown enough features for various sports activities and exercise.

This concept can integrate spa tourism in the Cankiri Kursunlu. There are no buildings within the protection zone. In this context, trekking and mountain biking have been determined in this zone. The flat sloping zones of the area are reserved for mini golf. The pond in the area offers an environment for sports-fishing and boating activities. Food and beverage and accomdation units for visitors in the area are planned.

A social centre unit will be opened to make the area active in the winter months, in which various indoors activities such as table tennis and chess can be performed.

Fig. 26.1. Camping and recreation tourism scenario, scale 1/25,000

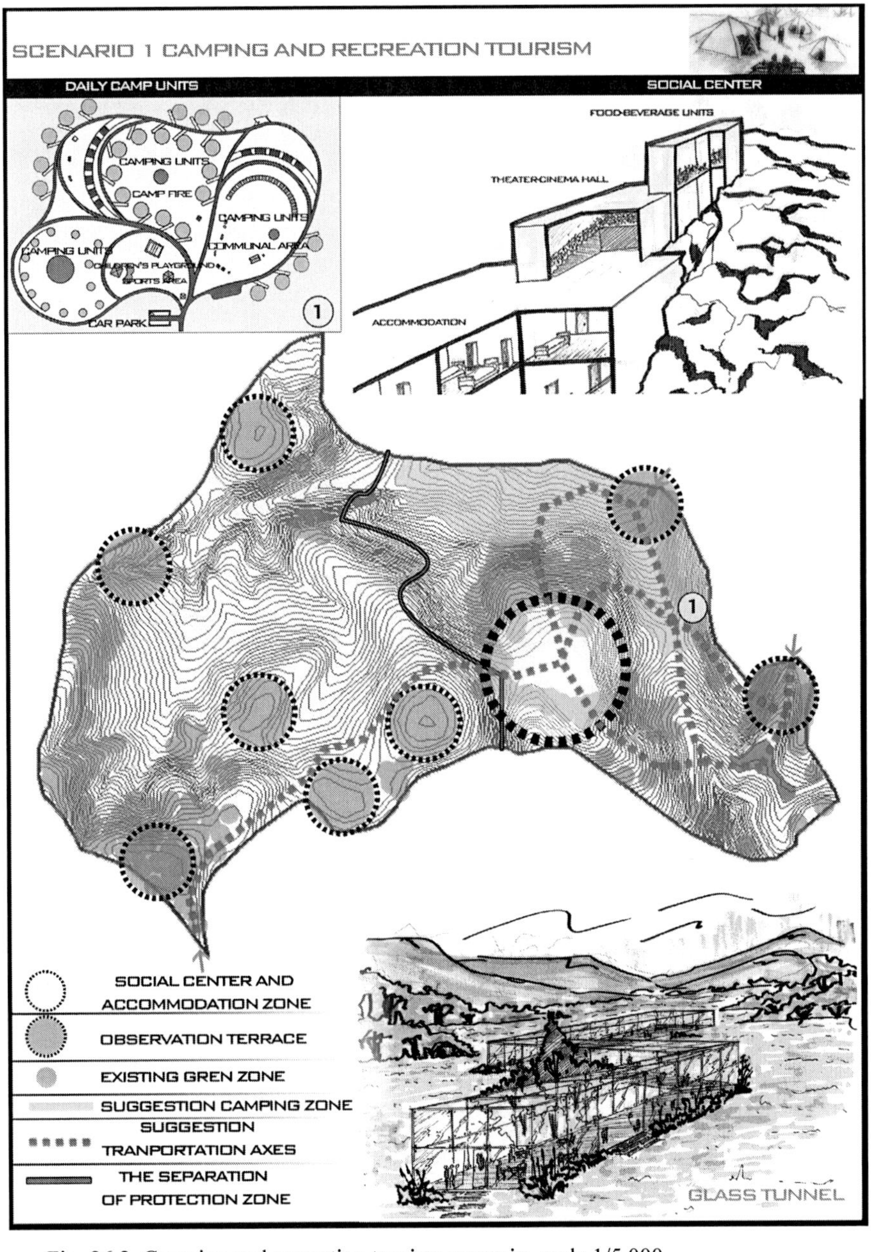

Fig. 26.2. Camping and recreation tourism scenario, scale 1/5,000

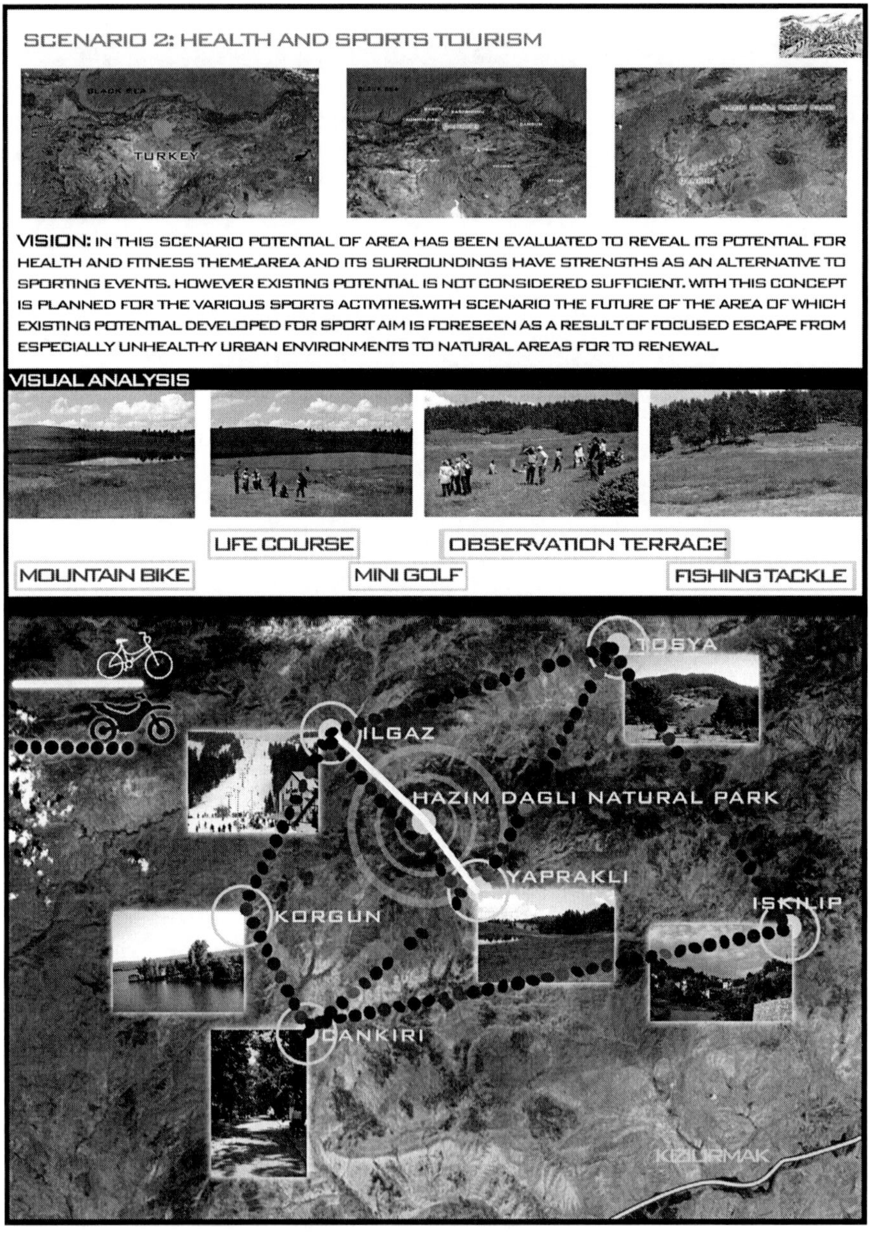

Fig. 26.3. Health and sports tourism scenario, scale 1/25,000

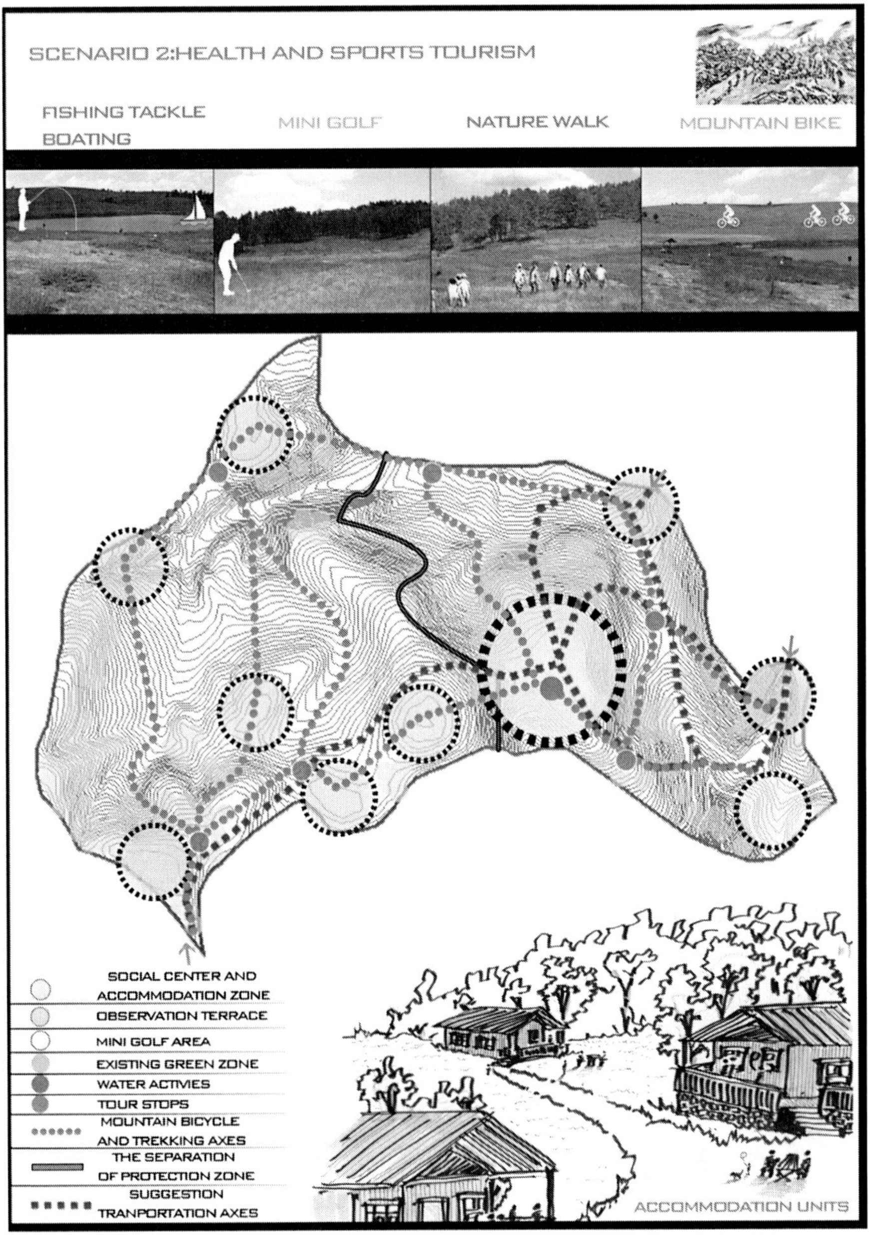

Fig. 26.4. Health and sports tourism scenario, scale 1/5,000

Fig. 26.5. Cultural and scientific tourism scenario, scale 1/25,000

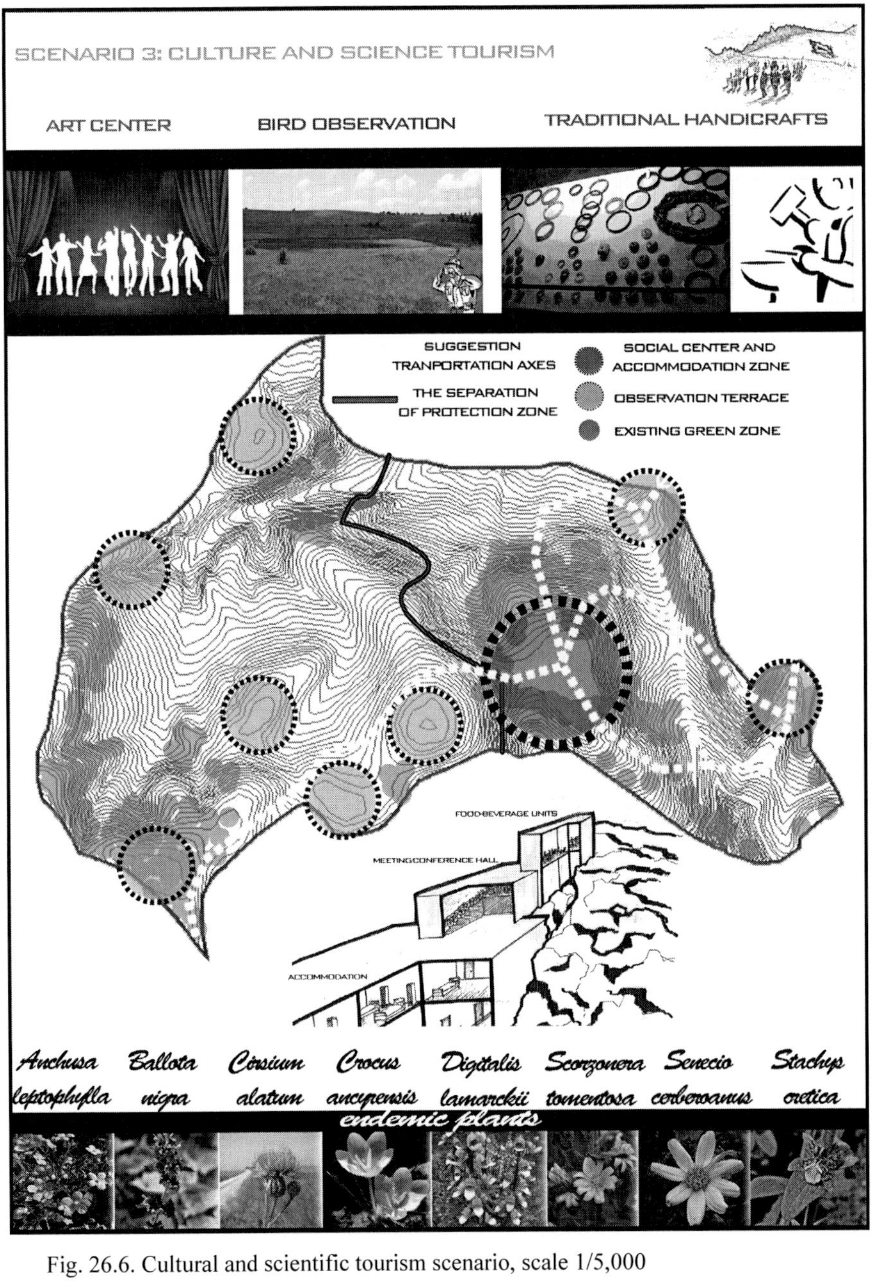

Fig. 26.6. Cultural and scientific tourism scenario, scale 1/5,000

Scenario 3: Culture and Science Tourism

Vision: This scenario aims to create a cultural and scientific direction of use in the area. Within the scope of this concept the spaces planned for culture, art and scientific activities can be performed. The future of the region can be considered as a centre for culture and science in this scenario. The concept elements of this scenario are a plant museum, accommodation, an observation terrace, an art centre, and a bird observation terrace.

Predicted time periods for the realization of this scenario:

Five years for the formation of the concept to cooperate with the municipality and relevant departments and mobilization of the project.

Twenty-five years for the creation of the art centre in the area and crafts linked to it, creating an environment for local food, art, music activities and the creation of a herbarium.

In the culture and science tourism scenario:

The area and its immediate surroundings are rich with regard to traditional cultural values. Also, natural values have great potential for scientific research and hobbies. An integrated tourism plan has been formed about the natural and cultural values of the area by integrating Cankiri's other districts and cultural activities in the surrounding provinces. In order to use the area in any season indoors, eating, drinking and accommodation areas have been considered.

In addition to meetings and conferences, traditional folk dances and cooking courses have been considered in the planned centre. The plant museum has been formed in the area for monitoring endemic and regional plants, and a bird observation terrace has been created on the high points of the area.

References

Calapkulu, F.; Shadaydeh, M.W. (1967). Çankırı-G31-d2 ve Çankırı-G31-c1 ve Çankırı G31-c2 Paftaları Jeolojisi. MTA. Ankara.

Dolarslan, M.; Gul, E. (2012). Cankırı-Yaprakli Hazim Daglı Tabiat Parkı'nın Dogal Sus Bitkileri ve Toprak Ozellikleri. Tarim Bil.Araşt. Dergisi 5(2):76-80.

Gunn, C.A. (1994). Tourism Planning, Basic, Concept, Cases. Publishing Office: Taylor and Francis, page:442, USA.

Meteoroloji Genel Mudurlugu. (2011). http://www.meteor.gov.tr

Milli Parklar (2010). T.C. Cevre ve Orman Bakanligi Doga Koruma ve Milli Parklar Genel Mudurlugu Hazim Dagli Tabiat Parki Uzun Devreli Gelisme Plani Analitik Etut Raporu. 92 sayfa. Ankara.

Weaver, D.B. (1999). Magnitude of Ecotourism in Costa Rica and Kenya, Annals of Tourism Research, 26 (4): 792-816.

CHAPTER TWENTY-SEVEN

FUNCTIONAL CHANGE AND ALTERNATIVE TOURISM POTENTIAL IN KAMIŞLI (POZANTI - ADANA), A RURAL SETTLEMENT

TÜLAY ÖCAL

Introduction

Urbanization activities that started with industrialization have accelerated with globalization. This was caused by migration from the countryside to cities due to population increase. Both the need for recreation and facilities created to address this need are the result of industrialization and urbanization. Even though it is not strictly defined, "recreation" is a basic concept in tourism and recreation geography (Özgüç 1998, 3). People living in towns and cities go up to the plateaus to relax during the summer months, staying there and engaging in various recreational activities (Somuncu 2005, 28).

In many countries of the world, people go to mountain areas for recreational purposes during the weekends and holidays to get away from the boring city life. All over the world, these houses people go to are named differently depending on the type of the house and how long people stay there. Those used during the summer are called summer houses, or second homes in general. Spending a portion of the year in these houses is very common in France, Scandinavia and North America (Özgüç 1997, 154). Social structure and the population of the cities and urban life experiences have created the need for rural areas (Müller 2007). According to a survey conducted in Switzerland, technological and global changes are the reasons behind the increased interest in houses in rural areas since the 1990s (Müller 2011). These second houses in rural areas are three to four hour drives from Oslo, Norway's capital, and outside the city, and to get to these houses people travel these distances on weekends (Overvåg 2011).

These activities we characterize as transhumance are recreational purposes in Turkey, such as cultural activities vineyards, gardening and husbandry. Although these are activities that were based on economy and livelihood in the past, they also have a social dimension. People living in the cities go up to the plateaus to live in a natural and cool environment, especially in the hot and humid summer months (Soydan 1994; Doğanay 1997). The plateaus used for this purpose are called "seyfiye" or "seyfiye yaylacılığı" (Tanoğlu 1966, 243; Tunçdilek 1967, 144; Emiroğlu 1997, 17). Using the Adana's plateaus as examples, Alagöz characterized this kind of transhumance activity as "urban transhumance," emphasizing that the cities are being depleted and everyone is going up to the plateaus (Alagöz 1993, 12–13).

Transhumance activities in rural areas are not only a way of life for Central Asian Turks but also a factor shaping Turkish culture from the past to the present. However, making nomadic tribes settle into sedentary life from the eighteenth century (Efe et al 2012, Halaçoğlu, 1997, 2006) became a turning point in the transformation of traditional transhumance in Turkish rural culture. Following this, increase in general welfare, change in socio-economic structure and urbanization laid the groundwork for transhumance (plateau tourism) done for recreational and leisure purposes (Koca 1995, 282–283). Therefore, it is possible to say that plateau and transhumance activities are in a process of change in terms of their purpose, settlement and use (Çetin 2012, 131).

The participants attending these transhumance activities want to get away from the extremely hot and stuffy environment of the coastal areas and rest in cooler areas, instead of engaging in any kind of economic activity. Today, plateaus like Tekir, Çamlıyayla, Gözne, Zorkun and Belen in the Adana Region of the Mediterranean have turned into plateau settlements with leisurely purposes, and even into seasonal or permanent settlements by going through a functional transformation (Koca 1995, 182–283). Çobanisa Village in Isparta of the Mediterrenean Region (Ceylan ve Demirkaya 2009) and Kirazlı Village in Aydın Kuşadası of the Aegean Region (Doğan Sertkaya 2009) are other examples. Also, plateaus like Bektaş, Kümbet, Uzungöl and Ayder have largely lost their traditional functions and gained new ones. Even regular settlements like the Hıdırdibi plateau city with tourism purposes have emerged (Zaman 2001, 217–2236). Given all these changes and developments in our country, the existence of a process of change from traditional transhumance activities to recreational activities and seasonal or permanent plateau settlements is evident (Çetin 2006, 178).

In this context, village settlements that are close to large metropolitan areas or in transit areas with specific historical and cultural values are witnessing an increasing number of tourism activities (Bakırcı 2012, 72). Although common in every region and area, transhumance activities differ from region to region and even differ within the same region. In the emergence of these transhumance activities, in addition to the economic and cultural differences, climatic and topographic conditions have important functions (Sevgi 1984, 177; Tıraş 2009, 189). Therefore, another region where annual vertical transhumance migration exists is the Mediterranean Region (Doğanay 1994, 268). The reason for this is the need for unspoiled and cool places in the hot and stuffy coastal parts of the Mediterranean region where the summers are really hot. The increase in urbanization, especially in Çukurova due to industrialization, the influx of people into the cities and people choosing plateaus because of their easy access on the road networks have led to a faster functional change. While the technological development brought about by globalization simplified life, it also made the cities uninhabitable. Affected by this, recreational transhumance intertwined the rural life with the urban life. In these rural areas, while all of the population was rural in the past, now most of the population is urban because of functional change. Moreover, people who left the rural areas during the 1980s and 1990s due to lack of adequate living conditions are returning to those rural areas. Retired in the city, the elderly population prefers the rural areas and settles there. Today's big technological developments make this a possibility. Appropriate working hours at the workplace make it possible for the working population to go up to the plateaus during the summer months. This leads to an increase in housing due to population increase in the villages and to wrong land usage. Pozantı and Akçatekir can be given as examples of this. Leading the way in this, the Kamışlı Village will cause untouched forest areas to disappear with its wrong land usage and disappearing vineyards and gardens.

Although many studies have been done about nearly every area suitable for plateau settlements, only the general conditions of the plateaus were studied. Looking at the Kamışlı Village of Adana's Pozantı district, this study analyzes the functional change that will happen in the future, the transformation process in transhumance related to this change, wrong land usage and the positive and negative effects of this for the rural population.

Material and Method

In this study, scientific studies written about transhumance and the research area, local sources about the area and 1:25,000 scale topographic map were used. For this purpose, fieldtrips were taken and observation was done by visiting the village continuously during the summer and fall of 2012, and data was collected through interviews. While the interviews were conducted with the village headman and the villagers, interviews were also conducted with the people who bought land for transhumance.

In this study, the functional change process of the Kamışlı Village in the past and the future, with its reasons and results, was examined, and the characteristics of the traditional transhumance in the village were determined especially. From the Republic days to the present, in addition to the geographic factors that led to functional change, the land usage, the purpose of transhumance, settlement and the effects of housing on the land were determined. The reasons for transhumance activities in the village, its present situation, its problems and its future were also examined.

Findings and Discussion

Location and natural geography characteristics of functional change

Kamışlı Village was founded in the Adana Section of the Mediterranean Region in the Middle Toros Mountains next to the Ecemiş fault line. It is the largest village settlement of Adana's Pozantı district. At 1,140 m above sea level, the village of Kamışlı is made up of mountainous slopes and uneven surfaces. The land is divided by the branches of the Körkün River, the most important river of Kamışlı, flowing from the Ecemiş River of Niğde's Çamardı district directly by the Kamışlı Village, cutting across the mountainous area of Adana's Karaisalı district and into the Mediterranean after joining the Seyhan River. The general slope of the land is towards Çukurova. The destruction of natural vegetation around the village to open up areas for vineyards and grazing animals has continued until today. Kamışlı Village is 16 km to Pozantı, 115 km to Adana, 60 km to Çamardı and 102 km to Niğde. It is on the edge of the Körkün River, which is in the Middle Toros Mountains and separates the Aladağlar with a valley, and is on the road that connects Adana and Pozantı to Çamardı (see Fig. 27.1 below).

Bordering the Aladağlar in the west, the fault line called the Ecemiş Corridor cuts the Toros Mountains in a NE-SE direction (Ketin 1983, 530). The faults that make the Ecemiş Depression were probably formed at the beginning of the Tertiary period, and over time a 45 km horizontal shift has happened due to the continuous movement of the faults. The settlements are arranged along the fault line to the western part of the mountain. The structure here made it possible for the fertile agricultural areas to form at the base of the depression where the alluvial and colluvial soil exist. Internal as well as external forces play an important role in the shaping of Aladağlar.

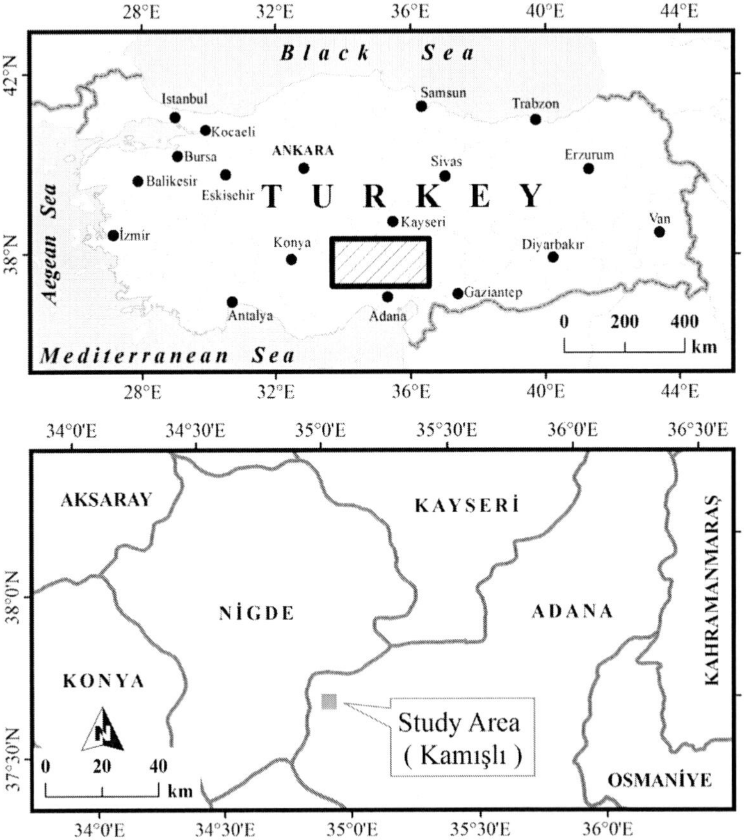

Fig. 27.1. Location of the Kamışlı Village

Polygenetic topography exists in the Aladağlar that is shaped by different processes and factors. Therefore, in the mountainous areas, fluvial, glacial and karstic landforms are intertwined. This diversity in the morphological structure in the mountainous areas creates a favourable environment for the transhumance activities. There are deeply incised valleys formed by rivers in the Aladağlar. The Ecemiş River to the west of the mountain, the Körkün River that follows the Ecemiş and many different valleys that join them form canyons in some places. Many of the plateaus are in the valleys of the branches of the main rivers that come from the higher grounds (Somuncu 2005) (Fig. 27.2).

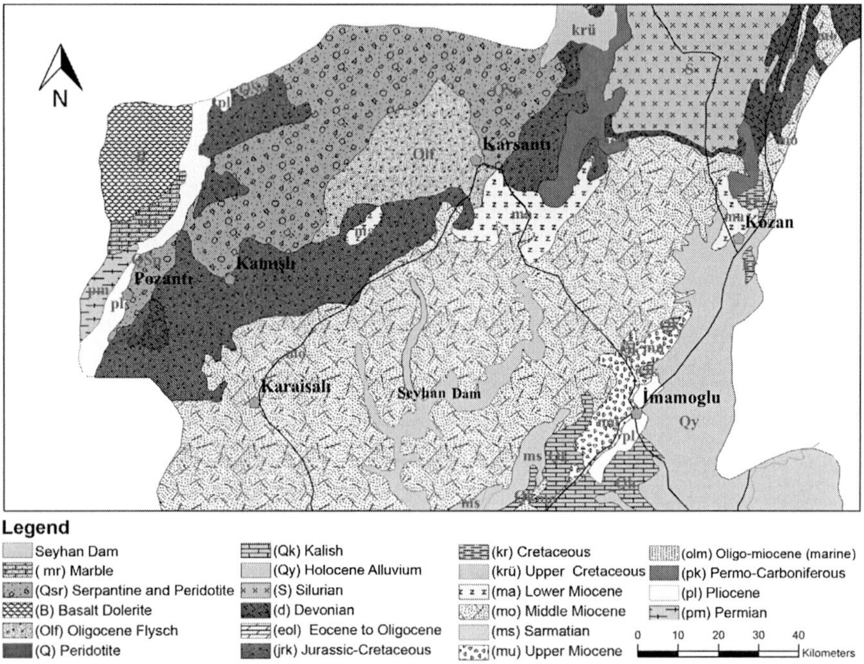

Fig. 27.2. Geological Map of the Kamışlı Village

Formed in the Alpin orogeny, the Ecemiş corridor uplifts like a large radius dome (anticlines) and subsequent undulations and folds occur within it. However, these movements become important coming to the holded area and also become main period characteristics of the folding (Blumenthal 1944). Severe and continuous orogenic movements of the Toros Mountains took place at the end of Cretaceous and during the Laramien phase. The lower Eocene exists in an uncomfortable discordance on older formations with an obvious basal conglomerate. Pebbles contain upper Cretaceous and older rock parts (Ketin 1965). These formations shape the heights around the Kamışlı Village founded along the Körkün River.

Among the hills surrounding the village, Küçük Yağmurca Hill to the north is 1,339 m high, the Kavlak Ridge to the northeast is 1,296 m, Kil Hill on the Hamidiye Ridge is 1,243 m, and Domuz Yatağı Hill to the east is 1,250 m. Among the other hills surrounding the village, Tosman Hill to the south is 1,249 m, Büyük Karahöyük Hill to the southwest is 1,573 m, Ballık Hill to the west is 1,563 m, Çatal Hill also to the west is 1,664 m, and Pandır Hill to the northwest is 1,763 m.

Table 27.1. Monthly and annual average temperatures in settlements around Kamışlı village

Months	Stations			
	Adana	Pozantı	Karaisalı	Çamardı
January	9,6	2,3	8,9	-0,5
February	10,4	3,4	9,5	1,1
March	13,4	7,8	12,4	5,0
April	17,5	12,5	16,4	10,2
May	21,8	16,6	20,6	14,7
June	25,6	21,5	24,7	18,7
July	28,2	25,4	27,6	22,0
August	28,5	25,3	27,8	22,2
September	26,1	20,4	25,4	17,3
October	21,5	14,8	21,0	11,5
November	15,3	8,4	15,0	6,1
December	11,0	3,9	10,4	1,8
Average Annual	19,3	13,5	18,3	10,9

The ones to the west, the Bolkar Mountains, consist of rocks with olistostrome characteristics and carbonate and clastic rocks formed during the Devonian Lower Tertiary period. The Aladağlar consist of shelf carbonate and clastic rocks that represent the Devonian-Cretaceous period (Özgül 1976). Surrounded by mountainous land, Körkün River in the valley is the most decisive factor for the Kamışlı village. Körkün River basin soil is zonal soil made from lime and rich rocks ranging from maroon-colored limestone to volcanic material. In addition to this, antizonal and azonal soil is also found. Soil types that can be found in the area are colluvial soils, brown soils, brown forest soils, alluvial soils, bare rocks and debris (Aladağ 1996). This valley land surrounding the Körkün

River, which flows along the Ecemiş fault, makes viticulture and gardening possible in Kamışlı Village.

Even though Pozantı's Kamışlı Village is in the Mediterranean Region, its climate shows significant differences due to it being 1,140 m above sea level. The 1,100 m difference between Kamışlı and Adana lowers Kamışlı's average temperature by 7–8°C, increases its snowy days and makes the snow stay longer. According to the Adana Meteorological Station records, the nearest station to the Kamışlı Village, the study area has an average annual temperature of 19.1°C, dropping to 13.5°C in Pozantı (780 m) due to its elevation and exposure. There are 300 m elevation difference between Kamışlı and Pozantı, causing the annual average temperature to be lower (Table 27.1).

Table 27.2. Monthly and annual average relative humudity in settlements around the Kamışlı village

Months	Stations		
	Adana	Pozantı	Çamardı
January	66	68	72,0
February	64	65	70,0
March	66	64	63,6
April	67	63	58,5
May	66	60	55,9
June	68	53	49,9
July	70	47	43,6
August	70	47	43,0
September	65	51	47,5
October	62	60	58,2
November	64	65	66,0
December	68	70	72,3
Average Annual	66	59	58,4

While the average temperature is 27.3°C in Adana during summer when transhumance is performed, this value is 23.73°C in Pozantı. This means that during the summer months the population in Adana goes up to the plateaus. The relative humidity, which makes people feel more hot than the actual temperature and makes the air stuffy, is around 66% in Adana during summer. On the other hand, at the Pozantı station this number is around 59% Table 27.2). Since the Kamışlı Village is 340 m higher than Pozantı, the relative humidity will be lower. Especially during the night, the relative humidity rises due to the decrease in temperature. The high temperatures in Adana and its surroundings and high levels of relative humidity during the summer negatively affect people's lives. While the hot and stuffy weather has adverse effects on people with heart conditions and high blood pressure, the work productivity of healthy people lessens. While the ideal temperature value for human activities in Turkey is determined to be 20.4°C, the ideal is actually is between 18.5 °C and 22.8°C (Sungur 1980). These values are even higher than the determined values around Adana during the summer. Therefore, during the summer people living around Adana go up to the village settlements in the Toros Mountains since they are cooler.

Since it is exposed to humid winds, the Kamışlı surroundings are rich in terms of vegetation. Apart from the areas used for grazing, it is covered with mixed forest trees. Above 1,300m these trees are red pine (*Pinus brutia*) and black pine (*Pinus nigra*). Some parts of the areas where people have destroyed the forest are covered with juniper (*Juniperus excelsa*) (Fig. 27.2).

On the Bolkar Mountains to the west of Kamışlı, red pine forests are common over 1,300 m, while up until a certain elevation on the Aladağlar red pine and black pine trees are common, and black pine forests are seen more often on higher elevations.

Fig. 27.2. Vegetation around Kamışlıı village

Water resources have been instrumental in selecting places to settle. The Kamışlı village, one of these settlements, was founded along the Körkün River but is also very rich in underground water sources. Also, being on the Adana-Çamardı road was instrumental in selecting the settlement place (Fig. 27.). These two factors come to the fore in the foundation and development of the village, attracting the people living in Adana.

Human factors causing functional change

During the Ottoman times, the Kamışlı village and its surroundings were the plateaus on which the nomads living in Adana's Karaisalı district wnet to graze their animals. The start date of transhumance activities in Kamışlı is not known. However, during the interviews, villagers explained their kinship with the people living in the Karaisalı district and stated that their relations still continued. With the population increase, Kamışlı's village population also increased and people who go up to Kamışlı settled here over time. The Ottoman Empire's policy of settling the nomads into sedentary lives became the beginning of a new period for the settlements. During this time, the Menemencioğulları lived a feudal life in the Karaisalı district, and the Karsantıoğulları in the Karsantı district on the Aladağlar rebelled (Baysun 1991, 150). Fırka-i İslahiye, making the nomads settle into sedentary life, affected all the settlements including the plateaus (Hallaçoğlu 1973, 1–2). During observations made in the field an old bridge on the Körkün River was observed, called Menemenci Bridge by the locals, interviews with the villagers confirm this. According to interviews with the elders in Kamışlı, the first settlements happened 150–200 years ago, coinciding with the time when the Ottoman Empire settled the nomads in Çukurova and on the Toros Mountains. Thus, permanent settlements emerged in Adana's Karaisalı-Pozantı and Aladağlar districts, including on the plateaus. Today, the Kamışlı village is a permanent settlement.

Fig. 27.3. Kamışlı Village Valley

The Kamışlı village was founded by the nomads settling to the northwest of the settlement. Because of this, the old name of the village is "Yukarıyirikler" (*yörük* is "nomad" in Turkish); over time, "yörükler" became "yirikler." According to the information received from the villagers, the first settlement is dated to around 1800. Later, other clans also settled in the village. The villagers in Yukarıyirikler move 3 km to the Pozantı-Çamardı road after 1960. Founded on the valley on flat land, the Kamışlı Village took the name Kamışlı in 1988. It is a significant settlement due to its proximity to other villages and being in the center of these villages. It is 16 km to the Pozantı district, and its population became

1,600 by 2013. Kamışlı Village started a vineyard culture and gardening making a functional change in the valley, since 1960. In addition to this, husbandry and forestry activities have continued until recently. However, the functional activity of vineyards and horticulture would eventually not meet demands due to population increase, and the young population of the village left to work and study after 1980–1990, living mainly in Adana.

During this time, there have been developments in the road network systems of Kamışlı. The Pozantı-Çamardı road was asphalted and this made the transportation easier, and which also intensified the relations between Adana and Kamışlı. With the road network system and more villagers buying cars due to the developments in Turkey, driving back and forth between Adana and Kamışlı became easier.

Now a village, Kamışlı was a township from the proclamation of the republic up to the 1960s. The villages of Aşçıbekirli, Çamlıbel, Hamidiye, Karakışlakçı, Yağlıtaş, Yazıcak and Yukarı Belemedik are linked to Kamışlı. Since the Kamışlı Village is at the centre, its social activities are abundant. The population in Adana began transhumance housing in the surrounding villages after 2000 because there was no land for housing in the centre of Pozantı and in Akçatekir. Pozantı's villages (the Alpu and Kamışlı plateaus) were packed with houses after 2000 and moved toward forest areas. In recent years, the Pozantı Forestry management has taken control of the construction of housing in the forests. Located in the vineyards and gardens of the valley, people started to take land from Kamışlı to make houses on the plateau. Since viticulture and gardening are labourious work, the villagers began to sell their vineyards-gardens.

Transhumance tourism in Kamışlı has shown rapid development after 2010. This big village settlement does not have a legal status or authority unit. Moreover, various problems such as uncontrolled housing and lack of control arising from the lack of governmental departments have started to emerge. Being at the centre of surrounding villages the crowded population also livens up the social structure (see Fig. 27.4 below). Every Friday people from the surrounding villages come to the bazaar/open market for shopping and this leads the population to rise twice or even thrice every Friday. There is one health centre, one post office and one elementary school serving the population. With these services, electricity came to the village after 1980. As drinking water, there is a spring and a well in the village. Sewage and a sewage treatment plant have been available since 2005, and the sewage water flows into the Körkün River after the treatment process. Since 2008, the waste has been collected and dumped 1 km away. As a result of these developments, the population of Adana have begun to escape the summer heat and to stay in the village. In the field studies, it is observed that construction will gain further momentum in the coming years because the villagers plan to make money by selling their vineyards and gardens.

During the observations, it was found that there are 20 grocery stores, one pharmacy, one jewellers, six carpenters, four butchers, two restaurants, one store, two coffee shops, two internet cafes, two car repair shops, two tire repair shops, two flour mills and one bakery in the village. Even though there are 1,775 houses, only 500 are used throughout the year. Out of these, 1,275 are transhumance houses and are not used during winter. Based on the interviews done with the village headman, it can be said that the village population is around 7,000–8,000. However, this number is expected to increase in the coming years because the Akçatekir public minibuses that run to Adana go through Kamışlı and serve the villages all the way to the Niğde border. This continuous transportation system attracts the Adana population. Kamışlı is expected to be one of the favorite plateaus in the coming years. However, even though it has 2,000 hectares of vineyards, it is also surrounded by forests. This brings forth the problems that the nature and wildlife will face in the future. Therefore, an area of 31,020.0 hectares was declared as a "Wildlife Development Area" by the goverment's September 13, 2006 decree, numbered 2006/10969. This was done to enforce the "Wildlife Conservation and Wildlife Development Regulations" that went into effect with the 25,637 Official Gazette based on the fouth article of the 4,915 Land Hunting Law. Since most of the Pozantı-Karanfil Mountain Wildlife Development area is within the borders of Pozantı, the administration of the forest areas was given to the Pozantı Directorate of Forestry (Ministry of Environment and Forestry 2009). In addition, Kamışlı Hydroelectric Power Plant I and Kamışlı Hydroelectric Power Plant II were planned on the Körkün River. Foundations for these plants were laid in 2010 and their construction is continuing.

Fig. 27.4. Kamışlı Village Center

The Kamışlı Village, with its settlement area, vineyards and orchards, comprises 2,000 hectares (see Fig. 27.5 below). To water these vineyards and gardens, irrigation canals were made to the east and west of the Körkün River on the arable land by Adana Soil and Water Management. Thus, on most of the land, irrigated agriculture and irrigated viticulture and gardening are performed. Grown in the Kamışlı Village, napoleon cherries are bought by the local traders and exported to France, England and Italy. Locally grown apples are sold in Çukurova. Other than these fruits, grape and peach orchards also exist. The products received from the grape gardens are turned into molasses and dried fruit pulp and the surplus is sold in Adana. The other economic activity is husbandry, and there are 100 cattle and around 2,000 sheep in the village.

The population structure of Kamışlı has a changeable structure. As of April 1, 2013, the population of the village was 1,600. Around 1,600,000 people from Adana go here for transhumance. Akçatekir public minibuses run between Adana and Kamışlı every day of the year, and 70 serve the village people and those who go up to the plateau.

Fig. 27.5. Vineyards and gardens of Kamışlı village

Kamışlı will be an important settlement for transhumance tourism in the coming years. Environmental conditions, infrastructure facilities like electricity, roads and water and availability of land for housing make transhumance with recreational purposes possible. Therefore, the local administration and private initiatives play important roles because the number of families going up to the plateau rises everyday due to increase in technological development and prosperity levels. The fact that the number of houses and population will increase in the near future should not be forgotten.

The Effect of Functional Change on Land Usage and Houses

Today, instead of traditional transhumance activities, recreational transhumance is done in Kamışlı Village. Çukurova's climatic conditions, increase in the number of apartments in Adana due to rapid population increase after 1980 and uninhabitable city conditions during the summer months have been effective in this. Our observations and the interviews with the people who come to the plateau to rest confirm this. Adana's development as a textile industry after 1980, and the hot, humid and stuffy city air have led people to find alternative solutions. The Pozantı district's village settlements have been popular because of their proximity to Adana. With technological developments, the road networks have shortened and every family has a car. The Çukurova-Ankara highway passing through Pozantı and the yearly public minibus schedule have made this place attractive.

During the 1980s, the opening of the irrigation canals by the Adana Directorate of Land and Water to supply the vineyards and gardens around the Körkün River made Kamışlı even more important, and the population that migrated to Adana has begun to return during the summer months. Today, those with ties to the village perform viticulture and gardening and continue to live in the city. In recent years, the Pozantı Directorate of Forest Management not allowing people to build houses in the forest areas led people with better economic conditions to buy land in the village and build a plateau house. Kamışlı, with its 2,000 hectares of land, is on its way to becoming a transhumance recreation centre.

Fig. 27.6. Housing in the vineyard and garden area of Kamışlı village

Today, plateau houses that are considered as second homes are being built on land ranging from 500 m to 2,000m. Out of 1,775 houses in Kamışlı, 500 are continuously inhabited. In recent years, infrastructure such as electricity and roads has caused a remarkable increase in the number of houses. Moreover, with increasing demand in home construction, the number of businesses that will sell construction materials will also increase. In 2004, zoning works were done in the land surrounding Kamışlı but did not bring any results. Creating a nature and wildlife protection region within the mountainous forest region of Kamışlı will act as a shield for the forest area. However, this led the population with better means to vineyards and gardens. The loss of vineyards and gardens means agricultural land becoming insufficient because of population increase, and this is especially detrimental for a country like Turkey where fertile land is limited (Fig. 27.6).

Almost all the houses on the plateau are built from concrete and have gardens. Even though materials like brick are used more often, stone-made houses can still be seen today. Despite the major change in the building material in the village, we cannot talk about a system of planned streets and avenues because the village's topographic structure does not allow for a systematic settlement. In addition to this, there are also problems in building the houses in gardens and how they are situated. Although multi-storey buildings are not wanted in a plateau village, houses' areas have widened and there are even 2, 3 and 4 storey houses. Many of the people who live in plateau houses also want to maintain their home life in the city. Because of this, all the houses in the village are identified by people's income.

Kamışlı Valley's problems and suggestions

Since recreational transhumance activities in the Pozantı's Kamışlı Village will continue in the coming years, they should be developed in a planned manner. Kamışlı Village and its surroundings have problems arising from lack of planning. The desire of Adana's population to have a second house up in the plateaus increases with each passing day, and even though it is forbidden to build houses for transhumance, it still continues. In 2004, zoning works happened in the village but brought about no results. Thus, in Kamışlı Village where there are no infrastructure and planning, there are problems due to too much housing. The potential of the destroyed area in the village's forest area should be determined and the number of houses and population should be limited as a measure of potential.

Local people's houses in Kamışlı Village are made of concrete and are adapted to today's conditions. However, since the people coming to the plateau have different education and income levels, they have begun to build houses not in harmony with nature, and the villa-style houses scattered irregularly in the vineyards and gardens especially corrupt the natural landscape. The building of these houses should be controlled and environmentally-friendly houses should be introduced instead.

Houses are being built by buying vineyards and gardens in the village. However, with time, lower-income people will begin to come to this area just like they did in Pozantı and Akçatekir, and house building in forest areas will begin. Even though the forest areas around Kamışlı Village are protected under the Wildlife Conservation of the Ministry of Agriculture and Forestry in 2006, human destruction of the forest area will be inevitable because the same problem is experienced in all the plateau areas of Turkey. To avoid this, the distribution of personal and public land should be clearly identified and the needed lands should be organized. Otherwise, it is possible to foresee illegal construction in forest areas in the coming years.

In the Kamışlı research area, the Adana, Pozantı and Çamardı asphalt roads should be improved and shortened. Since people come to this place using cars, traffic gets crowded in the village and parking problems emerge, and this will only increase in the future. Therefore, the streets in the village should be improved and parking spaces must be built.

Since 2008, refuse had been collected and dumped 1 km outside of the village. While collecting the trash is not a problem within the city, it is a problem for the 1,227 plateau houses. Since the houses are scattered, collecting the garbage during the summer months creates a problem. This problem will get bigger with the increase of house numbers in the coming years. Problems in the waste dump area are expected to happen in the future. The village's sewage and water treatment plant is sufficient for the population of 7,000 to 8,000 people. The sewage discharged into the Körkün River after being treated will create a hazard for the environment.

The population that had migrated to Adana from Kamışlı return to Kamışlı for agriculture and business since they own highland homes. Most of the people coming to the village for transhumance purposes are retired families; the others are tradesmen, farmers and civil servants. People go up to the plateau during the April to November period, and the June to September period is the most crowded since the schools are on holiday at that time.

The plateau houses in the research area form a different neighborhood. With each passing year, while the number of modern houses increases, changes are observed in the villagers' houses. These houses are built on the Ecemiş fault and great dangers await them in the future. Zoning works should start immediately in Kamışlı, and during such planning the fault line should be taken into consideration.

Conclusion

The nomadic lifestyle of the Mediterranean Region's Çukurova transformed over time due to husbandry and has become a permanent settlement. Kamışlı Village, in Adana's Pozantı district, is one of these settlements and it has begun to be used for transhumance purposes during the summer months, encouraged by the heat and humidity of the region. Kamışlı is a village settlement on 2,000 hectares of land that is engaged in viticulture and gardening. However, in recent years the population living in Adana have started to go up to the plateau and have built 1,227 houses to date. Therefore, in the study area the transhumance activity of people from Adana emerged and functional change has happened in permanent settlements.

Changing every year, the plateau population is made up of around 7,000 to 8,000 people. The dense transportation to the village during the summer did not disconnect the people's ties to Adana. On the contrary, the daily comings and goings from Adana to Kamışlı are continuous during the summer.

For the continuation of the current transhumance activities in Kamışlı and the efficient and sustainable use of land, it should be planned by taking the Ecemiş fault into consideration.

In fact, the main issue in all the plateau settlements in Turkey is the increase in the number of plateau houses aimed for recreational purposes and the transformation of these settlements after going through functional change. This means the pollution of the mountains and plateaus that encompass natural and biological richness, and water resources in settlements like Kamışlı Valley in the coming years. Also important for big settlements, polluting the water at its source will be the cause of unsolvable problems.

References

Aladağ, C. (1996). Physical Geography of Çamardı District in the City of Niğde Graduate Degree Thesis. Selçuk Univeristy Social Science Graduate School Department of Geography, Konya.

Alagöz, C.A. (1993). Transhumance Research in Turkey. Ankara Üniversitesi Türkiye Coğrafyası Araştırmaları ve Uygulama Merkezi, Türkiye Coğrafyası Dergisi, Ankara Üniversitesi Basım Evi, Ankara, 43.

Bakırcı, M. (2012). Tourism's Effect on Diversifying Economic Activities of Rural Settlements: Yayalar Village Case (Yusufeli/Artvin). Türk Coğrafya Dergisi, Sayı: 57, 71-85. http://www.tck.org.tr

Baysun, C. (1991). Cevdet Paşa Tezakir 21-39. Atatürk Kültür, Dil ve Tarih Yüksek Kurumu, Türk Tarih Kurumu Yayınevi, 150.

Blumenthal, M.M. (1944). Folding Unconformities of Çakıt River Strait, Türk Coğrafya Dergisi, Sayı: V-VI, Ankara, 118-132.

Ceylan, S.; Demirkaya, H. (2009). A Rural Settlement Developed due to Winter Tourism: Çobanisa Village (Isparta). Doğu Coğrafya Dergisi, 21, 79-94.

Çetin, B. (2012). Geographical Characteristics Functional Change of Alan Valley (Kırıkan-Hatay). Doğu Coğrafya Dergisi, Volume: 17, Sayı: 12, 129-150.

—. (2006). Recreational Activities that Emerge with Functional Change in Kefe Valley (Sefirhisar-Denizli). Doğu Coğrafya Dergisi, Sayı:15, 175-150.

Doğan, Sertkaya. Ö. (2009). Ecotourism Activities in Turkey and The Example of Kirazlı Village (Kuşadası-Aydın), Management and Education Academic Journal, 5(1): 45-48.

Doğanay, H. (1997). Turkish Human Geography. Milli Eğitim Bakanlığı Yayınları: 2982 Bilim ve Kültür Eserleri Dizisi: 877, Eğitim Dizisi: 10, İstanbul, s.276.

—. (1994), Turkish Human Geography, Gazi Büro Kitabevi, Ankara.

Efe, R., Soykan, A., Cürebal, İ., Sönmez, S. (2012). Madra Dağı'nın Kuzeybatısında Yaylacılık Faaliyetleri. Kazdağı ve Madra Dağı Belediyeler Birliği, Kaz Dağları Ulusal Çalıştayı. 12-14 Ekim 2012, Ayvalık-Balıkesir, Bildiriler Kitabı s. 107-124.

Emiroğlu, M. (1997). Plateaus and Transhumance in Bolu. Ankara Üniversitesi Dil ve Tarih Coğrafya Fakültesi Yayınları No: 273, Ankara, 17.

Halaçoğlu, Y. (1973). Fırka-i İslahiye ve Yapmış Olduğu İskan. Tarih Dergisi, sayı:27, İstanbul, 1-20.

—. (2006). Ottoman Empire's Settlement Policy and settlement of Nomadic Tribes. Türk Tarih Kurumu Yayınları, Ankara, 28-42.

Ketin, İ. (1983). A General Overview of Turkish Geology. İstanbul Teknik Üniversitesi Kütüphanesi, Sayı:1259, İstanbul.

—. (1965). Anatolia's Tectonic Units. MTA Dergisi, 66, Ankara, 20-37.

Koca, N. (1995). Transhumance Torism in Gözne. Doğu Coğrafya Dergisi, Sayı:1, Atatürk Üniversitesi Kazım Karabekir Eğitim Fakültesi Coğrafya Eğitimi Bölümü, Erzurum, 282-283.

Müller, D.K. (2007). Second Home in Nordic Countries Between Common Heritage and Enclusive Commodity. Scandinavian Journal of Hospitality and Turizm, Vol, No:3, 193/201, from http://www.tandfonline.com/page/terms-andconditions

—. (2011). Second Home in RuralAreas: Reflections on A Troubled History. Norsk Geografisk Tidsskrift Norweigian Journal of Geography, Vol: 65, 137/143. From http://www.tandfonline.com/foilsgeo20

Overvåg, K. (2011). Secondhome: Migration or Circulation. Norsk Geografisk Tidsskrift - Norwegion Journal of Geography, 65: 3, 154, 165. From http://www.tordfonline.com;loi/sgeo20

Özgüç, N. (1998). Characteristics of Torism Geography: Regions. Çantay Kitapevi, İstanbul, s.3.

—. (1997). Development and Characteristics of Seyfiye Settlements. İstanbul Üniversitesi Coğrafya Enstitüsü Dergisi, Sayı: 22, 145-162.

Özgül, N. (1976). Toros Mountains' Basic Geographical Characteristics. Türkiye Jeoloji Kurumu Bülteni, C:19. Sayı:65-78.

Paksoy, M. Y. (2006). The Flora of The Sarımsak Mountain and Körkün Valley (Pozantı-Adana). Niğde University, Graduate School of Natural and Applied Sciences Department of Biology, Master's Thesis, Niğde.

Sevgi, C. (1984). Adana İlinin Kuzeydoğu Kesiminde Yaylacılık. Ege Coğrafya Dergisi, 2: 177-197.

Somuncu, M. (2005). A Research on Transhumance in Aladağlar and Mountain Nomadism. Gündüz Eğitim ve Yayıncılık, Ankara, s.28.

Soydan, F. (1994). Recreational Transhumanism in Bozdağlar (Agean Region). Türkiye Kalkınma Bankası Turizm Yıllığı, Ankara, 293-306.

Sungur, K.A. (1980). An Essay on the Suitable and Non-suitable Temperature Distribution of Monthly Distribution in Turkey, İstanbul Üniversitesi Coğrafya Enstitüsü Dergisi, 23: 27-36.

Tanoğlu, A. (1966). Population and Settlement. İstanbul Üniversitesi Yayınları No: 1183, Edebiyat Fakültesi Coğrafya Enstitüsü Neşriyatı No: 48, İstanbul, 243.

Tıraş, M. (2009). Plateau Settlements at Yarpuz basin. Doğu Coğrafya Dergisi, Sayı:22, 18-27.

Tunçdilek, N. (1967). Turkiye Settlement geography (Köyaltı Settlement). İstanbul Üniversitesi Edebiyat Fakültesi Yayınları: 1283, Coğrafya Enstitüsü Yayınları No: 96, İstanbul, 147.

Zaman, M. (2001). An Example to Plateau Tourism Centers: Hıdırnebi Plateau City 1. Doğu Coğrafya Dergisi, Sayı:6, 217.

CHAPTER TWENTY-EIGHT

IDENTIFYING THE SOCIO-ECONOMIC FACTORS WITH THEIR IMPACT ON RURAL DEVELOPMENT (THE CASE OF FOREST VILLAGES IN THE WESTERN BLACK SEA REGION)

UFUK COŞGUN

Introduction

Development can be summarized as the positive changes in the structural features of a country or the regulation a nation as a whole in terms of economic, social or cultural aspects in a desired manner (Geray 1991). In other words, development is the act of mobilising the potential for increasing the monetary welfare level in underdeveloped countries (Gasper 2004). Rural area development policies have emerged as a result of the joint efforts of society and state so as to enhance economic, social and cultural opportunities of communities in rural areas, to help these communities reach national standards of living and enable them to participate in national development (DPT, ÖİK 2000). Social development is the process in which the efforts of small communities aimed at improving their economic, social and cultural conditions and the efforts of the state in this issue are combined and the process where these communities help integrate with the whole nation and make a contribution to national development efforts to the full extent (Geray 1981). Instead of dealing with local communities from only one aspect, overcoming their specific shortcomings and fulfilling their specific needs, it is envisaged to enable them to become integrated by establishing relationships among them (Geray 1982). Rural development has three main purposes:

1. Economic purposes which are intended for enabling people to spend more and to better meet their needs

2. Social purposes which render relationships more balanced and life more meaningful
3. Organizational purposes which help people act in a more planned way and enable them to display a specific behaviour (Yıldırak 1991).

The process of change constitutes the basis of social development. Indicators that can put the social change process forward should be made clear when determining the main principles of social development. Since its emergence up the present day, the concept of village development has been dealt with through different approaches. In time, these approaches have transformed and have begun to be grounded on more sound foundations.

The **Emotional (Romantic) Approach** is yielded in the literature in various different forms. In the beginning, it was not able to surpass the poetic romanticism (*güzelleme*) level which dealt with pastoral village life. The village and villagers' problems, pastures, the cold water of rivers, and mysterious lifestyles are praised, and the longing of the bourgeoisie for the pastoral life was given in the form of descriptions in a great number of poems, essays, stories and novels. Despite all these good intentions, this village movement discussed village life looking in from the outside and was sermonic and organised. This movement continued with the picnics organized by the intellectuals in the villages and did not lead to concrete results. An example for this type of village movement can be seen as in the 1940s to 1960s "community house" attempts by university students. In these examples of the emotional approach, the factors that led the change and development of economic and social life were not taken into consideration and were not understood, and sometimes only the situation was displayed (Çağlar 1986/b).

The **Social Services Based Approach** refers to village development, building village roads, community health care centres, schools and the provision of electricity. These approaches constitute the basis for the activities conducted by the Ministry of Village Affairs, founded in 1963. Despite the significance it has for the development of villages it has been considered that "all" people living in villages could reach a higher level of income with such kinds of services in the social service based approach (Çağlar 1986/b).

The **Independent approach** deals with village development independently from the development of other sections of society and aims to realize village development by preserving predominant production relationships in the villages without interfering with the restrictive aspects and by preserving the locations of villagers entirely. This approach includes the elements of the other two approaches. As a prerequisite for eliminating relative economic shortcomings, the importance of resource allocation that villagers need or might need for various economic activities is emphasised (Çağlar 1986/b).

The **Infrastructure Weighted Integrated approach** is the economic and social organization in non-agricultural sectors and the change and development in this organization, wherein the characteristic of production relationships in villages, the activities intended for meeting the social service requirements, and resource allocation for increasing the income level of village cooperatives and unions are considered as a whole. This approach attributes the reasons behind economic and social backwardness to the characteristics of predominant production relationships in Turkey in the general sense and to objective conditions in the production life in villages distinctively (Çağlar 1986/b).

In the 1960s, the Ministry of Village Affairs was among the ministries founded by 28th İnönü Government between 1963 to 1965. With the transition to a planned system, the ministry conducted studies for rural areas and forest villages. Within this scope the rural development projects conducted since the 1970s in Turkey are:

- Çorum-Çankırı Rural Development Project 1976–1981
- Erzurum Rural Development Project 1982–1989
- Muş-Bingöl Rural Development Project 1990–1998
- Yozgat Rural Development Project 1991–1999
- Eastern Anatolia Water Basin Rehabilitation Rural Development Project 1993–2001
- Ordu-Giresun Rural Development Project 1996–2004 (Coşgun 2005).

When the process regarding public relations and forest villagers development studies in Turkish forestry are evaluated, we can see that it goes back to the 1940s. In this period the perspective on forestry-human relations is of great importance regarding the future of forestry (Coşgun 2008). In addition to evaluating Turkish forestry in a general sense, there are some aspects concerning forest villagers within the scope of forest-human relations. According to Diker (1947), since the Ottoman Empire the forestry process is accepted as initiated with the foundation of the Forestry School in 1857 and is classified into four different periods: (a) 1857–1868 (ii) 1868–1893 (iii) 1892–1908 (iv) 1908–1920. The Republican period is classified into two (a) 1920–1924 (law on coppicing), (b) 1924–1937 (Forestry Law No. 3116).

In the period between 1950 and 1954, various special solutions for the reasons behind forest degradation witnessed in Seyhan and Ceyhan basins and solutions for the results concerning forest degradation were put forward (Peker 1958). Within this scope, opportunities for solutions concerning forest human relations and the development of forest villages have been discussed and put on the agenda. The study is regarded as important since it is a scientific research which analyzes human-forest relations at the local level.

The approaches towards forest villages in terms of forest-human relations are considered:

> … in the sense of development and the problems regarding the income levels of the people living in forest villages were brought to the agenda at the beginning of the 1950s. The discussion concerning the problem reached a legal conclusion with the "Forestry Law" numbered 6831 which was enacted in 1956. Article 13 of the code is interpreted as including both of the approaches regarding the "Protection of forests" …
>
> (Çağlar 1986)

The equivalent of rural development studies for forestry is social forestry. Social forestry can be described as the forestry concept intended for improving the life standards of the people living in a forest and at its edges, therefore regulating the forest and human relations (Özdönmez et al. 1996). According to Geray (1991), social forestry manages forest resources in a way which takes the poor rural section into account. "Forest Villages" which are generally located in mountainous and forest regions are the sections which benefit from national income at the lowest level (Esirci & Özcan 1993) and whose livelihood is through forests to a great extent. Therefore, the responsibility of increasing the welfare levels of the people living in these villages and reducing and/or eliminating the degradation they cause in the forest falls to the government. According to the applicable forestry code the applications regarding forest villagers can be summarized under three main headings:

- Direct resource transfer
- Providing employment in forestry activities
- Indirect resource transfer (Çağlar 1987).

Material and Method

This research covers forest villages in the Bolu, Karabük, Sinop, Zonguldak, Bartın and Kastamonu provinces in the Western Black Sea region, in which there are 972 forest villages in 6 provinces. The findings in this study were obtained from questionnaire data conducted face to face. Data was collected with a two phased sampling approach so as to administer the questionnaires.

A two phase method was used to determine the sample of the study. The first phase was conducted to determine the villages where the study is conducted, and the second was to determine how many enterprises (households) per province are included in the questionnaire study. Agriculture ranks first in the means of livelihood classification of forest villages. Therefore, in determining the villages sampled in the study, the average agricultural areas existing in these villages were used as data. In the first phase, villages were classified in three layers according to the average agricultural area size:

i. Villages with an average of 0–20 hectares of agricultural land
ii. Villages with an average of 26–50 hectares of agricultural land
iii. Villages with an average of more than 50 hectares of agricultural land

With formula (1), the number of villages included in the sample from each of the layers was determined (see Table 28.1 below). By using this data, it was determined that 113 sample villages from the region were used in the study. According to the number of forest villages in the provinces, it was determined in how many villages from each of the provinces were included in the study by taking the weighted averages into consideration.

$$n = \frac{S^2 * t^2 * N}{(N-1) * e^2 + S^2 * t^2} \quad \text{formula (1)}$$

N : Main mass volume = 972

t : Standard normal distribution value, (for P=95% probability value 1.96).

S^2 : Absolute main mass standard deviation.

: Sub main mass average

e : Acceptable maximum error

The second phase was to determine how many enterprises were included in the questionnaire from each province. In the first phase, data regarding agricultural area sizes of households/enterprises in the villages were obtained. In choosing the households in villages, formula (1) was used to determine how many households/enterprises were used for sampling in each layer:

i. 117 enterprises for villages with an average of 0–25 hectares of agricultural land
ii. 142 enterprises for villages with an average of 26–50 hectares of agricultural land
iii. 125 enterprises for villages with an average of more than 50 hectares of agricultural land

Table 28.1. Distribution of sampled villages according to province

PROVINCES	I. Layer		II. Layer		III. Layer		General Total	
	No. of villages in the layer	No. of villages to be sampled from the province	No. of villages in the layer	No. of villages to be sampled from the province	No. of villages in the layer	No. of villages to be sampled from the province	Total no. of villages in the provinces	Total no. of villages to be sampled from the provinces
Bartın	23	3	60	7	43	5	126	15
Bolu	10	1	45	5	48	5	103	11
Karabük	27	3	39	5	23	3	89	11
Kastamonu	72	9	124	15	203	23	399	47
Sinop	25	3	111	13	70	8	206	24
Zonguldak	29	3	12	1	8	1	49	5
Western Black Sea Region	186	22	391	46	395	45	972	113

It was determined that 384 households/enterprises in total were used in the questionnaire. The number of forest villages in each layer were distributed according to the weighted averages (see Table 28.2 below).

Sixty-two variables were investigated as the variables assumed to affect rural development, and were reduced to 39 during the factor analysis phase. The analysis was performed with 39 variables. Socio-economic factors which are important for rural development were determined with factor analysis. In the factor analysis, the Principal Component Analysis was used as an Extraction Method and Promax with Kaiser Normalization method was used as a Rotation Method. The KMO value of the study was found to be 0.719 (see Table 28.3 below).

Cluster analysis was performed in the study as the second analysis. The purpose of this analysis is to identify whether they show a different grouping in terms of 62 different socio-economic variables which are identified in forest villages and sampled by being classified under different natural conditions according to various agricultural areas. According to Çağlar (1986), it should not be disregarded that micro level studies are very important for the studies intended for the development of forest villages. Further, it is necessary to reveal the general living conditions of forest villages at the regional and basin level, because living conditions of forest villages and the general living conditions of the regions where these forest villages are located show parallels with each other (Çağlar 1986).

Table 28.2. Distribution of the number of enterprises to be sampled from provinces

PROVINCES	I. Layer		II. Layer		III. Layer		General Total	
	No. of villages from province to be sampled	No. of agricultural enterprises to be sampled from province	No. of villages from province to be sampled	No. of agricultural enterprises to be sampled from province	No. of villages from province to be sampled	No. of agricultural enterprises to be sampled from Province	Total no. of villages to be sampled from province	Total no. of enterprises to be sampled from province
Bartın	3	**16**	3	**22**	3	**14**	3	**52**
Bolu	1	**5**	5	**15**	5	**14**	11	**34**
Karabük	3	**16**	5	**15**	3	**8**	11	**39**
Kastamonu	9	**48**	15	**46**	23	**64**	47	**158**
Sinop	3	**16**	13	**40**	8	**22**	24	**78**
Zonguldak	3	**16**	1	**4**	1	**3**	5	**23**
Western Blacksea Region	22	**117**	46	**142**	45	**125**	113	**384**

Table 28.3. KMO and Bartlett's Test

Kaiser-Meyer-Olkin Measure of Sampling Adequacy.		.719
Bartlett's Test of Sphericity	Approx. Chi-Square	4507.612
	df	741.000
	Sig.	.000

Results

Various research projects and implementation projects were performed for achieving the development of the population living in rural areas. The factors which can affect rural development were not emphasised. Therefore, it is important to identify the criteria in the beginning, and the studies conducted by taking these criteria into account will be more fruitful. Sixty-two different socio-economic factors which can be important for rural development were used in the analysis of variables. With this analysis they were reduced to 39 variables in the factor analysis process. The "Agricultural Area" factor ranks first in the socio-economic parameters group which can be efficient for the development of forest villages in the analyses conducted. "Degradation" ranks second, "Fruit Yield" ranks third and "Animal Husbandry" ranks fourth. It can be seen that the Western Black Sea region forest villages have similar structures with other agricultural structures in terms of animal presence. In the period from 1980 to the present day, the presence of sheep and goats has decreased gradually. Cattle breeding was replaced by sheep and goat breeding (Coşgun 1999). However, cattle breeding is not performed in a systematic manner. Animals graze outdoors for long hours, there is no informed pasturing system and the fodder needed for the animals is not available. Furthermore, the fifth factor is "Migration," the sixth is "Agricultural Mechanisation," the seventh is "Corn Area Yield," the eighth is "Population," the ninth is "Animal Husbandry Problems," the tenth is "Organisation," the eleventh is "Village Leadership," the twelfth is "Deciding the Use of Income in the Family," the thirteenth is "Desire for Development Projects," the fourteenth is "Satisfaction" and the last is "Utilisation of Firewood."

According to Coşgun (2005), the first factor is "Agricultural Areas," the second is "Forest Crimes," the third is "Fruit Yield" and the fourth is "Animal Husbandry." It can be seen that in both studies, the factors which rank at the top are similar. Among the factors intended for the development of the community agricultural activities rank first and this is an expected development because it represents the main resource of the livelihood of forest villages.

In the study including eighty socio–economic variables which impact rural development, the first factor is "Agricultural Areas," the second is "Degradation," the third is "Animal Husbandry," the fourth is "Forest Crimes," the fifth is "Housing Conditions," the sixth is "Migration," the seventh is "Agricultural Mechanisation," the eighth and ninth are "Village

Solidarity" and "Succeeding Together" and the last is "Commercial Relations" (Coşgun 2005).

In the analyses performed, not only factors consisting of economic variables but also factors consisting of social variables are underlined, especially significant is the "being organized" factor, which manifests itself within the scope of cooperative membership. An organized structure displayed by local people in rural development necessitates unanimity in the decisions to be made and joint responsibility. An organized rural sector is significant in terms of displaying the ability to work jointly. On the other hand, the "Village Leadership" factor is very important data for identifying the priority target group, as it is a priority for leaders in the villages to take ownership and believe in rural development. In addition, the "Deciding the Use of Income in the Family" factor is important for revealing how the income in the family is spent is a social variable which will determine how enterprises and households will be affected. Yet another social variable "Desire for Development Projects" is important in terms of determining which rural development studies will be implemented in the region. The "Satisfaction" factor presents the level of satisfaction of persons living in the village and to what extent they accept the conditions. Being satisfied with the living conditions in the village means a desire for a future in different environments, cities, etc. From this point of view it can be seen that local people display a structure adopting the environment they live in. The last factor is the "Firewood" factor, and it is impossible for a community living within forests to not make use of the forests. However, in order to ensure that people make use of these forests within legal frameworks and in a sustainable manner, rural development activities are needed; ranked at the top are "forest crimes." The assumptions for forests and the people living in them included in forestry laws and the Turkish Constitution are not enacted by the bureaucrats (Gökçe 1992). For this reason, these two factors show themselves clearly.

When we look at the socio-economic variables important for the development of forest villages in the Western Black Sea provinces, we can see that some similarities are observed . Accordingly, the first effective factor for the development of the enterprises in forest villages in the Bartın province is "Agricultural Areas," the second is "Degradation," the third is "Forest Crimes," the fourth is "Decision making in the Enterprise," the fifth is "Grazing and Vegetable Cultivation Area," the sixth is "Economic Deprivation," the seventh is "Presence of Cattle," the eighth is "Use of Wood," the ninth factor is "Population" and the tenth is "Social Awareness" (Coşgun 2005).

The first factor effective in the development of forest villages in Bolu is “Agricultural Area,” the second is “Degradation,” the third is “Forest Crimes,” the fourth is “Migration,” the fifth is “Fruit-Vineyards and Orchard Areas,” the sixth is “Village Leadership and Social Life Leadership” and the seventh is “Property” (Coşgun 2005).

The first factor effective in the development of forest villages in Karabük is “Agricultural Areas,” the second is “Forest Crimes,” the third is “Degradation Factor,” the fourth is “Employable Work Force,” the fifth is “Economic Deprivation,” the sixth is “Use of Wood” and the seventh is “Village Solidarity” (Coşgun 2005).

The first factor effective in the development of forest villages in Kastamonu is “Agricultural Areas,” the second is “Forest Crimes,” the third is “Degradation Factor,” the fourth is “Animal Husbandry,” the fifth is “Vegetable Yield,” the sixth is “Corn Yield,” the seventh is “Development of Enterprise Commercial Structure,” the eighth is “Economic Deprivation and Lack of Employable Work Force,” the ninth is “Making Use of forest for economic purposes or Illegally” and the tenth is “Village Leadership and Social Solidarity” (Coşgun 2005).

The first factor effective in the development of forest villages in Sinop is “Agricultural Areas,” the second is “Degradation Factor,” the third is “Forest Crimes,” the fourth is “Animal Husbandry,” the fifth is “Unused Agricultural Areas that Can Be Used” and the sixth is “Low Vegetative Yield” (Coşgun 2005).

The first factor effective in the development of forest villages in Zonguldak is “Agricultural Areas,” the second is “Forest Crimes,” the third is “Input problems in Animal Husbandry,” the fourth is “Barley Yield Efficiency,” the fifth is “Use of Forest Areas,” the sixth is “Low Vegetative Yield,” the seventh is “Animal Husbandry,” the eighth is “Agricultural Mechanisation and State of Housing,” the ninth is “Area for Vegetables,” the tenth is “Degradation,” the eleventh is “Grazing” and the twelfth is “Economic Deprivation” (Coşgun 2005).

It is seen that socio-economic variables important for rural development bear great similarities among the provinces forming this region. According to Çağlar (1986), the living conditions of forest villages show parallels with the general living standards of the region in which they are located. With this approach, according to Coşgun (2005), the data presented above correspond to each other.

A cluster analysis performed in this study has revealed that forest villages that constitute the region show unity. The average agricultural area sizes of the villages have been identified with stratification. It is observed that villages with different average agricultural area sizes are grouped in the same clusters. When we look at it from this point of view it supports the idea that “the living conditions of forest villages show parallels with the general living standards of the regions they are located in.”

Conclusions

Determining the socio-economic data effective in rural development is an approach that will ensure studies are permanent and effective. There are not enough resources available in Turkey. However, the approaches concerning rural development have continued since the foundation of the Republic.

Experiences in the development of rural development projects and project implementation in Turkey can be traced back to half a century ago. Forest villages are included in rural development projects directly or indirectly. However, in the implementation process of these projects no effort has been made towards finding out which subjects should be taken into account for the region and area. In fact, the efforts made to this end will ensure activities are implemented in an easier manner. In other words, it would be effective for obtaining lasting results and ensuring sustainability.

In the study in which eighty socio-economic variables important for rural development are included it is observed that there are great similarities between the factors created by the variables coming to the forefront in development and the factors that have come to the fore during the evaluation performed for the provinces. It is seen that agricultural areas, forest crimes and degradation, animal husbandry and efficiency in agricultural production are at the top of the list. Similar factors also rank at the top of the lists in this study. It is revealed that the first intervention should be made towards agricultural production among the studies regarding rural development. Heading towards an agricultural production where modern techniques are used and irrigation and infrastructure problems will be solved from conventional agricultural production should be the main purpose. It is possible for forest villages very close to forest areas which are their natural living environment to make use of forests and to achieve a planned and sustainable structure through rural development studies.

The cluster analysis performed reveals that classification of agricultural area sizes of forest villages is not a significant element for rural development at regional and local levels. It is necessary to perform new and different studies by working with similar data in other regions in Turkey and make interregional comparisons.

For the development of forest villages it is necessary to put internal and external dynamics into practice. Internal dynamics cover improving conventional agricultural and husbandry practices in agricultural enterprises. Forest villages have serious production losses as a result of their settlements. Insufficiencies of infrastructure are an indicator of the transportation of forest villages to district centres. Unity of a great number of institutions and agencies that carry out rural development studies under an organization as a "Centre/Institution for Natural Resource Management and Use" is a leading factor for external dynamics. Institutions and agencies that conduct rural development studies must review their structures and should ground their purpose, targets and investment criteria on objective principles on the basis of basins unity and turn them into rural industrial investments.

Mountain villages (mostly located inside the forests) conduct their main livelihood activities under more limited circumstances compared to villages located on the plains. Their average agricultural areas are smaller and more dispersed, and their efficiency levels are considerably limited. Thus, when introducing their products to the market they are subject to the same market conditions and contradictions emerge.

The development of forest villages which constitute the poorest part in the rural section is very important. Sustaining life in rural areas will ensure prevention of migrations from rural areas to urban areas, and the factors identified in this study reveal which variables should be taken into account in rural development. It is natural to include agricultural activities in the studies to be conducted for forest villages because the main sources of livelihood for these villages are agriculture and animal husbandry. The development of agriculture can be possible through the development of such infrastructure.

References

Coşgun, U. (1999). Sinop İli Orman İçi Köylerinin Sosyo-ekonomik Yapısı, Kalkındırılma Koşul ve Olanakları Üzerine Bir İnceleme, İzmir, s32, Ege Üniversitesi Fen Bilimleri Enstitüsü Doktora Seminer Notları (yayınlanmamış).

—. (2005). Batı Karadeniz Bölgesi Orman İçi Köylerin Sosyo-Ekonomik Yapısı ve Bu Köylerin Kalkındırılmasında Etkili Olan Sosyo-Ekonomik Faktörlerin Çoğul Sayısal Analiz Yöntemleriyle Belirlenmesi, Batı Karadeniz Ormancılık Araştırma Enstitüsü Müdürlüğü Teknik Bülten No: 11, 2005, Bolu.

—. (2008). Batı Akdeniz Ormancılık Araştırma Müdürlüğü'nün 50. Yılında Orman-Halk İlişkileri ve Sosyal Ormancılık Araştırmaları, Batı Akdeniz Ormancılık Araştırma Müdürlüğü 50. Yıl Etkinlikler "Bildiriler", Antalya.

—. (2011). ORKÖY Üzerine Değrlendirmeler, TOD Orman ve Av Dergisi, Sayı:1, Ocak-Şubat, sayfa: 10-14, Ankara.

Çağlar, Y. (1986a). Türkiye'de "Orman Köyleri" Ve Kalkındırılmasına Yönelik Etkinlikler, Ankara, s. 216, MPM Yayınları No:340.

—. (1986b). Köy, Köylülük Ve Türkiye'de Köy Kalkınması Sorunu, Türkiye Ziraatçılar Derneği Yayınları, Ankara, s. 218.

—. (1987). Orman Köylüleri Ve Küçük Üreticilik Üzerine, 11. Tez Kitap Dizisi No: 7, İstanbul, s. 59-83.

—. (1999). Meşeler Artık Gövermeli, Ankara, s. 56.

Diker, M. (1947). Türkiye'de Ormancılık, Dün-Bugün-Yarın, Tarım Bakanlığı Orman Genel Müdürlüğü Yayın No: 61, Akın Matbaası, Ankara.

DPT (2000). Kırsal Kalkınma Özel İhtisas Komisyonu Raporu, VIII. Beş Yıllık Kalkınma Planı, Ankara, s.79.

Esirci, Y.; Özcan, Y. (1993). Orman ve Orman Köylerinin Kalkındırılması, I. Ormancılık Şurası Tebliğler ve On Çalışma Grubu Raporları, Cilt II, Seri No:13, Yayın: 006, Ankara.

Gasper, D. (2004). The Ethics of Development, Edinburg University Pree Ltd

Geray, C. (1981). Kırsal Türkiye'de Toplum Kalkınması ve Kooperatifçilik (Örnek Olaylar), Tuna Yayıncılık ve Dağıtım, Ankara, s: 10-45.

Geray, A.U. (1982). Toplum Ormancılığı, Türkiye'de Toplum Ormancılığı Geliştirme Projesi, Vezirköprü, Kunduz Eğitim Merkezi Seminer Notları,Ankara, s. 35-48.

—. (1991). Toplum, Orman Kaynakları, Kültür İlişkileri. İ.U. Orman Fakültesi Dergisi, Cilt:41,Sayı:1-2, Sayfa: 57-72, İstanbul

Gökçe, O. (1992), Türkiye Ormancılık Politikası, Türkiye Sosyal Ekonomik Siyasal Araştırmalar Vakfı, İstanbul, s.330.

Özdönmez, M.; İstanbullu, T.; Akesen, A. (1996). Ormancılık Politikası, İ.U. Orman Fakültesi Yayınları, İ.U. Yayın No: 3968, Orm. Fak. Yayın No:435, I.U. Basımevi ve Film Merkezi, İstanbul.

Peker, H. (1958). Seyhan ve Ceyhan Havzalarında Orman Tahripleri, Sebepleri Neticeleri ve Bunlara Karşı Alınacak Tedbirler Hakkında Araştırmalar, Ziraat Vekaleti Orman Umum Müdürlüğü Yayınları, Ormancılık Politikaları Serisi, Sıra No: 281, Seri No: 16, Sinan Matbaası, Ankara.

Yıldırak, N. (1991). Kısal Kalkınma, A.Ü. Fen bilimleri Enstitüsü Ders Notları, Ankara, s: 18-24.

CHAPTER TWENTY-NINE

ANALYSIS OF LANDSCAPE ELEMENTS IMPORTANT FOR COASTAL GEOMORPHOLOGY IN AMASRA TOURISTIC SETTLEMENT AND EVALUATIONS ON THEIR SUSTAINABILITY

YELIZ SARI NAYIM

Introduction

The approach of various different disciplines was used in scientifically shaping and grasping the concept of landscape. The most important among these is geography that accepts landscape as a basic unit. Geographers have focused on natural landforms or physiographic units along with the dynamic relationships between human-cultural groups when defining landscape (Forman & Godron 1986; Ayaşlıgil 2008). Today, one of the pioneers of landscape ecology that has developed as a separate scientific area is geographer Carl Troll (Thomas 2012), who defined landscape ecology as the "scientific discipline focused on the study of the physico-biological relationships active in the management of the various environmental units in a region" (Forman & Godron 1986; Ayaşlıgil 2008). The different land units and the study of the physico-biological relationships stated in Troll's definition are very important for the understanding of landscape. Here, the term "various environmental units" stands for different landscape units that are different landscape elements, whereas when we refer to the study of the physico-biological relationships we mostly understand it as geographical and biological knowledge of landscape.

Landscape ecologists and geographers study the landscape as a whole (Zonneveld 1989). Here, landscape is thought of as a complex system and is assumed to be the collection of the many heterogeneous parts that interact with one another (Hall et al. 2004; Garcia Aguirre et al. 2010; Forman & Godron 1986; Soykan et al. 2012).

Heterogeneous landscapes have been shaped as a result of geomorphological and pedological processes (Conacher 2002; Anonymous 2013a), subject to the process that occurs along the surface between the lithosphere, atmosphere and hydrosphere, and resulting in the appearance of various area forms on the world's surface. Geomorphological diversity is the result of various different landscape elements that are sensitive to change. Geomorphological and pedological diversity, focusing on landscape sensitivity with respect to change, are important for the understanding of geodiversity (Thomas 2012) (Fig. 29.1).

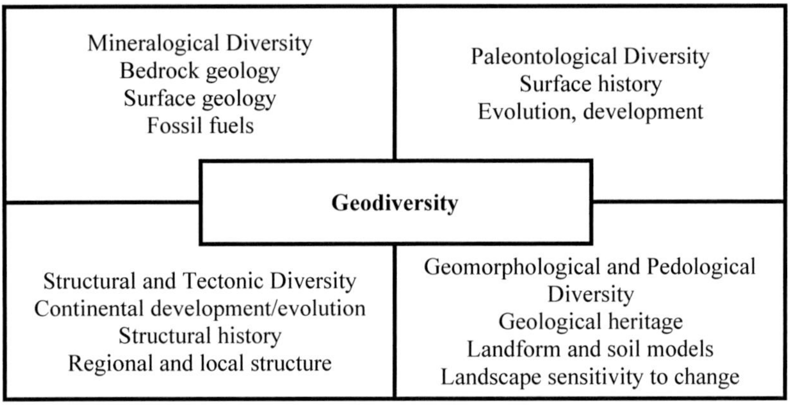

Fig. 29.1. Basic elements of geodiversity (Thomas 2012)

Results of recent studies indicate distinct relationships between geomorphological heterogeneity and biological diversity (Burnett et al. 1998; Nichols et al. 1998; Parks & Mulligan 2010; Thomas 2012; Tasser et al., 2009). For this reason, geodiversity and biological diversity are now taken into account together in nature protection studies. These explanations indicate that geodiversity should be taken into account for the protection and management of ecosystems that make up the landscape (Gordon et al. 2011; Thomas 2012).

Landscape can be defined in terms of various bio-geomorphologic land units consisting of specific biophysical elements such as litology, landform and land cover (Naveh 2007; Garcia-Aguirre et al. 2010). Bio-geomorphological land units make it easier to understand the holistic system of landscape (Zonneveld 1995; Garcia-Aguirre et al. 2012). The term "Land" mentioned here has the same meaning as "Landscape," whereas "Land unit" corresponds to the landscape element which is a basic concept in landscape ecology. This term represents the homogeneous part in landscape with regard to ecology. When landscape elements (land units) are classified, along with landform, soil and vegetation properties, human-based changes that are effective in all three are also taken into account during analysis (Zonneveld 1989).

Special geomorphological forms are important parameters in defining ecologically important landscape elements during landscape planning and management studies that play an important role in environmental protection.

The definition of landscape in terms of geomorphology not only ensures the understanding of the vegetation models in an area (Garcia-Aguirre et al. 2010) but also enables more holistic and effective decisions for the protection and designation of the landscape.

The objectives of this study are:

- to define the special geomorphological elements that contribute to the understanding of landscape in and around the Amasra settlement
- to evaluate the current protection status for the sustainability of geomorphological elements within the scope of urban land use planning and management
- to state the need for taking active protective measures in leading geomorphological elements at the coastal region of Bartın and the local scale of Amasra to attract geotourism.

Material and Method

The study was carried out in the city of Amasra, which is a historical and touristic settlement of the Western Black Sea region. Amasra is located within the Bartın city limits and its northern neighbour is the Black Sea, whereas Bartın is located to the southwest and Kurucaşile is located to the east (Sarı 2001) (Fig. 29.2).

The focus of this study has been on the analysis of significant landscape elements in terms of geomorphology. A five-stage method developed by Anonymous (2013b) has been used for the analysis of landscape elements important for coastal geomorphology. Regarding area, these are:

- understanding of the geomorphological process and properties
- reading and interpreting topographic and geologic maps
- determining the landform from the maps
- carrying out site studies (photos of the site, keeping track of site notes, determination of different landforms)
- presenting suggestions for the sustainability of the current geomorphological structure.

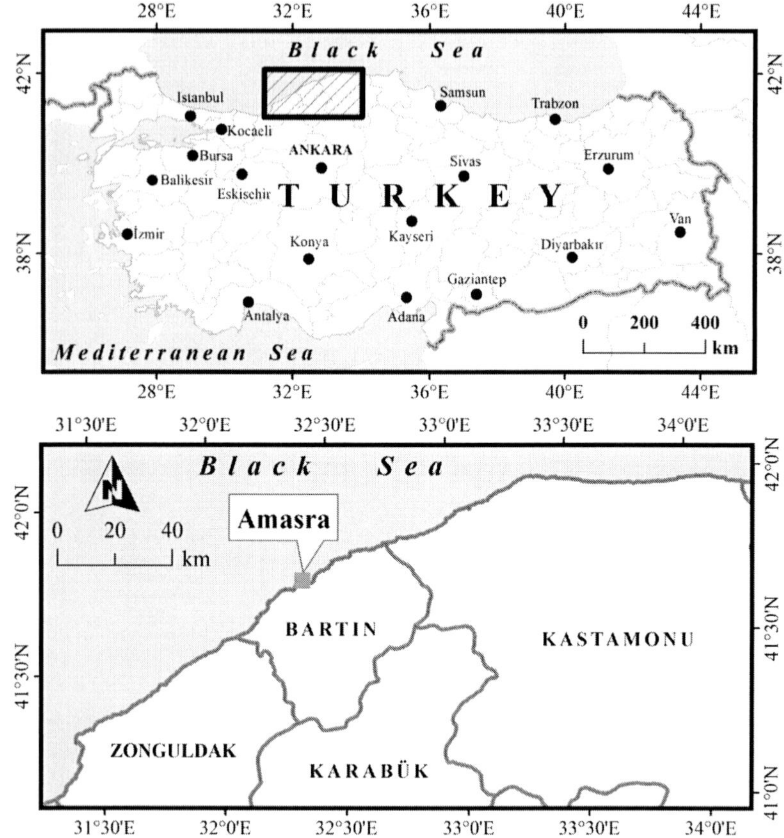

Fig. 29.2. Geographical position of the study area

Initially, various written and verbal resources were evaluated in order to understand the geomorphological process and features of the area. Afterwards, the landscape area was read via overlapping the three dimensional topographic maps obtained via geographical information systems with the geology map. The special geomorphological forms determined were supported with field studies and details recorded in the field. Photographs shot from various angles helped put forth the landforms specific to the coast. Field studies contributed to the three dimensional perception of landforms and the understanding of the entire landscape. Landscape information obtained from a geomorphological perspective specific to Amasra was evaluated with the current protection status and suggestions were developed for the study area.

Results and Discussion

The understanding of the landscape of the study area in terms of geomorphology was actualized through the findings for the analysis stages specified in the method and their evaluation. The findings of this analysis are as follows.

Understanding the geomorphological process and properties: Amasra, defined by Sakaoğlu (1999) as "the peninsula with five islands and two coves" was formed as a small peninsula along the Black Sea coast. The Amasra town centre and its environs have a different look in terms of coastal geomorphology and were formed on the foothills of the Black Sea coastal mountains. The islands that make up the backbone of the Amasra town centre are Boztepe, Zindan Island, Küçük Island and Tekke Hill. Büyük Island and Tavşan Island are in a separate location away from the peninsula formed by the merging of four islands (Sakaoğlu 1999; Sarı 2001) (Fig. 29.3).

The second largest tombolo of the Black Sea coast of Turkey is in Amasra. Various geomorphological processes have caused a cove to form along with several islands in front of it prior to its formation (Demirca 1999; Ceylan 2011). A sandbar formed with the wave effect of detritic material brought about by Hisarönü Creek first connected Tekke (Kumbahçe) Hill, which was once an island along with the old Zindan (Kale) Island with the Anatolian coast. Thus, with this tombolo the former island has transformed into a peninsula (Ceylan 2011). Natural coves, known today as Küçük and Büyük Port, also formed during the formation process of the Amasra Peninsula.

Sheer cliffs can frequently be seen in the coastal geomorphology of the city of Amasra. Limestone has prevented the formation of beaches, instead making for rocky coasts. The existing beaches are generally narrow and on the foothills of sheer cliffs (ATI 2001; Sarı 2001).

Reading and interpreting topographic and geologic maps: Topographically, Amasra can be defined as:

- Mountainous areas with slopes bordered by Kuşkaya Hill (443 m), Gölyanı Hill (415 m), Tavacık Hill (393 m), Çingen Hill (411 m) and Doruk Hill (386 m)
- Hilly areas with lower heights such as Büyükilit Hill (211 m), ending with the lower slopes and valley extensions of the mountainous ranges and Zeytin Hill (92 m)
- Alluvial flats extending from the foothills of mountainous and hilly areas

- A peninsula formed by the merging of an island with four other islands.

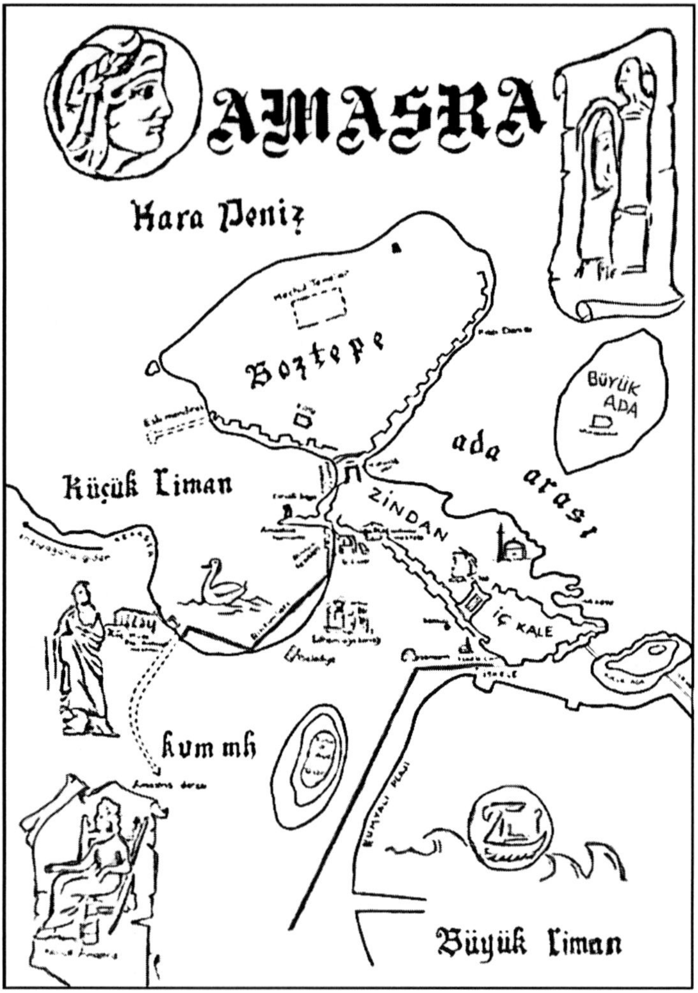

Fig.29.3. Historical map of Amasra Peninsula (modified after Anonymous 2013c)

Hisarönü Creek flows from the mountainous slopes of Amasra into the Küçük Harbor after mixing with the Kestaneli Kuru Creek somewhere close to the urban settlement. Many materials are carried to the lower parts and the Black Sea via this spring due to the steep slope of the area (Sarı Nayim 2010) (Fig. 29.4).

When the area is examined from a geological perspective it is observed that the İnaltı Formation (JKi), composed of neritic limestone, is dominant along the coast that encompasses the Amasra urban settlement. The Kilimli Formation (Kk), containing sandstone and shale, developed during the same period as the İnaltı Formation, is the best example around Amasra. Pelagic-semipelagic limestone Kapanboğazı member (Kyk) and volcanogenic sandstone, shale and limestone containing the Yemişliçay Formation (Ky) is observed around the mountainous region surrounding the Amasra settlement. The Kapanboğazı member is quite distinct at the region of the Kuşkayası Road Monument (Tüysüz et al. 2001). Quaternary aged alluvion (Qal) units appear on the flats of the Amasra settlement (Sarı Nayim 2010) (Fig. 29.5).

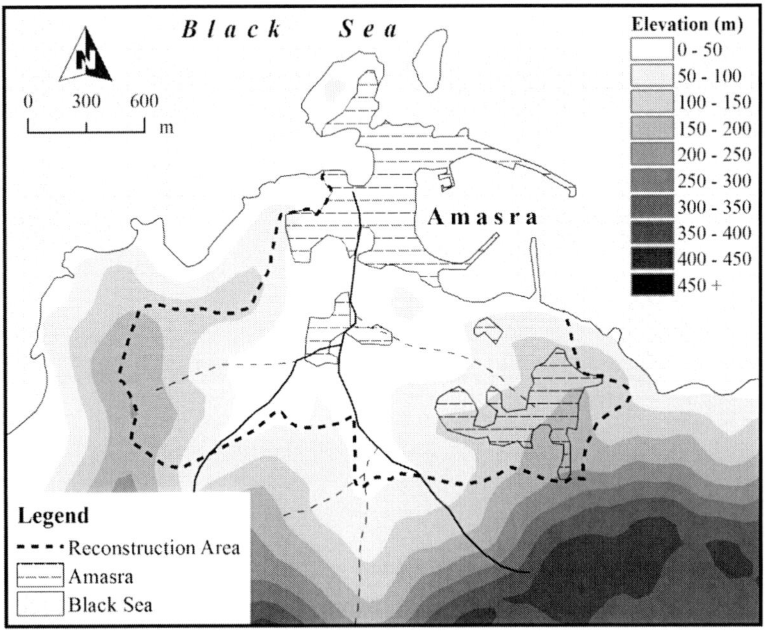

Fig. 29.4. Elevation levels of the study region.

Putting forth the landform from maps and field studies: The results acquired during the first three stages of the study were combined with the field studies carried out during the fourth stage. Accordingly, it is observed that the coastal region of Amasra has a unique geomorphological structure among the coastal ecosystems of the city of Bartın. Landscape elements that are unique in terms of geomorphology determined in this study are as follows:

- Amasra/Boztepe Peninsula and Tavşan Island
- Büyük Harbor and Küçük Harbor natural coves
- Calcareous cliffs close to the sea
- Rock blocks inside the sea with varying dimensions
- Alluvial flats
- Rocky and mountainous terrains.

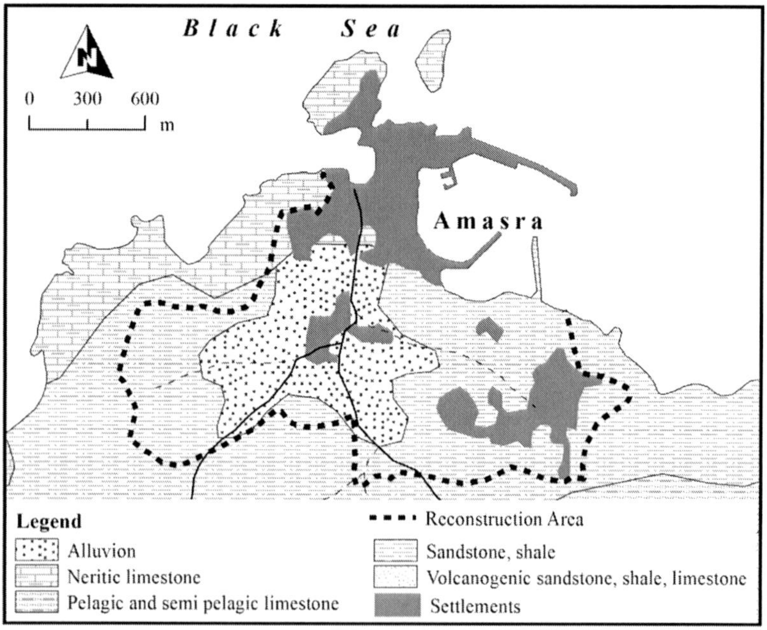

Fig. 29.5. Geological map of the studied region.

The neritic limestone coastal geomorphology of Amasra has especially caused unique landscape forms to appear. These have been determined as the Amasra/Boztepe Peninsula and Tavşan Island, Büyük Harbour and Küçük Harbour natural coves, sheer calcareous cliffs by the sea, and rock blocks inside the sea with varying dimensions. The elements that make up the Amasra/Boztepe Peninsula are Tekke Hill, Küçük Island, Eski Zindan Island and Boztepe. These similar forms have the same landscape characteristics with the Tavşan Island in terms of topography, geology and geomorphology (Fig. 29.6).

Fig. 29.6. Boztepe and Tavşan Island.

The Büyük Harbour and Küçük Harbour natural coves located on both sides of the Amasra/Boztepe Peninsula are coasts with narrow beaches that have formed at the foothills of sheer cliffs (Fig. 29.7).

The calcareous sheer cliffs close to the sea with neritic limestone properties cover the Amasra/Boztepe Peninsula, Tavşan Island and all the coves neighbouring the Black Sea at the western and eastern tips of the city. Again, the rock blocks inside the sea with varying dimensions observed at the sides of calcareous cliffs near the coast have been determined as landscape elements contributing to the geomorphological landscape integrity of the study area (Fig. 29.8).

Fig. 29.7. Amasra Peninsula and natural coves on both sides.

Hisarönü Creek, springing up from the mountains of Amasra and flowing into the Black Sea at the Küçük Harbor Cove after passing through the Amasra/Boztepe Peninsula and its environs, appears as a different landform. The areas specified as alluvial flats continue all the way up to the foothills of the hilly and mountainous areas to the south of the settlement (Fig. 29.9).

The hilly and mountainous areas surround the basin containing the Amasra/Boztepe Peninsula, and the alluvial flats that are its continuation have a unique geomorphological structure (Fig. 29.10). When the three dimensional terrain model analysis and the geological map are overlapped and examined together, it has been put forth that;

- hilly and mountainous terrains, including sandstone and shale, border the alluvial flats of the Amasra settlement from the west, south and east
- the Pelagic-semipelagic limestone and volcanogenic sandstone, shale and limestone containing mountainous terrains that border the Amasra settlement from the south contribute significantly to the geodiversity of the mountainous areas.

Current protection status of the study region: Half of Boztepe and all of Tavşan Island are under protection as 1st Degree Archeological and Natural Protected Areas. The rock block to the coast of the Old Zindan Island looking north, known as the Kuşna Cliff, is a 1st Degree Protected Area, whereas Küçük Harbor Cove is a 1st Degree Archeological Protected Area. In addition, the hilly areas and sheer cliffs that extend all the way to the sea located to the north of the Bartın-Amasra highway are 3rd Degree Protected Areas (Ministry of Culture and Tourism 2007). Amasra/Boztepe Peninsula, the two natural coves, alluvial flats, sheer cliffs and mountainous terrains are under intense pressure due to commercial settlement, transportation, tourism and recreation.

Results and Conclusion

In this study, landscape elements around the historical and touristic city of Amasra and its environs that are striking have been mentioned and suggestions have been made for the sustainability of the current geomorphological structure.

Landscape elements or biotopes that are important in nature protection and landscape planning are determined using landscape ecology research methods. Geomorphological diversity appears as a significant parameter in the determination of different landscape elements. The relationships between geomorphological diversity and vegetation distributions are especially important for understanding and defining the landscape.

This study emphasizes the importance of geomorphology in landscape ecology studies. The island, peninsula and tombolo characteristics of

Amasra have a unique geology and geomorphology with its calcareous cliffs close to the coast. Various geomorphological features of limestone have added a characteristic diversity to the landscape.

It was observed during the study that geomorphological integrity was not taken into account during the determination of natural protected areas.

Fig. 29.8. Steep calcareous coasts of the Old Zindan Island looking north

Fig. 29.9. Alluvial flats to the south of the Amasra settlement where structures abound

Fig. 29.10. Hilly and mountainous terrains bordering the Amasra Peninsula

Landscape elements that contribute to geomorphological diversity should be taken into protection not as a single unit but within a holistic system. In this scope, it will be suitable to protect:

- the Boztepe Island, Old Zindan Island, Küçük Island and Tekke Hill form the Amasra/Boztepe Peninsula with Tavşan Island
- the natural cove formations on both sides of the Amasra Tombolo
- all the sheer calcareous cliffs close to the coast
- the whole morphological structure of the rock blocks inside the sea together.

This requires the reevaluation of the current protected area decisions:

- The forest and pseudo-maquis communities surrounding the Amasra Peninsula along with the covered hilly and mountainous areas should be protected within the integrity of the landscape and should be handed down to the next generations.
- The landscape forms encountered in the study area also serve as a biotope for many different kinds of plant covers and wildlife. Significant relations between the geomorphologic and biological diversity have been put forth in detail during the biotope mapping study carried out in the region by Sarı Nayim (2010). The sustainability of geomorphological diversity in Amasra should be taken into account not only as geo-preservation but also within a wide landscape perspective that will preserve its biodiversity.
- It is observed that some of the landscape forms in Amasra that have significant contributions to the urban ecology and aesthetic of Amasra are under pressure due to residential and commercial settlement, transportation, tourism and recreation, even though they are under protection as 1st and 3rd Degree Protected Areas. This increasing pressure has resulted in the removal of the natural protection area properties of Tekke Hill, the urban settlement area of Boztepe along with Old Zindan Island and various parts to the south of the city with the latest decision about Amasra taken by the preservation board. Environmental values lost due to wrong decisions will in time result in a decrease of the expectations of Amasra regarding tourism.
- It is important that the sustainable decisions stated in this article are included in the development plan for landscaping and preservation, and the landscape elements defined in this study that have geomorphological significance should be brought into geotourism at the Bartın coastal region and the local Amasra scale.

References

Anonymous (2013a). Victoria's Geomorphological Framework. Victorian Resources Online Statewide, Australia, http://wro.dpi.vic.gov.au/dpi/vro/vrosite.nsf/pages/landform geomorphology.

—. (2013b). Geomorphic Landscape Anaylsis Project. Carleton University, Canada, http://serc.carleton.edu/NAGTWorkshops/geomorph/activities/23482.html.

—.(2013c). Amasra map. Amasra District Governorship, Amasra Museum Management web page, http://www.amasra.gov.tr/muze.html.

Ayaşlıgil Y. (2008). Landscape ecology course notes, IU Faculty of Forestry, Department of Landscape Architecture, İstanbul.

ATİ (2001). Notes of the interview carried out with Amasra Hard Coal Facilities Plan Project Department Geology Engineer, Amasra.

Burnett, M.R.; August, P.V.; Brown Jr. J.H. and Killingbeck, K.T. (1998). The Influence of Geomorphological Heterogeneity on Biodiversity:I.A Patch–Scale Perspective.Conservation Biology, 12:363-370.

Ceylan M.A. (2011). General Overview of the Tombolos with Urban Settlements on Turkey's Coastlines. Marmara Geography Journal, 23, 352-372, ISSN:1303-2429, İstanbul.

Conacher, A. (2002). A Role for Geomorphology in Integrated Catchment Management. Australian Geographical Studies 40(**2**):179-195.

Demirca, A. (1999). Coastal Geomorphology Between Bartın Creek Estuary-Amasra-Çakraz. İstanbul University Institute of Social Sciences, Master Thesis, (unpublished), Istanbul.

Forman, R.T.T and Godron, M. (1986). Landscape ecology. John Wiley, New York.

Garcia-Aguirre, M.C.; Alvarez, R.; Dirzo, R.; Ortiz, M.A. and Eng, M.M. (2010). Delineation of Biogeomorphic Land Units Across A Tropical Natural and Humanized Terrain in Los Tuxtlas. Veracruz, Mexico, Geomorphology, 121, 245-256.

Garcia-Aguirre, M.C.; Alvarez, R. and Aceves, F. (2012). Geology and Geomorphology in Landscape Ecological Analysis for Forest Conservation and Hazard and Risk Assessment. Illustrated with Mexican Case Histories, Earth Sciences, Books, InTech, www.intechopen.com, ISBN:978-953-307-861-8.

Gordon, J.E.; Barron, H.F.; Hansom, J.D. and Thomas, M.F. (2011). Engaging with Geodiversity-Why It Matters. Proceedings of the Geologists Association, 122.

Hall, O.; Hay, G.J.; Bouchard, A. and Marceau, D.J. (2004). Detecting Dominant Landscape Objects Through Multiple Scales: An Integration of Object Specific Methods and Watershed Segmentation. Landscape Ecology 19, 59-76.

Harita Genel Komutanlığı (2001). Zonguldak E28-d2, E28-c1, E28-c2, E28-b3 colored topographic maps, 1:25.000, Ankara.

Ministry of Culture and Tourism (2007). Amasra (Bartın) Protected Area Borders and Appendix. Karabük Culture and Natural Assets Protection Regional Council, Date of Decision:28.09.2007, Karar No:674, Ankara.

Naveh, Z. (2007). Landscape Ecology and Sustainability. Landscape Ecology 22, 1437-1440.

Nichols, W.F.; Killingbeck, K.T. and August, P.V. (1998). The Influence of Geomorphological Heterogeneity on Biodiversity: II. A Landscape Perspective. Conservation Biology, 12:371-379.

Parks, K.E. and Mulligan, M. (2010). On the Relationship Between A Resource-Based Measure of Geodiversity and Broad Scale Biodiversity Patterns. Biodiversity and Conservation, 19: 2751-2766.

Sakaoğlu, N. (1999). Çeşm-i Cihan Amasra, Publication of Ministry of Culture Turkey Economic and Social History Association, Ankara.

Sarı, Y. (2001). A Study of the Natural and Cultural Landscape Values of Amasra With Regard to Sustainable Tourism. ZKÜ Institute of Science, MSc. Thesis, Zonguldak.

Sarı Nayim, Y. (2010). Mapping of the Important Biotopes Located Between Amasra-İnkum (Bartın). İstanbul University Institute of Science, Doctorate Thesis, İstanbul.

Sharp, R.P. (1982). Landscape Evolution (A Review). JSTOR:Proceedings of the National Academy of Sciences of the United States of America, Vol: 79, No:14, Part 2: Physical Sciences, 4477-4486, National Academy of Sciences, USA.

Soykan, A.; Sönmez, S., Efe, R.; Cürebal, İ. (2012). Türkiye Turizm Politikalarına ve Ayvalık Adaları Tabiat Parkı Uygulamalarına Eleştirisel Bakış. Kazdağı ve Madra Dağı Belediyeler Birliği, Madra Dağı Ulusal Çalıştayı. 12-14 Ekim 2012, Ayvalık-Balıkesir, Bildiriler Kitabı s. 82-90.

Tasser, E.; Ruffini F.V. and Tappeiner, U. (2009). An Integrative Approach for Analysing Landscape Dynamics in Diverse Cultivated and Natural Mountain Areas. Landscape Ecology 24, 581-713.

Thomas, M.F. (2012). A Geomorphological Approach to Geodiversity–Its Applications to Geoconservation and Geotourism, Quaestiones Geographicae, 31(**1**), UK.

Tüysüz, O.; Genç. Ş.C.; Tar,ı U. (2001). Geological and Morphological Features of Bartın and its Proximity. İTÜ Development Foundation Avrasya Institute of Geology Technical Report, 2000/11/003, İstanbul.

Zonneveld, I.S. (1989). The Land Unit-A Fundamental Concept in Landscape Ecology and Its Applications, Landscape Ecology, Vol:3, No:2, 67-86, SPB Academic Publishing.

—. (1995). Land Ecology. SPB Academic Publishing, 90 p., Amsterdam.

CHAPTER THIRTY

ALTERNATIVE TOURISM ACTIVITIES AND ECOTOURISM ROUTES IN THE ZAMANTI STREAM BASIN

HAKKI YAZICI, NUSRET KOCA
AND SULEYMAN HILMI ŞAHIN

Introduction

The Zamantı Stream, fed by several springs in Uzunyayla, runs through the counties of Tomarza, Develi, Yahyalı and forms the Seyhan River, joining the Göksu Stream in the vicinity of Aladağ County, which is 80 km to the north of Adana. It is located in the vicinity of the County of Pınarbaşı in Kayseri at an elevation of almost 1,500 m. The stream covers all the western side of the Tahtalı Range and runs in narrow and deep valleys, joined by a great number by streams and tributaries around this area. The area is approximately 8,820 km^2, supplying 46% of the water to the Seyhan River. The Çamlıca III dams are located on the course of the Zamantı Stream on the point in the east of Bahçelik and Yahyalı, and south of Pınarbaşı. Natural attractions within the canyon formed in the lower basin of the Zamantı Stream and the species of endemic plants and fauna found in the basin provide important resources for ecotourism in the region.

The current situation of the ecotourism activities which can be carried out in the light of natural, historical and cultural resources and the alternative routes which can be exploited in ecotourism in the Zamantı Stream as a largest tributary of the Seyhan River, with a drainage area located in different geographical regions, have been investigated here. Issues such as identifying the ways to use the resources in the basin, and using and planning the tourism resources of the basin without compromising on the sustainability have been studied. The aim was to contribute to the planning conducted in the area in the future by analyzing the data gathered and using the findings which have emerged during this analysis.

Although the Zamantı Stream Basin has great potential in terms of ecotourism it has not attained the desired level in this aspect. Identifying the causes for this and providing solutions is crucial to the systematics of our study. As a result, it is important to assess the current ecotourism resources in the basin from a geographical point of view and to propose solutions for the future.

Materials and Methods

During the first phase of this study, we surveyed the literature, observations and CBS. National and international essays, declarations, books and theses related to the utilization of natural, historical and cultural resources of the Zamantı Stream Basin in terms of sustainable tourism. In addition, we assessed the surveys carried out in the basin through scientific programs. The raw data on the field studies were used with numerical input into the CBS system. The potential cultural and natural ecotourism values of the basin were transferred to the map.

The Current Use and the Potential of the Resources in the Zamantı Stream Basin in Terms of Ecotourism

Zamantı Stream Basin is an area with great potential for ecotourism activities. These activities are explained below with the current assets and are shown in detail in Fig. 30.1.

Utilization of the River: River tourism which doesn't necessitate large amounts of investment unifies the historical, archaeological, cultural and authentic values of a region with other tourism types. Therefore, while studying the tourism potential of our rivers, we should also know the other tourism values that are located around the river which can be developed in an integrated way. In our opinion, the Zamantı Stream is convenient for river tourism. A "National Rafting Course" has been drawn within the borders of Yahyalı County in the basin for the rafting activities performed between April and September in the area. A Rafting Festival is organized by the municipality of Yahyalı in June each year and these activities are promoted. In addition, the tour companies which generally operate in Avanos, Ürgüp and Göreme in Cappadocia occasionally organize tours in this area. However, these tours have a negative impact on the economic development of the local residents because they are organized as excursions and the needs of the tourists are met in Cappadocia (The Municipality of Yahyalı 2011). If the shores of the dam and ponds are arranged for the recreational use of the local residents as parks, these areas can also be used for domestic tourism as well. Particularly, the arrangements to be made around the Bahçelik Dam in the north will facilitate both trout sales and serve for entertainment purposes. There have been some attempts to offer rafting activities in the basin in the past, but these have failed due to lack of demand. Two tour companies applied to do rafting from Israel and the Netherlands in 2004 and 2005, and one-day rafting tours were organized in the region in these two years. However, the tours failed to continue in the following years due to the high level of difficulty in the Zamantı Stream lap and lack of demand. Furthermore, rock falls due to the fault line concentrations on the course of the stream and the existence of steep rocks in the stream bed pose dangers.

The Utilization of the Caves: The Zamantı Stream Basin includes the cave systems in the carstic structure in the Mid-Taurus Mountain Range. If the physical attractions in these caves (Zindan-I Köşk I, II, Perili Cave) are utilized, it will be to the advantage of the ecotourism in the region. To this end, promotion of the caves, betterment of the transportation facilities and their opening to tourism are needed. Some protective measures should be taken for their sustainability.

Nature and Bird Watching: Accepting the fact that the best teacher is nature, it has something to say to those who know how and where to look. One of these observations is nature and bird watching, a kind of observation sport which enables one to get to know nature and the world of birds. The Aladağlar National Park, located within the Zamantı Stream Basin, is an important place among the bird habitats with regard to its population of Bearded Vultures, Griffins, Golden Eagles and Caspian Snowcocks. This important bird habitat is known as one of the areas where there are some species whose protection is in an adverse status in the assessment made by the Association of Nature Conservation.

Paragliding: Paragliding is done by means of running on a high hill before jumping off with a parachute. It is an activity which can be done in the suitable spots on the Tahtalı Range, which covers the whole east side of the Zamantı Stream Basin, and on the eastern side of the Aladağlar in the south after the completion of the required planning and preparation processes. If the necessary education and opportunities are provided, the region can compete against the other paragliding areas in Turkey.

Utilization of the Campsites: It is known that people resort to different ways for spending their holidays. Today, it is harder to find a clean environment to spend time. One of the ways the tourists want to spend time is on the mountains away from their everyday concerns. There are many areas in the Zamantı Stream Basin which are convenient for camping during the summer months, and especially areas in the Aladağlar National Park which offer attractive sites for camping activities. The natural beauty of this area, its cold waters and its clean air can attract tourists.

Mountainaring, Nature Trekking and Horse Riding: The canyon in the lower basin of the Zamantı Stream around Yedigöller (Seven Lakes) area is located in the Aladağlar National Park. The summits and the plateaus have a great potential to allow tourists to carry out various activities (Fig. 30.2). The Aladağlar National Park contains the most suitable spots to take photos, view the scenery and climb the summits with its mountain lakes and plateaus.

Fig. 30.1. Group of foreign tourists on the entranceway of Karpuzbaşı Falls in Aladağlar National Park

By educating the people who live in the area, we can generate revenue for the regional and national economy and if the opportunities are available and used properly, we can both develop tourism and protect the environment.

Horse-riding is one of the activities that can be done in the Zamantı Stream Basin, especially in the lower parts where the valley gets narrower and paths extend along the stream. The ways to Karpuzbaşı Falls and Uzunyayla, located in the north within the borders of Pınarbaşı County, are suitable places for such activities. Uzunyayla is one of the best recreational areas with its cool weather and ample waters during the summer. The Jockey Club of Turkey established Uzunyayla Horse Riding and Horse-riding Training Facilities in the county of Pınarbaşı in 2001.

Activities of Transhumance: Recreational activities in the study area, such as observing transhumance tourism as well as transhumance culture, its historical places, areas of natural beauty, flora, fauna, forests, streams, canyons, breathing in its clean air and drinking its water pave the way for diversifying the tourism in the region. Aladağlar (Range) has become the site where domestic and foreign tourists visit and take part in recreational activities in increasing numbers in recent years due to its natural and cultural resources in addition to its transhumance activities (Somuncu 2005, 243; Efe et al. 2012). The transhumance activities which can be carried out in the areas where plateau and mountain climate conditions prevail in the northern side consist of activities based on stockbreeding. The transhumans, especially those from Çukurova, live on the plateaus to carry out activities of stockbreeding as well as spending time in their summer homes.

The Utilization of Hunting Areas: The area has a natural environment for game animals known to exist such as Caspian snowcocks (wild turkey), wolves, wild boars, foxes and lynx. There are also other game animals like speckled partridges, gray partridges, quails and turtledoves. Those who want to hunt in the area can do so in the region after they get the required permissions from the competent authorities. However, it is also a fact that some species are facing the threat of extinction as a result of reckless hunting. For hunting tourism in the area, wild goats are hunted in WLDA according to the quota imposed by the Ministry of Environment and Forestry (Aladağlar Plan for the Management and Improvement of Wild Life Development Areeas 2009). The fish found in the tributaries of the Zamantı Stream in the county of Pınarbaşı in the north of the basin represent a remarkable place for fishing, such as the tributaries of Örenşehir, Karagöz, Gümüşgün, Kabaktepe and Kırkgeçit every day during the hunting season.

The Utilization of Endemic Plant Species: Turkey has a wide range of climates due to its geographical variations, and it is a natural land bridge between three continents, which adds to its plant diversity. Here, we can see plants rarely found in other parts of the world (www.kultur.gov.tr). The endemic species in the Zamantı Stream Basin should be identified and integrated into ecotourism. However, a balance for the protection of these species should be set while promoting them and any harm posed by the local residents, and foreign tourists should be prevented to maintain sustainability.

The Ecotourism route of the Zamantı Stream Basin

Ecotourism has emerged because people want to get away from the chaotic environment created by the economic and technological advances in today's world, and it is growing rapidly each year. The Zamantı Stream Basin has enough potential for the growth of ecotourism with its natural, historical and cultural assets. Therefore, we tried to determine the usability route in the basin with regard to ecotourism. An ecotourism route within the Zamantı Stream basin needs to be drawn that will allow the domestic and foreign tourists visiting the region to travel more easily and to determine their own itinerary. It must give information as to which events will be done, where, when and for how long they will be done, and where the tourists can find accommodation. In addition to the recreational activities, one can also hunt, fish, go trekking, go on photo safari, ride horses, do transhumance activities, and go climbing and rafting. The routes denoting these activities are divided into exploring nature and exploring history and culture.

Exploring Nature Routes

- These are the routes which are shown within the walking trails located in the recreational area of Karpuzbaşı Falls in Aladağlar National Park in accordance with the instructions of the Provincial Directorate of Environment and Forestry. One of these is the route that starts from the entranceway of the national park and ends in the northwestern side of Elif Falls. The second starts from the Kapuz River, on which Elif Falls are located, extends in a northwest then southwest direction, goes over the hill on which the falls are located and down to the Bozarmut Stream, passing Güney Falls, ending where the Aksu Stream starts. This route can be used in all seasons except for the winter.
- The route within the 15.500 m^2 area of the Zamantı Stream lying within the borders of Yahyalı County is defined as a "National Rafting Course," located in the South Zamantı canyon. Rafting Festivals are organized by the municipality of Yahyalı in June every year. In addition, the tour companies normally located in Avanos, Ürgüp and Göreme in the Cappadocia area occasionally organize tours to this area.
- There is a route called the Aladağlar National Park Route. The area includes tarns, plateaus, summits (3725m) and the Yedigöller (The Seven Lakes) which have very beautiful landscapes. This area is a

perfect route to take photos and enjoy the scenery and the summits. The best time for these activities is the period between May and September.

- The route that starts from the Hacer Pass and extends through the tarns, falls and spectacular rock formations goes through the deep Hacer Valley, onto the Ulupınar Plateau and the Soğukpınar area of the Hacer Forest, an important route ending in the Karpuzpaşı Falls, one of the most magnificent waterfalls in Turkey (Kayseri Provincial Culture and Tourism Assets 2012). The best times to visit the route are late spring and early autumn.
- There is a route along Gökoluk, Akçay, Aksu Valley, Aksu Canyon and Karpuzbaşı Falls. This route is suitable for camping, canyon trekking and canyon tours by carrying the loads on mules and by means of motor vehicles as three-day tours (Çukurova University 2000).
- This is the route along which one can see the natural bridge formations in the vicinity of Yeşilköy. Access to the area is possible by arriving in Yeşilköy by motor vehicle and then by walking for the rest of the trip. The natural bridges in the area can be considered within the scope of geoparks.
- This route also includes the recreational activities which can be done around Homurlu Falls, located within the borders of Develi County. The local residents can go for recreational activities as well as cures in the summer months in this area, as it is believed to be a cure for some health problems, especially rheumatic diseases.
- The areas of the entranceway of Hacer Forest in the village of Ulupınar, the entranceway of Aksu Canyon, the Zamantı Stream in the village of Çamlıca and the area around the Bozarmut Stream and Güney Falls can be designated as camping areas.
- It is possible to conserve the 26 species of endemic plants in the Aladağlar National Park and to exhibit these to visitors during spring and summer.
- It is possible to take photographs of the endemic animal species in appropriate seasons which live in the basin by obtaining necessary permissions from the Provincial Directorates of Environment and Forestry.
- River fishing can be carried out within the borders of the Pınarbaşı and Tomarza Counties in the north, paying attention to the hunting season.

The Routes for Exploring History and Culture

- It is possible to visit the area in the vicinity of the village of Tomarza Kömür, where there are remnants of sarcophagi as the area was a sarcophagus production centre in Roman times. One can also see the complete stages by which a sarcophagus is made (separation from the main rock, transportation, raw product and the end-product). The area is already open to visitors all year round.

- It is possible to visit the areas where examples of religions before Islam in Anatolia can be found. The rock temples and altars of the Hittite era, the rock churches of the Roman era and the churches belonging to the Byzantines and the Armenians can be visited all year round.
- It is possible to see the cuisine, carpet and rug motifs, and homemade artifacts particular to the region, which are all symbols of the ethnographic culture in the basin in all seasons of the year. Visitors can easily access these products which form an attractive part of the ethnographic culture.

We can ensure the sustainability of the ecotourism activities in the basin if the routes mentioned above can be maintained throughout the year. The spots which have been selected in the basin can be used as laboratories where extracurricular activities in the open air can be performed and where natural beauties can be seen on the spot, remains of various cultures can be explained and sports and recreational activities can be followed.

Conclusion

In this study we have identified the ecotourism activities which can be practised with the natural, historical and cultural assets of the Zamantı Stream Basin, located in the east of the Central Anatolian Region. The findings obtained from the field study, the interviews made with the persons of interest in the areas covered, the geographical information systems evaluated, the analysis based on remote sensing and the statistical data obtained from the related organizations have been used in this study. Each of the ecotourism assets in the study area has been assessed separately, the potential of the area has been evaluated, alternative activities which can support the current ecotourism activities have been identified and certain suggestions have been made for the future. All these studies have been mapped with the help of geographical information systems and an ecotourism route has been drawn.

We can conclude that the physical factors such as geological, geomorphological and climatic features, and cultural and historical factors can play a great role in the formation of current potential ecotourism assets of the Zamantı Stream Basin. The main ecotourism assets in the area are the Karpuzbaşı Falls, the Aladağlar National Park, and historical remains and artifacts. There are ecotourism assets that can be utilized in historical and cultural tourism in the Zamantı Stream Basin as well as activities such as rafting, river fishing, trekking, climbing, hunting, cave tours and transhumance. It is necessary to improve the current facilities and the quality of the service to increase the number of the accommodation facilities, to improve the transportation facilities and to create awareness about tourism in the local residents as a result of interviews we have conducted with the local residents. In addition, the sustainability of these assets is of great importance for these resources to be passed down to the next generations and the conservation of ecotourism assets.

Acknowledgement: This study was supported by the BAP (Scientific Study Projects) unit of Afyon Kocatepe University as the project numbered 10-EGT-03.

References

Aladağlar Yaban Hayatı Geliştirme Sahası Yönetim ve Gelişme Planı (2009). Kayseri il Çevre ve Orman Müdürlüğü

Atalay, İ. (1994). Türkiye Coğrafyası, İzmir.

—. (2004). Genel Fiziki Coğrafya. Ege Üniversitesi Basımevi, İzmir.

Blamey, R.K. (2001). Principles of Ecotourism, D.B. Weaver, (Ed.) The Encyclopedia of Ecotourism, Australia: CABI International.

Çukurova Üniversitesi, (2000). Aladağlar Milli Parkı Uzun Devreli Gelişme Planı (Mastır Planı), Çukurova Üniversitesi,Adana.

Değirmencioğlu, A.Ö. ve Ahipaşaoğlu, S. (1999). Anadolu'da Turizm Rehberliği, Detay Yayınevi. İstanbul.

Doğanay, H. (1997). Türkiye Beşeri Coğrafyası.Milli Eğitim Bakanlığı Yayınları: 2982, Eğitim Dizisi: 10. İstanbul.

—. (2001). Türkiye Turizm Coğrafyası.Çizgi Kitabevi Yayınları, Konya.

Doğaner, S. (2001). Türkiye Turizm Coğrafyası.Çantay Kitabevi, İstanbul.

Efe, R., Soykan, A., Cürebal, İ., Sönmez, S. (2012). Madra Dağı'nın Kuzeybatısında Yaylacılık Faaliyetleri. Kazdağı ve Madra Dağı Belediyeler Birliği, Kaz Dağları Ulusal Çalıştayı. 12-14 Ekim 2012, Ayvalık-Balıkesir, Bildiriler Kitabı s. 107-124.

Erdoğan, N. (2003). Çevre ve Eko-turizm. Erk Yayıncılık, Ankara.

http//www.kultur.gov.tr Erişim Tarihi: 20.10.2011.

Kayseri Kültür ve Turizm Envanteri (KKT Envanteri), 2011 (2012?)

Kayseri Yahyalı Belediyesi 2011 Yılı Basın ve Halkla İlişkiler Verileri

Kopar, İ. (2009). Aladağlarda 2 Fosil Mağara: Zindanı Köşk-1 ve Zindanı Köşk-2 Mağaraları, Türkiye Coğrafya Dergisi, Sayı 53, 69-83, İstanbul.

Somuncu, M. (2005). Aladağlar Yaylacılık ve Dağ Göçebeliği Konusunda Bir Araştırma, Gündüz Yayıncılık, Ankara.

Özgüç, N. (1984). Beşeri Coğrafya'da Veri Toplama ve Değerlendirme Yöntemleri. İstanbul Üniversitesi Yayın No: 3849, Fakülte Yayın No: 2511. İstanbul.

—. (2003). Turizm Coğrafyası. Çantay Kitabevi, İstanbul.

Yeşil Atlas Dergisi, Kasım, (2003), İstanbul.

Weaver, D.B. (2001). Ecotourism in the Context of Other Tourism Types, D.B.Weaver (Ed.), The Encyclopedia of Ecotourism (73-83), CABI International, Australia.

Wood, M.E. (2002). Ecotourism, Principles, Practices and Policies for Sustainability United Nations Publication, UNEP.

CHAPTER THIRTY-ONE

CLIMATE CHANGE AND THE POSSIBLE MIGRATION OF THE POPULATION IN GEORGIA

NODAR ELIZBARASHVILI, GIORGI MELADZE AND ZURAB SEPERTELADZE

Introduction

Actuality of the problem

Discussion about the climate problem is not going away. On the contrary, it has become a subject of interest not only to scientists but also politicians. It is known that the expected consequences of climate change are discussed not only global and environmental but also social and economic problems.

Some scientists see nothing alarming in global climate change. They compared this phenomenon to the scales of (especially an increase carbon and its high concentration) some natural events and processes (for example, volcanic activity, biosphere waste decay, and increase in the temperatures of oceans) For example, an erupting volcano releases as much carbon as all the industrial companies around the world in a hundred years. The scientists argue that the climate change has happened many times in the history of the geographical formation. Global warming was especially prominent in the Middle Ages to the extent that a viticulture was developed in England at that time. Such considerations have solid foundations, but cannot be explained by citing the natural processes and rate of increase of atmospheric temperature and the amount of carbon at the global level during the last century (Davitashvili and Elizbarashvili 2011).

The reader's attention will focuse on the brief description of some of results, already evident as a result of climate change in certain regions of the world as well as in Georgia.

Climate variability in the Northern Hemisphere became obvious at the beginning of the twentieth century. Warming was recorded in the 1930s (the average air temperature increased by 0.6^{0}C), but the 1950s then noted a cooling before recording another warming in the 1960s. There are two kinds of explanations for global warming. The first is associated with the greenhouse effect, but the second is related to the 11 and 80 year cycles of solar activity. It is established that in Georgia the average air temperature increased by 0.6^{0}C in the twentieth century.

- The signs of global warming became alarming from 1980s. It was fixed that on a number of islands in the Northern Arctic Ocean (Greenland, Spitsbergen and New Land) and on the coastal strip; the average annual air temperature has increased by 2-3^{0}C during the century. The territories which were used for growing annual crops 600 hundred years ago were freed from ice in Iceland. The rapid decrease in ice cover on Greenland and Alaska was mentioned, as well as in all moderate zones of mountains. In Georgia, in the glacier-nivalis zone of the Caucasus, the increase in the number of glaciers is obvious at the expense of great glacier fragmentation and the reduction of total area.
- In addition to reducing the ice-covered area, mentioned in the Northern Hemisphere is the process of displacement of the lower border to the North of the Permafrost, Taiga and Tundra. Similar signs have been observed in the mountainous Caucasus region, which are reflected in the rise of the subalpine forests upper border.
- There was a further decline in atmospheric precipitation in arid regions of the Northern Hemisphere, especially during the cold period of the year. These developments led to a decrease in river discharge and caused loss of water level in internal reservoirs. The droughts magnitude, the intensity and a number were increased, what became a reason for vital environmental and economic activity degradation in many regions. In recent decades the most obvious decrease in sediments is (10%) in North Africa and South-East Asia's tropical belt. Abnormalities were observed if the amount of sediments increases on the shores of the Western Australia. On the contrary, we are dealing with another event on the Eastern Shores. It is estimated that the eastern steppe landscape has increased the share of a semi-desert area which is covered with plants.
- In almost all regions of the world, a characteristic thermal regime shows that global warming is observed almost everywhere, however with various degrees. Maximum climate change was mentioned in the land of the Northern Hemisphere, between N.L. 40-70^{0}C. Here, with the increase of the average annual temperature, a frequency of extreme weather indices is mentioned. For example, in the Scandinavian countries where almost the entire twentieth century did not display significant temperature fluctuation, an increase in winter temperature is

observed. In Georgia, in recent years (2011-2012), extreme and high temperature has been noted in places where they have never been noted.

- Characteristics of the oceans in climate variability are still less, most even the negative abnormalities are observed in some of the water. Although the results of the entire world ocean warming are obvious, its level increased by 10-25 cm in the last 100 years. Such a situation is primarily associated with the global trends in melting ice. Similar is the case of the Black Sea coast of Georgia.
- Changes in distribution of sediment variance are evident due to the changes in characteristics of the river runoff, groundwater and soil moisture. In Eastern Europe, it has been observed almost everywhere the rise of the soil humidity, water-layers, and groundwater resource (stocks).
- Global warming is also reflected in the duration of snow cover. Its index decreased in Eurasia and North America continents. The situation is similar in Georgia, where during the last half century, almost the entire snow distribution rate changed.
- As a result of global warming in the equatorial belt, the activation of evaporation process is underway. The Increase in the salt level of the equatorial and tropical belt waters is expected, this will lead this type of water to be complicated and sink. The Gulf Stream's heavy, warm and salt waters, will give "some places'' to the cold arctic light and freshwaters. Thus, it is expected not only to moderate subtropical coastal belt or climate change in the characteristics, but also the living environment for residents in the ocean. Hints are obvious in Eastern coasts of North America and in various regions of Europe is mentioned annually the increase in the snow cover sickness and frosts intensity.
- Degradation of the ozone layer and its protective qualities pose a serious threat to mankind and the natural environment. Mankind is threatened with impairment of the immune system, increased skin and eye diseases and complicating impairment of the body's resilience against infectious diseases and deterioration of mental attitude.
- Degradation of the ozone layer and ultraviolet radiation increase will significantly reduce pumpkin and pea type plants, vegetation productivity, much more sunflower, cotton, and some types of grain (maize and rice) and their productivity will be decreased by 5%. The trend of decreasing grain harvest is also obvious in Georgia.
- Variability of ozone layer affects soil fertility. Bacteria that are able to use nitrogen from the atmosphere, which is then used to develop the growth of plants, lives in the upper layers of soil. It is known that bacteria processed by as much nitrogen (35b. tone) as much nitrogen

fertilizer is produced in the world. In case of ultraviolet radiation increase, microorganisms' life and soil fertility will be significantly reduced.

- It is observed in the Northern-Hemisphere atmospheric circulation, from ozone depleted air masses high longitude (for example Greenland), the frequent moving process towards Europe. Hydrometeorological services of Russia have already fixed the ozone layer's decreasing fact by 40% in Ural and Siberia. It is known as well that the variability of the ozone layer in the Northern Hemisphere is characterized by well expressed the seasonal mood. If the layer thickness is normal in spring, in late autumn and winter, the same cannot be said.

The Climate Change Trends in Georgia

The climate change trends are set out in the article ''Georgia's Second National Communication to the United Nations Framework Convention Climate Change'', which was prepared in the framework of a joint project by the Ministry of Environmental Protection of Georgia and the United Nations Development Program (Tbilisi, 2009). According to the project processes and trends connected to the climate change got in touch with some of the geographical settings among which are: air temperature, sediments and moist quality. On this basis, the trends of climate change have been estimated, as well as quantitative, space and time characteristics for the development of extreme natural events. Analysis was conducted by using statistical methods on the basis of the data that holds the National Environmental Agency of Georgia, which covers the period from the middle of the twentieth century to the period of 2007 - 2010. In addition, by 2010, a global climate model assessed the main climate parameters in Georgia for the three priority regions (Black Sea Coastal-Poti Weather Station, The Caucasus Mountains-Lentekhi Weather Station and the Lowland between the Mountains-Dedoflistskaro Weather Station.) For each parameter, the trends for changes are assessed in accordance with the current year and seasons.

According to the national post, during the last half-century, on the territory of Georgia is mentioned the average annual increase in air temperature and sediments (especially, in recent decades), as well as the absolute minimum (except, Dedoplistskaro) and maximum warming in the three priority regions (Poti-sediments 13%, t-0,2^{0}C, Lentekhi-sediments 8%, t-0,4^{0}C, Dedoplistskaro-sediments 6%, t-0.6^{0}C). Comparatively, a somewhat different picture gives an analysis of changes in sediments for the Eastern and Western Georgia. In this case the decreasing annual trend is observed in Western Georgia (by 27mm), while the same figure in Eastern Georgia by 41mm higher.

An interesting and somewhat different picture of global climate models will be manifested by the year 2100.

According to the first model (HadAM3P) the average air temperature will rise in Western Georgia from +1.8 degrees and reach 12.7 degrees, but the amount of sediments will be reduced by 30mm and slightly exceed 1500mm. In Eastern Georgia, the expected increase in average air temperature is two times more than that of Western Georgia (+3.5 degrees), as the decrease in sediments is more than twice expected (70mm). Thus, for the existing semi-arid and semi-humid forest landscapes in the eastern lowlands of the mountains, the critical ecological situation created what will be related to moisture deficit. It should be noted that the decrease in rainfall in Eastern Georgia against the background of increasing air temperature, will be the characteristic of all seasons of the year, especially in the summer.

According to the second model (ECHAM4) the average air temperature will rise in Western Georgia until +3.4 degrees and reach 14.3 degrees, the amount of sediments will be increased by 36mm and reach 1600mm. In Western Georgia, a distinct decrease is in summer precipitation (by 50mm), but the average seasonal air temperature increases by 5 degrees. According to the model, the average air temperature and expected change in sediments in Eastern Georgia is almost the same as Western Georgia (+3.9 degrees increase and 72mm decrease). Thus, according to the first model, the second model forecasts critical ecological situation for the semi-arid and semi-humid forests existing in the lowlands between the mountains of Eastern Georgia.

The Landscape- the basic unit of measurement for natural environmental conditions

Forecasting the consequences of climate change is beyond any doubt, however, it is arguable what refers to the approaches to the analyses and evaluation of various environmental responses. For ecologists these are ecosystems and biocenosis, for hydrologists they are river basins, for climatologists they are temperature and synoptic field, for agrarists they are agrobiocenosis and so on. The fact is, climate change is simultaneously reflected in the elements of climate, also in the waters, relief, soil, plants, productivity, demographic behavior of the population and economic activity. Thus, the predicting focus appears as natural so also the socioeconomic environmental elements. Because of the natural environment forecasting, only integrated approach should be considered essential (Elizbarashvili 2005).

The constituent elements are connected to each other with the cause-effect relationship. Such links are created for ecological systems or landscapes, which are characterized by uniform relief, geological structure, climate, water regime and the soil-vegetation cover. Given that the landscape scale is a key step in the challenging hierarchical ladder that the whole geographical shell (the earth) or its main constituent units are (for example, Continents, Oceans and Natural Areas) submitted in the middle part of Large geographical regions (for example, the Caucasus, the Himalayas, the Andes, etc). The landscapes (ecosystems), on the lower level are elementary eco or geosystems.

The integrity of geographical shell and its compound landscapes is caused by the constant energy and substance transformation (by the structure and function). Due to the formation history of the landscape, it characterizes natural, natural anthropogenic mode, ethocycle and the dynamic development tendency. The landscape is connected with each other's anthropogenic-technological elements caused by the natural and public demands. Despite the modern scales of anthropogenic transformation, the landscape is mainly developed to characterize the natural processes. Modern landscape is the natural-historical phenomenon, in which is represented in the face of territorial systems, the ecological variety of natural and socioeconomic processes' results undergoing in the geographic shell.

Landscape state and functioning can be considered as the unity of several natural and anthropogenic processes, among which the most important are: physical, (mechanical) chemical, biological and socioeconomic processes. The intensity of the each is determined by the specificity of substance and energy converting by the potential of natural resource, traditions, forms of economic activity, population density, the structure and the nature of the transport infrastructure and others (Georgia's Second National Communication 2009, Nikolaishvili 2009).

For the prediction reliability of the condition and functioning of the landscape, it is important to identify which component of the landscape is experiencing direct or indirect impact, let's define the function and stability of the component in a given system. In the knowledge of the landscape, taking into account the known sequence, according to the functions in the natural landscapes, elements are located in the following sequence:

L (Relief and Geological Structure) ⟶ **A** (aeromases or the climate elements) ⟶ **H** (hydromasses or hydrosphere elements) ⟶ **S**

(pedomasa or the soil) ⟶ **B** (biomasses or the live world). In the current scheme, an arrow shows a major interconnections sequence. According to the scheme, the right locates an environmental element that is much more dependent on the left, but less is dependent on the element to the left. For example, the soil transformation leads to a significant change of biological component, partially hydromass, even slightly aeromass transformation.

Georgian Landscape Zonality and Geographical Peculiarities

The main geographical peculiarities of Georgian landscapes spreading are associated with a number of factors, among them are: altitude of the place, terrain shape, exposure, geological structure, rocky surface peculiarities, its proximity to the Black Sea and etc. (Georgia's Second National Communication 2009, Kvachakidze 2001).

On the plain and hilly parts of Georgia are the nearly 30% of the Georgia territory, 6 types of landscapes are represented. They are united into the Colchis humid - subtropical and moderately humid and semi-humid, sub-tropical climate regions of Iveria. From their geographical peculiarities must be the following: Plain and hilly forms of terrain; Accumulative; karstic; Erosive and Arid–denudation geo-morphological processes. They are hypsometric represented 800m above sea level in western Georgia and 200m -1000m in Eastern Georgia all in the range of altitude. There is represented sub-tropical, humid, semi-arid and arid, also sub-tropical, semi-arid and temperate warm arid climate types; the average annual sunshine duration is 2000 - 2400 hours, the average temperature of January is from $-2^{0}C$ to $+5^{0}C$, and the average temperature of July is from $+22^{0}C$ to $+25^{0}C$, the average amplitude of temperature is 17 - 24%. Sum of active temperatures is 3000 - 4000 degrees, average annual wind speed is 2.0 - 6.0 m / sec, sediments are 400mm - 2000 mm, moisture coefficient is 0.6 - 2.0, the snow cover is formed during 3-30 days, and so on. Forest vegetation is presented of marsh, grove, poly-dominant, oaks, hornbeams, chestnuts, pine relict forests, arid and sparse forests etc. The quantity of phytomass ranges from 5 to 300 t/ha. Major parts of the plain and hilly landscapes are characterized by their factually transformed or changed horizontal structure; naturally functioned landscapes are represented in the protected areas, on the separate river valleys and wetland plains.

Presented are (5 types) Georgian mountain landscapes of the Caucasus, Minor Caucasus and Meskhet - Javakheti volcanic physical - geographical constituencies. From the peculiarities of Georgian mountain landscapes geographical area mentioned the following: low, medium, high mountains, mountain caves, paleoglacial and glacial relief forms, which range above the sea level from 300 – 5201 meters.

The climate is mostly temperate and warm, temperate humid, semi-arid, semi- humid and arid with growing damp forms. Sunshine duration is from 2200 to 2500 (in Southern Georgian Highland) and from 1900 to 2000 h/year (in Caucasus). The average temperature in January is -2^{0}C - 12^{0}C, and in July +4^{0}C - +20^{0}C, the temperature amplitude is 18^{0}C - 25^{0}C, the total sum of the active temperature is 200 - 3500°, Average wind speed 1.0 - 5.0 m/sec, precipitations ranges within 600mm - 2800mm. Evaporation 300mm - 800mm.

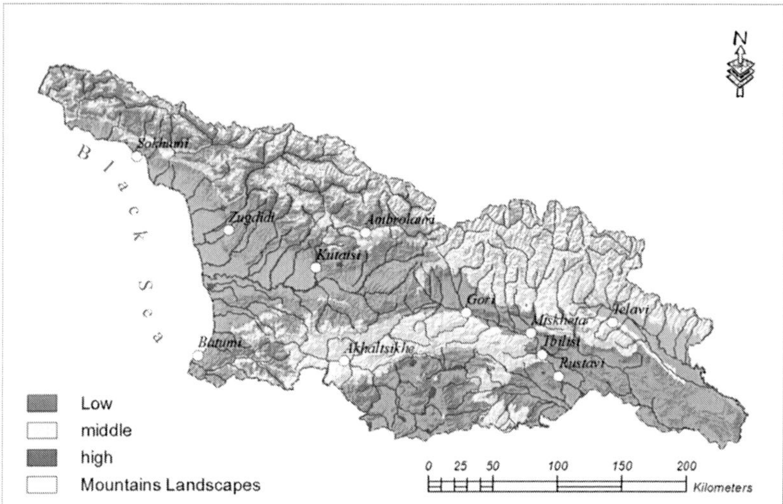

Fig. 31.1. Degree of Possible Change of Landscapes of Georgia

Fig.31.2. Arid Landscapes of Georgia

Fig.31.3. Semi-Arid landscapes of Georgia

Fig. 31.4. Transformation of Semi-Humid Landscapes of Georgia

Snowy days are 30-150 days; growing damp coefficient is 0.6-1.5. Forest vegetation is represented; Colchis hemi hills, mainly beech, oaks, beech-dark coniferous, spruce and birch tree forests, arid sparsely forests. Phytomass number ranges from 10 to 800 t/ha anthropogenic transformation is of a medium quality; some minor fragments of a landscape are found that factually is unchanged (Elizbarashvili 2009, Kvachakidze 2001, Piloting Landscape Planning 2009).

Climate Changes and the Possible Migration of Population

Global climate changes will affect landscapes and geographical features of Georgia that will ultimately affect water and biological resources, population and a distribution of agricultural specialization. In accordance with the second and the third National Notification of 2100 years, the most notable trends of climate changes will be obviously seen on the eastern lowland landscapes, currently almost half of the population, two million inhabitants (the capital of Georgia - Tbilisi included). Population density is highest in the Shida Kartli plains, in Kvemo Kartli plains, and in the hilly part of Kakheti. The global climate changes will negatively impact the following landscapes:

- **Kvemo Kartli plain semi-desert landscapes:** An area of 370 square kilometers has a population of more than 500 thousand inhabitants. Without the development of an irrigation system, desertification becomes a threat.
- **Shida Kartli plain steppe landscapes**: An area of 560 square kilometers has a population of over 150 thousand inhabitants. Without the development of an irrigation system, it will become a semi-desert.
- **Iori Upland steppe landscapes**: An area of 1140 square kilometers has a population of more than 50 thousand inhabitants. Without the development of an irrigation system it will become a semi-desert;
- **The extreme southern part of the Iori Upland steppe landscape:** An area of 170 square kilometers has a population of approximately 5 thousand permanent and seasonal residents. Without the development of an irrigation system it will become a semi-desert;

- Georgian extreme south-eastern part of the **semi-desert landscapes:** An area of 90 square kilometers, precipitates totals of 350, there are no permanent population. It is already a desert.

Conclusion

Thus, according to global climate change and its expected results, ecosystem changes or degradation problems threatens almost 2,330 square kilometres, or 3.5% of Georgian territory. For such processes preventive actions will need to be carried out otherwise, over 800 thousand people olmost 20% of the Georgian population will become ecological migrants.

References

Davitashvili, Z.; Elizbarashvili, N. (2011). Geography of Global Problems. Tbilisi, (in Georgian).

Elizbarashvili, N. (2005). Geo-ecological Basics of Landscape Planning - Tbilisi, (in Georgian).

Elizbarashvili, N.; Meladze, G.; Svanadze, D. (2009). Landscape Planning: Methodology and Experience - Tbilisi, (in Georgian).

Georgia's Second National Communication to the United Nations Framework Convention on Climate Change. Tbilisi, 2009 (in Georgian).

Kvachakidze, R. (2001). Georgian Forests, Tbilisi, (in Georgian).

Nikolaishvili, D. (2009). Space - Time Analysis of Georgian Landscape - Tbilisi, (in Georgian).

Piloting Landscape Planning in the Countries of the Southern Caucasus-Baku, Tbilisi, Erevan, 2009.

CONTRIBUTING AUTHORS

Abdurrahman Dinç
Necmettin Erbakan University, Faculty of Tourism, Konya – Turkey

Abdurrahman Kanabakan
Akdeniz University, Antalya - Turkey

Alaattin Kızılçaoğlu
Balıkesir University, Necatibey Education Faculty, Balıkesir - Turkey

Alejandra Meda
"G. d'Annunzio" University, Faculty of Economy, Department of Economics, Pescara - Italy

Ayla Bozdağ
Selçuk University, Faculty of Engeneering, Department of Geological Engineering, Konya - Turkey

Burçin Kırlar
Pamukkale University, Tourism Faculty, Department of Tourism Management, Denizli - Turkey

Carmen Bizzarri
European University of Rome, Department of Human Sciences, Rome - Italy

Cemali Sarı
Akdeniz University, Education Faculty, Antalya – Turkey

Christian Hoffmann
EURAC Research, Institute for Regional Development and Location Management, Bolzano- Italy

Daniela La Foresta
Federico II University, Faculty of Political Sciences, Naples - Italy

Döndü Üçeçam Karagel
Mustafa Kemal Univesity, Faculty of Arts and Sciences, Department of Geography, Antakya, Hatay – Turkey

Elisa Ravazzoli
EURAC Research, Institute for Regional Development and Location Management, Bolzano-Italy

Elsayed Abbas Zaghloul
National Authority for Remote Sensing and Space Sciences (NARSS), Cairo-Egypt

Filippo Favilli
EURAC Research, Institute for Regional Development and Location Management, Bolzano- Italy

Fizyon Sönmez
Çukurova University, Faculty of Agriculture, Department of Landscape Architecture, Balcalı, Sarıçam, Adana - Turkey

Franca Miani
University of Parma, Department of Economics, Parma - Italy

Giorgi Meladze
Institute of Geography, Ivane Javakhishvili Tbilisi State University, Tbilisi- Georgia

Güler Göçmez
Selçuk University, Department of Geological Engineering, Konya - Turkey

Hakan Alphan
Çukurova University, Faculty of Agriculture, Department of Landscape Architecture, Balcalı, Sarıçam, Adana - Turkey

Hakkı Yazıcı
Afyon Kocatepe University, Faculty of Education, Afyonkarahisar - Turkey

Hasan Sayılan
Ankara University, Institute of Social Sciences, Ankara - Turkey

Hilmi Demirkaya
Akdeniz University, Education Faculty, Antalya – Turkey

Hulusi Karagel
Mustafa Kemal Univesity, Faculty of Education, Antakya, Hatay - Turkey

Ibrahim Atalay
Mehmet Akif Ersoy University, Faculty of Arts and Sciences, Department of Geography, Burdur - Turkey

Isa Cürebal
Balıkesir University, Faculty of Arts and Sciences, Department of Geography, Balıkesir - Turkey

Margherita Pedrana
European University of Rome, Department of Human Science, Rome - Italy

Maria V. Shishmanova
South-West University "Neofit Rilski", Faculty of Mathematics and Natural Science, Departmment of Geography, Ecology and Environment Protection, Blagoevgrad – Bulgaria

Marina Fuschi
G. d'Annunzio University, Departmnet of Economic, Pescara - Italy

Mehmet Ali Özdemir
Afyon Kocatepe University, Facutly of Arts and Sciences, Department of Geography, Afyonkarahisar - Turkey

Mehmet Emin Sönmez
Kilis 7 Aralık University, Faculty of Art and Science, Department of Geography, Kilis - Turkey

Meryem Atik
Akdeniz University, Faculty of Agriculture, Department of Landscape Architecture, Antalya-Turkey

Mimoza Dushi
University of Prishtina, Faculty of Mathematical and Natural Sciences, Department of Geography, Prishtina-Kosovo

Murat Bayram
Pamukkale University, Faculty of Tourism, Department of Tourism Management, Denizli - Turkey

Neşe Yücesoy
Ege University, Faculty of Agriculture, Department of Landscape Architecture, Izmir - Turkey

Nodar Elizbarashvili
Tbilisi State University, Department of Regional Geography and Landscape Planning, Tbilisi - Georgia

Nurhan Koçan
Ege University, Faculty of Agriculture, Department of Landscape Architecture, Izmir - Turkey

Nusret Drešković
University of Sarajevo, Faculty of Science, Department of Geography, Sarajevo – Bosnia and Herzegovina

Nusret Koca
Afyon Kocatepe University, Faculty of Education, Afyonkarahisar - Turkey

Osman Yılmaz
Mehmet Akif Ersoy University, Faculty of Education, Burdur - Turkey

Paola Mucciante
University "G. d'Annunzio" of Chieti Pescara, Economic Department, Pescara - Italy

Rahman Nurković
University of Sarajevo, Faculty of Science, Department of Geography, Sarajevo – Bosnia and Herzegovina

Recep Efe
Balikesir University, Faculty of Arts and Sciences, Department of Geography, Balikesir - Turkey

Suleyman Hilmi Şahin
Ministry of Education, Geography Teacher, Afyonkarahisar – Turkey

Salwa Elbeih
National Authority for Remote Sensing and Space Sciences (NARSS), Cairo- Egypt

Serkan Bertan
Pamukkale University, Faculty of Tourism, Department of Tourism Management, Denizli – Turkey

Sevgi Tatar (Dönmez)
Afyon Kocatepe University, Institute of Social Sciences, Afyonkarahisar - Turkey

Šemsudin Džeko
Tourism Association of Bosnia and Herzegovina, Sarajevo - Bosnia and Herzegovina

Tahsin Tapur
Necmettin Erbakan University, Ahmet Keleşoğlu Education Faculty, Meram, Konya - Turkey

Thomas Streifeneder
EURAC Research, Institute for Regional Development and Location Management, Bolzano-Italy

Tülay Öcal
Niğde University, Facutly of Education, Niğde - Turkey

Ufuk Coşgun
Southwest Anatolia Forest Research Institute, Antalya - Turkey

Valentina Albanese
University of Parma, Department of Economics, Parma - Italy

Vedat Karadeniz
Erzincan University, Faculty of Education, Erzincan - Turkey

Veli Ortaçeşme
Akdeniz University Faculty of Agriculture Department of Landscape Architecture, Antalya-Turkey

Yeliz Sarı Nayim
Bartın Üniversity, Faculty of Forestry, Department of Landscape Architecture, Bartın – Turkey

Zurab Seperteladze
Ivane Javakhishvili Tbilisi State University, Department of Geography, Tbilisi - Georgia